Springer Theses

Recognizing Outstanding Ph.D. Research

Aims and Scope

The series "Springer Theses" brings together a selection of the very best Ph.D. theses from around the world and across the physical sciences. Nominated and endorsed by two recognized specialists, each published volume has been selected for its scientific excellence and the high impact of its contents for the pertinent field of research. For greater accessibility to non-specialists, the published versions include an extended introduction, as well as a foreword by the student's supervisor explaining the special relevance of the work for the field. As a whole, the series will provide a valuable resource both for newcomers to the research fields described, and for other scientists seeking detailed background information on special questions. Finally, it provides an accredited documentation of the valuable contributions made by today's younger generation of scientists.

Theses are accepted into the series by invited nomination only and must fulfill all of the following criteria

- They must be written in good English.
- The topic should fall within the confines of Chemistry, Physics, Earth Sciences, Engineering and related interdisciplinary fields such as Materials, Nanoscience, Chemical Engineering, Complex Systems and Biophysics.
- The work reported in the thesis must represent a significant scientific advance.
- If the thesis includes previously published material, permission to reproduce this must be gained from the respective copyright holder.
- They must have been examined and passed during the 12 months prior to nomination.
- Each thesis should include a foreword by the supervisor outlining the significance of its content.
- The theses should have a clearly defined structure including an introduction accessible to scientists not expert in that particular field.

More information about this series at http://www.springer.com/series/8790

Nicolas Ackermann

Growing Stock Volume Estimation in Temperate Forested Areas Using a Fusion Approach with SAR Satellites Imagery

Doctoral Thesis accepted by
Friedrich-Schiller University of Jena, Germany

Springer

Author
Dr. Nicolas Ackermann
Science, Applications and Future Technologies Department
European Space Agency ESA-ESRIN
Frascati
Italy

Supervisors
Prof. Christiane Schmullius
Department of Earth Observation
Institute for Geography
Friedrich-Schiller University of Jena
Jena
Germany

PD Dr. Christian Thiel
Head of Radar Science and Application Section
Department of Earth Observation
Institute for Geography
Friedrich-Schiller University of Jena
Jena
Germany

Dr. Maurice Borgeaud
Head of the Earth Observation Science, Applications, and Future Technologies Department (EOP-S)
European Space Agency (ESA), ESRIN
Frascati
Italy

ISSN 2190-5053 ISSN 2190-5061 (electronic)
Springer Theses
ISBN 978-3-319-13137-5 ISBN 978-3-319-13138-2 (eBook)
DOI 10.1007/978-3-319-13138-2

Library of Congress Control Number: 2014954607

Springer Cham Heidelberg New York Dordrecht London

Printed on acid-free paper

Springer International Publishing AG Switzerland is part of Springer Science+Business Media (www.springer.com)

'Fais de ta vie un rêve, et d'un rêve, une réalité.'

Antoine de Saint-Exupéry

Supervisors' Foreword

Temperate forests hold several vital functions for the biosphere and the well-being of humans. They form an important terrestrial sink for carbon and play a significant role in the economy of many countries, both in the industrialised and developing world. Furthermore, forest degradation and deforestation are linked to a significant increase in carbon dioxide concentration in the atmosphere. The measurement of forest stand properties, especially biomass and stem volume, is therefore essential for mitigation of the actual global warming and the assessment of the forest resources exploitation. In recent decades, spaceborne remote sensing data have shown to be a promising tool to monitor forest biomass at a global scale, particularly data collected by SAR systems. In this context, the research undertaken by Dr. Ackermann is relevant and timely as he has been through a meticulous and thorough analysis of a large test site in the Thuringian forests, located in Central Germany.

This thesis presents an in-depth study of L- and X-band radar backscatter and interferometric coherence for the derivation of Growing Stock Volume (GSV) in the temperate zone. As part of the data analysis, a deep assessment of the influence of different weather conditions, forest structure and characteristics of the SAR data is proposed. Among other things, it is the first time that with the use of TanDEM-X coherence no saturation was found between X-band data and GSV. These detailed investigations are significant contributions for future scientific studies focusing on the estimation of forest GSV. As one of the highlights of this thesis, the reader is introduced to the role that topography plays in the scattering of electromagnetic waves transmitted by the radar. Though it is known that topography affects the retrieval of biomass from SAR data over hilly terrains, the work presented a thorough and meticulous analysis of this effect by comparing SAR backscatter and interferometric coherence with ground measurements. Moreover, this thesis proposes an innovative radar remote sensing method for large-scale mapping of temperate forest biomass. For this purpose, a comprehensive dataset of 222 radar scenes acquired by the German TerraSAR-X and TanDEM-X satellites, the Italian satellite constellation Cosmo-SkyMed, and the Japanese ALOS PALSAR sensor were considered. The proposed algorithm is ingenious and includes two approaches based on the derivation of forest/non-forest as well as tree species composition masks and the fusion of biophysical parameters. The developed methodology integrates L- and X-band data in an innovative synergy process, which may be very relevant for the operational mapping of Essential Climates Variables (ECVs) such as biomass.

This thesis describes complex and technically very challenging topics in a simple way. The content of the manuscript is well structured, precise and clear. It includes an excellent scientific style and high-quality graphics that are proposed in a proportionate and precise manner, describing the effects linked to topography, scattering mechanisms and spatial/temporal de-correlation for the usefulness of SAR data to retrieve forest parameters. From beginners to experts, all types of readers interested in SAR and forestry should find their way through this thesis.

University of Jena, October 2014 Prof. Christiane Schmullius
University of Jena PD Dr. Christian Thiel
European Space Agency (ESA) Dr. Maurice Borgeaud

Acknowledgments

I would like to first acknowledge the different space agencies for providing the SAR data used in this thesis. I am very thankful to the Japan Aerospace Exploration Agency (JAXA) for providing the ALOS PALSAR images, the Italian Space Agency (ASI) for providing the Cosmo-SkyMed data and the German Aerospace Center (DLR) for delivering TerraSAR-X (TSX) and TanDEM-X (TDX) acquisitions as well as the RapidEye (RE) data through the RapidEye Science Archive (RESA) service. This Ph.D. was sponsored by the DLR under the radar and optical satellite data for ENVIronment and LAND use applications (ENVILAND2) project with federal funds provided by the German Federal Ministry of Economics and Technology on the basis of legislation by the German Parliament (Grant no. 50 EE 0844-50 EE 0847). With reference to the ancillary datasets used in this thesis, I thank the TLVermGEO for providing the digital orthophotos and LIght Detection And Ranging (LiDAR) data and the 'Deutscher Wetterdienst' (DWD) for delivering weather data.

This Ph.D. was carried out over a 4.5-year period with 3 years spent at the Department of Earth observation of the Institute of Geography at the FSU Jena and 1.5 years spent at the Chair of Earth Observation and Remote Sensing of the Institute of Environmental Engineering at the Swiss Federal Institute of Technology Zürich (ETHZ). For the duration of my time at these two institutions, I had the chance to meet exceptional persons.

First, I thank Prof. Christiane Schmullius, Head of the Earth Observation Department at the FSU Jena, for having given me the opportunity to study within a single project three exciting sub-topics, namely radar, optics and forestry. Prof. Schmullius has been more than a supervisor. With her eternal good humour, kindness and very professional approach, she encouraged me to give my best and inculcated in me a real passion for my work. Second, I acknowledge my supervisor for thesis, Dr. Christian Thiel. Dr. Thiel has shown a great deal of professionalism and has guided me in my choices while giving me the freedom to make decisions. This helped me to develop a certain degree of independence and maturity in my work. Third, I am particularly grateful to Dr. Maurice Borgeaud, Head of the Earth Observation Science, Applications and Future Technologies Department (EOP-S) of

ESA. Dr. Borgeaud introduced me to the Ph.D. position, and as my co-supervisor, supported me with decisions made regarding the important guidelines of my work.

On my first arrival at the 'Jena Paradies' train station, I was immediately charmed by the small city of Jena and the University's campus. The Earth Observation Department, which is located just in front of the Paradies station, housed several persons to whom I would like to express some words of gratitude. First, I thank my dearest friend Ralf Knuth for always welcoming me to his home and for the open discussions which we had concerning our respective topics. Ralf has been an exceptional person, whether during or outside working hours. His efficient and structured way of working has been a tremendous example to me. Then, I acknowledge Claudia Hilbert and Tanvir Ahmed Chowdhury. By working together at the Thuringian Forest with different datasets and techniques, the three of us were able to have useful discussions regarding our different observations. Our friendly collaboration was ideal for me, even in difficult times. Lastly, I thank all the other people at the Geography Institute who contributed to making my stay in Jena a tremendous experience. In particular, I thank Dr. Valerio Avitabile for the lovely hikes and 'Stilbruch' dinners, Dr. Oliver Cartus and Reik Leiterer for their valuable advices, Dr. Martin Herold for the Pétanque' games, Roman Gerlach for the bike tours and billiards, Jiazhou Zhu for our long discussions in the Kartoffelhaus restaurant and his wonderful support during the forest campaigns, Michael Schultz and Mikhail Urbazaev for the unforgettable time spent in the Thuringian Forest, the Geography football team for the many great times we played football, Johannes Reiche for the fitness and jogging tours as well as for pre-processing some of my SAR data, Martyna Stelmaszczuk for returning my bike to Switzerland, Dr. Sören Hese, Marc Bennemann and Rainer Hoffmann for solving system administration issues and Annett Habenstein for her tremendous support when I needed to solve different administrative matters.

As explained in Sect. 1.2, this thesis was a part of the ENVILAND2 project. In this framework, I express special thanks to Christian Berger, Marcus Bindel and Tanja Riedel for correcting my reports written in German and to Jonas Eberle and Olivier Ballereau for their support in solving programming issues. I also express my sincere gratitude to the entire ENVILAND2 team for the great working atmosphere which made our meetings very pleasant.

While this Ph.D. was an individual work, it benefitted from the insights of external people with expertise in the remote sensing domain as well as in other related fields of interest such as forestry and hydrology. Referring first to the remote sensing domain, I thank Dr. Maurizio Santoro from GAMMA Remote Sensing AG for his support in pre-processing the SAR data and answering several of my questions. Solving the 'radcal' command problem was like a Christmas present to me. I also acknowledge Dr. Thuy Le Toan from the Centre d'Etudes Spatiales de la BIOsphère (CESBIO) and Prof. Eric Pottier from the Institute of Electronics and Telecommunications of Rennes (IETR) for their valuable discussions concerning some of the important issues pertaining to my thesis and for welcoming me at their respective institutes. With respect to the forestry domain, I first acknowledge Dr. Sergej Chmara, head of the 'GIS-Developments' of the TLWJF, for providing me with a large sample of the Thuringian Forest inventory. I also express special thanks to Prof. Anka Nicke, who teaches at the FHE, for all her support. Even though she was very busy, Anka always answered my different questions willingly and helped me to organize the forest campaigns. In addition, the discussions we had during our hikes in the Thuringian Forest helped me to discover a new passion for forest ecosystems and provided me with a great understanding of how to interpret the remote sensing data. Thank you again so much Anka! The execution of the forest campaigns required the involvement of a number of experienced people. In this regard, I also acknowledge Ralf Götze and Jörn Grassmann for performing the measurements. Lastly, I thank M. Andreas Schiene, who introduced me to the Thuringian Forests. Concerning the hydrology modelling, I extend my sincere gratitude to Dr. Sven Kralisch for providing me with the JAMS modelling system and to Daniel Varga for all the time that he invested to solve problems for pre-processing the weather data.

As mentioned previously, a part of this Ph.D. thesis was performed at the Chair of Earth Observation from the ETHZ. This research group, which was founded in November 2009, is led by Prof. Irena Hajnsek, who is also the Head of the 'Pol-InSAR' group at the DLR Hochfrequenztechnik und Radarsysteme (HR) Institute. I express my deepest gratitude to Prof. Hajnsek for welcoming me to her ETH group and for her financial support. Without this tremendous opportunity, I believe that the achievement of my Ph.D. would not have been possible. I also thank her for the great inputs she provided me regarding my work. Although the research group at the ETHZ was formed relatively recently, there are a number of outstanding people who I would like to acknowledge. First of all, I thank Dr. Armando Marino for his great support in the form of daily advice and long discussions. Then, I thank Silvan Leinss for his IDL routines and his pertinent suggestions. In the ETH group, my thanks are extended to Dr. Othmar Frey for his many words of advice, Manuele Piechierri and Vincenzo Di Fiore for the great times spent on the 'vita parcours', Haijing Wang for her unforgettable good humour, Simon Zwieback for the bike tours and Rita Wicki for her support in administrative and financial issues.

During the final stage of this thesis, proofreading was performed by the editing agency EnagoTM. EnagoTM has been very serious and professional in their approach to fine-tuning my document. I thank EnagoTM and the proofreader who worked on this manuscript for the remarkable job done. I also thank my friend Charlie for performing a final check of my work.

Finally, I would like to thank my aunt Maureen for her great support and my family: Louis, Janet, Claire, May and my love Silvia. There are no words to say how grateful I am to you all for your encouragement and sacrifices. You are my inspiration, my joy and my life. Thank you again so much. In writing these last lines, I have a special thought for my dearest sister Claire Jane Ackermann, who left us prematurely on 14 August 2010. I dedicate this Ph.D. thesis to you Claire, because you loved nature as much as I do and because I could never take out time to see the world with you. Rest in peace my dearest sister; you will always be in my heart.

Contents

Acronyms and Symbols

AGB	Above ground biomass
ALOS	Advanced Land Observing Satellite
ALTM	Airbone Laser Terrain Mappers
ANN	Artificial neural network
ASAR	Advanced Synthetic Aperture Radar
ASI	Italian Space Agency
ASL	Above seal level
ATI	Along-track interferometry
BMBF	German Ministry of Education and Science
BSA	Backward scatter alignment
C	Carbon
CESBIO	Centre d'Etudes Spatiales de la BIOsphère
CO_2	Carbon dioxide
CSK	Cosmo-SkyMed
CTI	Cross-Track Interferometry
CV	Coefficient of variation
dB	Decibel
DBA	Distorted born approximation
DBF	Digital beamforming
DEM	Digital elevation model
DHDN	Deutsches Hauptdreiecksnetz
DBH	Diameter at breast-height
DLR	German Space Agency
DSM	Digital surface model
DSW	Datenspeicher Wald
DWD	Deutscher Wetterdienst
ENL	Equivalent number of looks
ENVISAT	ENVIronment SATellite
EO	Earth observation
ERS	European remote sensing satellite
ESA	European Space Agency

ETHZ	Swiss Federal Institute of Technology Zürich
FBD	Fine beam dual mode
FBS	Fine beam single mode
FDD	Freeman–Durden model
FHE	Fachhochschule Erfurt
FSA	Forward scatter alignment
FSU	Friedrich-Schiller University of Jena
GIS	Geographic information system
GLCM	Gray level co-occurrence matrices
GSD	Grid sample dimension
GSV	Growing stock volume
H	Hydrogen
HH	Horizontal sending and Horizontal receiving (like-polarisation)
HIS	Intensity-hue-saturation
HR	DLR Hochfrequenztechnik und Radarsysteme
HS	High resolution Spotlight mode
HV	Horizontal sending and Vertical receiving (cross-polarisation)
IDW	Inverse distance weight
IETR	Institute of Electronics and Telecommunications of Rennes
InSAR	Interferometric SAR
IPCC	Intergovernmental Panel on Climate Change
IWCM	Interferometric water cloud model
JAMS	Jena Adaptable Modelling System
JAXA	Japan Aerospace Exploration Agency
JERS	Japanese Earth resources satellite
k-NN	k-nearest neighbour
LiDAR	Light detection and ranging
LOOCV	Leave-one out cross-validation
MAE	Mean absolute error
MIMICS	Michigan Microwave Canopy Scattering Model
ML	Maximum likelihood
MLI	Multi-look intensity
MoM	Method of moments
NED	National Elevation Dataset
NESZ	Noise-equivalent sigma zero
O	Oxygen
OA	Orientation angle
PALSAR	Phased Array type L-band Synthetic Aperture Radar
PDF	Probability distribution functions
PLR	Polarimetric mode
PolInSAR	Polarimetric SAR interferometry
PolSAR	SAR polarimetry
PRF	Pulse repetition frequency
RADAR	Radio detection and ranging
RAR	Real aperture radar

RE	RapidEye
RGB	Red, Green and Blue
RMSE	Root-mean-square error
ROI	Regions of interest
rRMSE	Relative RMSE
RT	Radiative transfer
RVoG	Random volume over ground model
SAR	Synthetic-aperture radar
SC	ScanSAR
SIBERIA	SAR Imaging for Boreal Ecology and Radar Interferometry Applications
SL	Spotlight
SLC	Single look complex
SM	Stripmap
SNR	Signal-to-noise ratio
SRTM	Shuttle Radar Topography Mission
TDX	TanDEM-X
TLVermGEO	Landesamt für Vermessung und Geoinformation
TLWJF	Thüringer Landesanstalt für Wald, Jagd und Fischerei
TSX	TerraSAR-X
VH	Vertical sending and Horizontal receiving (cross-polarisation)
VHF	Very high frequency
VV	Vertical sending and Vertical receiving (like-polarisation)
WCM	Water cloud model

Forestry

B	Biomass
GSV	Growing stock volume
GSV_{tree}	GSV for a single tree
GSV_{stand}	GSV for a forest stand
ρ	Wood density
h	Tree height
h_{g}	Average forest stand height
h_{100}	Average height of the 100 tallest trees ha^{-1}
DBH	Diameter at breast-height
G	Basal area
f_z	Form factor
N_{trees}	Number of trees per hectare
ABT	Abteilung
ALT	Age
$AREA$	Area
$BAAN$	Dominant species percentage
$BEART$	Number of understory layers
ETF	Eigentumsart

FOA	Forstamt
HAVOR_GES	Growing stock volume (inventory attribute)
HBA_S	Tree species
HÖHE	Height
N	Stand density
REV	Revier
VG	Relative stocking
TFL	Teilfläche
UABT	Unterabteilung
p_{d}	Daily precipitations
p_{h}	Hourly precipitations
t_{d}	Daily air temperature
t_{h}	Hourly air temperature
w_{d}	Daily wind
w_{h}	Hourly wind
h_{d}	Daily air humidity
h_{h}	Hourly air humidity

Electromagnetic Waves

$\vec{E}$	Electric field
$\vec{H}$	Magnetic field
$\vec{D}$	Electric flux density
$\vec{B}$	Magnetic flux density
$\vec{J}$	Electric current density
ρ	Electric charge density
ε	Relative permittivity (dielectric constant) of the medium
μ	Permeability of the medium
$\vec{k}$	Wave vector
t	Time
$\vec{r}$	Position vector
v	Wave speed
ω	Angular frequency
T_{p}	Wave period
f	Carrier frequency
φ_0	Initial phase angle
φ	Phase angle
E_0	Electric field amplitude
I	Intensity
A_0	Amplitude
c	Speed of light
λ	Wavelength

SAR Fundamentals

δ_g	Ground range resolution
δ_a	Azimuth resolution
ρ_r	Pixel spacing in range
ρ_a	Pixel spacing in azimuth
L_r	Number of looks in range
L_a	Number of looks in azimuth
θ	Incidence angle
θ_1	local incidence angle
R	Slant range distance
$\varphi_{\text{scatterer}}$	Phase due to distributed scatterers
φ_{delay}	Phase due to atmospheric and ionospheric effects
σ	Radar cross-section
σ^0	Radar backscatter coefficient (sigma nought projection)
β^0	Beta nought projection
γ^0	Gamma nought projection
W	Bandwidth
PRF	Pulse repetition frequency
τ	Pulse duration
L	Antenna length
P_r	Received power
P_t	Transmitted power
A_r	Effective area
G_t	Antenna gain at the transmitter
G_r	Antenna gain at the receiver
A_σ	Reference area
k	Boltzmann constant
NF	System noise figure
L_{tot}	Total signal losses
T_0	Equivalent noise temperature
V_s	SAR platform flight speed

Statistical Variables

SNR	Signal-to-noise ratio
$NESZ$	Noise-equivalent sigma zero
P_I	Probability density function of the intensity
L	Number of looks
Γ	Gamma distribution
ENL_t	Theoretical ENL
ENL_s	Statistical ENL

N_s	Number of samples
CV	Coefficient of variation

Physics

h_{smooth}	Surface height variation
a	Sphere radius
R_{HH}	Fresnel reflection coefficient in HH polarisation
R_{VV}	Fresnel reflection coefficient in VV polarisation
N_b	Number of scatterers
σ_v	Volume backscatter
α_{ext}	Extinction coefficient
h_V	Forest layer height
σ_{veg}	Vertical reflectivity
α	Terrain slope
A_{flat}	Ground scattering area for flat terrain
A_{slope}	Ground scattering area for inclined terrain
τ_x	Optical path length
K_e	Coefficient of extinction
H_c	Crown height

SAR Interferometry

$\Delta\varphi$	Interferometric difference
$\Delta\varphi_{scatterer}$	Phase difference due to distributed scatterers
$\Delta\varphi_{delay}$	Phase difference due to atmospheric and ionospheric effects
$B_{\|}$	Spatial baseline
$B_{\|}$	Parallel baseline
$B_{\perp}$	Perpendicular baseline
B_n	Normal baseline
ϕ	Interferometric difference (developed)
ΔR	Slant range difference
Δz	Altitude difference
$\Delta\eta$	Coherent displacement
$\Delta\rho$	Path delays due to atmospheric and ionospheric effects
φ_{noise}	Interferometric phase noise contribution
n	Waves cycles factor
g	Complex image
γ	Interferometric complex coherence
$\lvert\gamma\rvert$	Interferometric coherence (complex coherence amplitude)
$\lvert\hat{\gamma}_{ML}\rvert$	Maximum Likelihood estimate coherence
γ_{proc}	Processing decorrelation
γ_{SNR}	Noise decorrelation

γ_{az}	Spectral-shift decorrelation in azimuth
γ_{rg}	Spectral-shift decorrelation in range
γ_{vol}	Volume decorrelation
γ_{temp}	Temporal decorrelation
$\gamma_{\mathrm{spatial}}$	Spatial decorrelation
σ_{γ}	Coherence bias
σ_{ϕ}	Phase bias
k_{z}	Vertical wavenumber
Δf_r	Frequency shift

SAR Polarimetry

E_{0x}	Component x of E_0
E_{0y}	Component y of E_0
δ	Polarisation state
δ_x	Component x of δ
δ_y	Component y of δ
$\vec{e}_x$	Component x of the Euclidean space
$\vec{e}_y$	Component y of the Euclidean space
z	Wave position along z-axis
χ	Ellipticity angle
ψ	Orientation angle
$\vec{E}^{\mathrm{r}}$	Received electric field
$\vec{E}^{\mathrm{t}}$	Transmitted electric field
$[S]$	Scattering matrix (Sinclair matrix)
S_{HH}	HH component of $[S]$
S_{HV}	HV component of $[S]$
S_{VV}	VV component of $[S]$
S_{VH}	VH component of $[S]$
$E_{\mathrm{h}}^{\mathrm{t}}$	Horizontal component of $\vec{E}^{\mathrm{t}}$
$\vec{E}_{\mathrm{v}}^{\mathrm{t}}$	Vertical component of $\vec{E}^{\mathrm{t}}$
σ_{pq}	Backscatter cross-section
$\lvert S_{pq}\rvert$	Generalized scattering matrix
$\vec{k}_{\mathrm{L}}$	Lexicographic vector
$\vec{k}_{\mathrm{P}}$	Pauli vector
$[C]$	Covariance matrix
$[T]$	Coherency matrix
f_{s}	Surface expansion coefficient
f_{d}	Double bounce expansion coefficient
f_{v}	Volume expansion coefficient
f_{c}	Helix expansion coefficient
P_{s}	Surface power

P_{d}	Double bounce power
P_{v}	Volume power
P_{c}	Helix power
P_{t}	Total power
η	Orientation angle
v	Orientation angle ($\pi/2$ corrected)
$[S(\eta)]_{\mathrm{corr}}$	Scattering matrix corrected for azimuth slopes
$[R_S(\eta)]$	Rotation matrix for $[S]$
$[T(\eta)]_{\mathrm{corr}}$	Scattering matrix corrected for azimuth slopes
$[R_T(\eta)]$	Rotation matrix for $[T]$

Polarimetric SAR Interferometry

γ_{u}	Complex polarimetric interferometric coherence
$\vec{w}_1$	Unitary filter vector image 1
$\vec{w}_2$	Unitary filter vector image 2
$[\Omega_{12}]$	Joint image complex coherency matrix
$\tilde{\gamma}_{\mathrm{m}}(\vec{w})$	RVoG model
$\tilde{\gamma}_{\mathrm{vol}}$	Approximated volume decorrelation
$M(\vec{w})$	Ground-to-volume ratio
$m(\vec{w})$	Ground-to-volume amplitude ratio
$m_G(\vec{w})$	Ground scattering
$m_V(\vec{w})$	Volume scattering
h_{V}	Volume height
σ_v	Mean extinction coefficient
h_{a}	Height of ambiguity
σ_v	Mean extinction coefficient

Modelling

$G\hat{S}V$	Estimated *GSV*
N_{train}	Number of training samples
$N_{\mathrm{test}}^{\mathrm{model}}$	Number of testing samples for model selection
$N_{\mathrm{test}}^{\mathrm{accuracy}}$	Number of testing samples for accuracy evaluation
A	Regression parameter A
B	Regression parameter B
C	Regression parameter C
R^2	Coefficient of determination
$RMSE$	Root-mean-square error
k-NN	k-nearest neighbour
w	Weighting factor
d	Spectral distance
s	Spectral distance exponent

$R(h)$	Cross-correlation function
$H_{\text{norm}}(x)$	Normalized histogram difference function
$R_{\text{norm}}(h)$	Normalized cross-correlation function
T_r	Threshold
g_{ij}	Masked imaged
S	Score

Chapter 1
Introduction

This study aims to estimate the growing stock volume (*GSV*) in temperate forest regions using a fusion approach based on synthetic-aperture radar (SAR) satellite imagery. This introductory chapter will present the general elements associated with this work in three different sections. The first section will briefly relate the context of the thesis. In particular, this section will provide the key issues and their potential solutions obtained by using SAR satellite systems. The second section will introduce the background of this work by discussing three topics: (1) briefly characterising the temperate forests and the global carbon cycle, (2) providing a broad overview of specifications and technical characteristics of the satellites addressed in this thesis and (3) briefly describing the objectives and the different entities and institutions involved in the project radar and optical satellite data for ENVIronment and LAND use applications (ENVILAND2) which is linked to this work. In the final section of this chapter, the structure of the thesis is presented.

1.1 Context

Temperate forests hold several vital functions for humans and the biosphere. They are a large terrestrial carbon sink, which is important for maintaining the global carbon cycle. They also play an important role in the economic sustainable development of industrialised and developing countries. For example, they supply fuel, food, fibre, building materials and medicinal plants, which constitute some of the main sources of basic human needs. In addition to being a substantial carbon sink and having a high commercial timber value, temperate forests are also known as complex ecosystems which are essential for wildlife habitats and for the maintenance of biological diversity at local, national and international levels.

For several centuries, owing to deforestation by humans for timber industry and construction activities, forests in the Northern temperate zones have suffered significant losses. Nowadays, tree planting, landscape restoration and natural expansion

N. Ackermann, *Growing Stock Volume Estimation in Temperate Forested Areas Using a Fusion Approach with SAR Satellites Imagery*,
Springer Theses, DOI 10.1007/978-3-319-13138-2_1

of conifers and deciduous trees reduced the net loss of the temperate forest regions worldwide. However, these environments are currently urbanised, so naturally occurring temperate forests remain in a few regions globally.

With reference to global warming, recent reports from the Intergovernmental Panel on Climate Change (IPCC) showed a significant increase in the Earth's surface temperature in the past decades (Solomon et al. 2007). It has been agreed that this general trend is highly correlated to the concentration of carbon dioxide (CO_2) released into the tropospheric atmosphere by human activities. Industry, traffic and exhaustive agriculture activities have been shown to be the primary sources of anthropogenic greenhouse gases. However, forests degradation and deforestations have also been shown to be responsible for the significant increase in CO_2 concentration in the atmosphere.

In the present context, to mitigate global warming, governments have been urged under the Kyoto protocol to significantly reduce their emissions. To do so, a system of emission rights has been defined for the regulation of carbon into the atmosphere. However, an important issue which affects the establishment of this new system is the quantification of the pool of carbon on a national scale. In fact, owing to the rapid forest changes and their high spatial extension, ground-based measurements are often limited to expensive, inaccurate and long-term forest cover updates.

With the development of remote sensing techniques over the last decade, there is a great desire to use SAR satellite systems in order to overcome these limitations. Spaceborne systems have the ability to cover large regions at a reasonable cost. Moreover, SAR microwave signals can interact with plants' structures, allowing the potential retrieval of different biophysical forest parameters, and the SAR sensors are operational at night and in bad weather conditions. Therefore, SAR remote sensing systems have the ability to produce reliable and up-to-date temperate forest biomass maps, which may be used to not only implement carbon-emission trading but also improve the predictions of climatic models and support activities of the forest industry.

1.2 Background

1.2.1 Temperate Forests and the Global Carbon Cycle

Temperate forests cover 13 % of the Earth's land surfaces (FAO 2010) and span five continents in large discontinuous blocs. They extend beyond the tropic of Cancer in the North and the tropic of Capricorn in the South, and are by far one of the most extensive forest groups worldwide. As the latitude increases, the temperate forests gradually disappear, giving way to the boreal forest and the polar regions (Fig. 1.1).

The distribution of temperate forests is far from being even. They principally appear in the Northern Hemisphere, forming a continuous belt across North America, Europe and Asia. The presence of temperate forests is also noticed in some regions

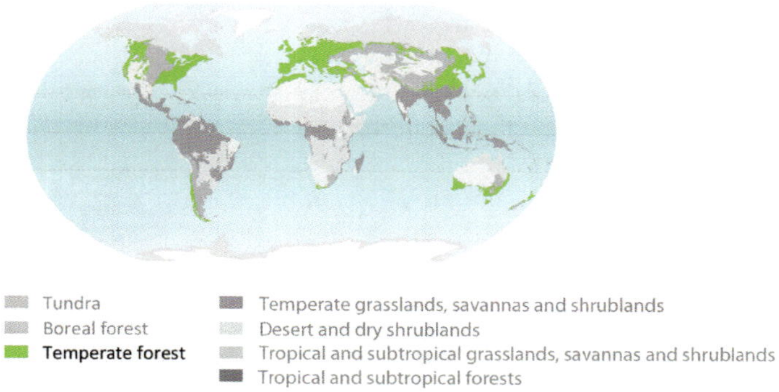

Fig. 1.1 Temperate forests worldwide. Adapted from Pravettoni (2009)

of the Southern Hemisphere. Small pockets particularly appear in South America, Australia and New Zealand. Despite the widespread extension of temperate forests globally and the wide variety of trees in these forests which differ from region to region, temperate biomes are commonly classified in three different families: (1) conifers, which are also widely represented in boreal (Northern) forests; (2) broad-leaved deciduous forests, which are typically from central Europe, North-East America and East Asia and (3) the denominated broad-leaved evergreen forests, which currently appear in only a few regions worldwide such as the North-West regions along the US coast and Canada, the South-East region of Australia and the South of New Zealand. The distribution of the trees over the temperate regions primarily relies on abiotic factors such as the soil conditions, air temperature and frequency of precipitation. These factors are closely linked to the climate types of the regions which are characterised by well-defined seasons in temperate areas.

The actual state of the temperate forests results from several thousands of years of existence, during which time they have provided land and natural resources for different civilisations. Historically, after the last glaciation (15,000 years BP), the temperate forests spread all over the Northern Hemisphere and cover more than 80 % of the Earth's land area in this period (Reich and Frelich 2002). However, with the advent of agriculture, temperate forests have been increasingly felled to clear lands for farming and to subsequently meet the increasing demand of wood for heating, construction materials and paper industry materials. The currently on-going reforestation process was started in the eighteenth century and aimed to reach a peak at the end of the nineteenth and early twentieth centuries (Nabuurs et al. 2007). As a result, the existing temperate forests consist of mostly secondary forests, aged and often monospecies plantations (Nabuurs et al. 2007).

Temperate forests belong to the terrestrial ecosystems which constitute one of the reservoirs of the carbon cycle. Referring to the global carbon cycle, the terrestrial ecosystems represent a total storage of about 2,190 Gt C of carbon, with approximately 610 Gt C stored in vegetation and 1,580 Gt C present in soils and organic matter. About two-thirds of the forests' carbon is thus stored in the soils. With the vegetation photosynthesis process, the terrestrial ecosystems are also characterised by a net land-to-atmosphere flux of about 121 Gt C year^{-1}. This flux is neutralised by the respiration of vegetation and soils, which together release about 60 Gt C year^{-1} into the atmosphere. Although the scientific community appears to agree with the existing carbon reservoirs and flux, the accuracy of the aforementioned quantities remain very approximate. The approximations of these values were essential owing to the complex interactions between the different carbon pools, the large heterogeneity of different biomes and the recent action of humans on land use and land management (Houghton 2007).

The current carbon stocks reported for temperate, boreal and tropical forests are estimated to be 119 ± 6, 272 ± 23 and 471 ± 93 Gt C, respectively (Table 1.1). The amounts of carbon stored in boreal and tropical forests are thus much higher than that in temperate forest ecosystems. This difference is mainly explained by the climatic conditions of these ecosystems. In the case of tropical forests, the humidity enhances the growing conditions, allowing the rapid development of vegetation. In the case of boreal forests, the frozen temperatures reduce the rate of the organic matter decomposition, allowing the high accumulation of matter and carbon in the soils. Therefore, the carbon storage is mostly located in the above ground biomass (*AGB*) of tropical forests and in the soils of boreal forests. In the case of temperate forests, modest carbon levels are located in both soils and plants. With reference to the carbon sinks, the total averaged annual change of the forest ecosystems reaches for the period 2000–2007 about 2.3 Gt C year^{-1} (Table 1.1). With the sequestration of 0.78 Gt C year^{-1}, temperate forests comprise the second most influential carbon sink, after tropical forests. In addition, within the last 2 decades, the carbon sinks in temperate forests have increased, while the uptake from tropical forests has declined. This difference is primarily owing to human activities. As mentioned above, regarding the temperate forests, reforestation and afforestation programmes have been developed over the past few decades and centuries, especially in China and the US. However, for tropical forests, the last few decades have consisted mainly of deforestations, thereby converting forests into sources of carbon.

Table 1.1 Comparisons of carbon stocks and sinks for different forest ecosystems (Pan et al. 2011)

Designation	C Stock (Pg C)	C Sink (Pg C year^{-1})	
	2007	1990–1999	2000–2007
Temperate	119 ± 6	0.67 ± 0.08	0.78 ± 0.09
Boreal	272 ± 23	0.50 ± 0.08	0.50 ± 0.08
Tropical	471 ± 93	1.33 ± 0.35	1.02 ± 0.47
Total	861 ± 66	2.50 ± 0.36	2.30 ± 0.49

1.2.2 SAR Satellite Missions

As part of this thesis, four recent SAR spaceborne systems were investigated, namely ALOS PALSAR, TerraSAR-X (TSX), TanDEM-X (TDX) and Cosmo-SkyMed (CSK). Table 1.2 briefly summarises the characteristics of these satellites. Additional characteristics are provided in JAXA (2012), Eineder et al. (2009), Fritz et al. (2012) and ASI (2007).

The satellite system Advanced Land Observing Satellite (ALOS) was launched on 24th January 2006 by the Japan Aerospace Exploration Agency (JAXA). This spacecraft has three main instruments onboard, which include the Phased Array L-band Synthetic-Aperture Radar (PALSAR) sensor. The PALSAR sensor was designed to follow a sun-synchronous orbit, with a 46-day repeated cycle at an altitude of 691 km. The active system uses an L-band frequency (23.5 cm, 1.27 GHz) and integrates an off-nadir pointing function which permits it to aim between 18° and 55°. The sensor allows four different operation modes: fine beam, scanSAR (SC), Direct downlink and polarimetric (PLR). Data acquired in fine beam (high resolution) can be measured over a 55 km × 65 km frame and delivered in either single polarisation (FBS, HH or VV) or dual polarisations (FBD, HH/HV or VV/VH). The SC mode offers a larger swath width (250 km), but at the cost of lowering the spatial resolution (100 m). For reduced spatial resolution capabilities, a direct downlink mode allows the direct transmission of data to the X-band ground stations. Finally, the PLR mode offers a complete polarisation scheme (HH/HV/VV/VH) for 30 m range resolution on a 30 km swath. The bandwidth of the sensor varies with the acquisition modes. In general, the bandwidth is 28 MHz in FBS mode and 14 MHz in the FBD, PLR and SC modes. On Thursday, April 21 2011, at 22:30 GMT, owing to a power generation anomaly, the three observation instruments onboard ALOS were automatically shut down to a power-saving mode. On the next day, ALOS 'Daichi' was declared to be lost.

Table 1.2 Principal characteristics of ALOS PALSAR, TSX, TDX and CSK instruments

Designation	ALOS PALSAR	TerraSAR-X	TanDEM-X	Cosmo-SkyMed
Space agency	JAXA	DLR	DLR	ASI
Number of instruments	1	1	1	4
Launch date	24th January 2006	15th June 2007	21st June 2010	8th June 2007
Band name	L-band	X-band	X-band	X-band
Frequency (GHz)	1.27	9.65	9.65	9.6
Bandwidth (MHz)	14–28	150–300	150–300	400
Orbit geometry	Sun-synchronous	Sun-synchronous	Sun-synchronous	Sun-synchronous
Orbit altitude (km)	691.6	514.0	514.0	619.6
Orbit inclination (°)	98.16	97.44	97.44	97.86
Orbit period (min)	98.7	94.92	94.92	97.2
Revisit time (day)	46	11	11	1

TSX belongs to the new generation of high-resolution satellites which are capable of measuring spatial resolutions up to 1 m. Initiated by the close cooperation between the German Ministry of Education and Science (BMBF), the German Aerospace Centre (DLR) and the Astrium GmbH, the instrument was successfully launched on June 15, 2007 and was operational seven months later. This satellite was conceived for scientific and commercial utilisation, and the SAR sensor consists of an active phased array X-band system operating at 9.65 GHz (3.1 cm) with nominal and maximal range bandwidths at 150 and 300 MHz, respectively. The spacecraft follows a sun-synchronous orbit with a nominal orbit height at the equator of 514 km. Its orbital velocity approaches 7.6 km s^{-1}, resulting in about 15 orbits per day and a revisit time of 11 days. TSX was designed for four different imaging modes: High-resolution Spotlight (HS), SpotLight (SL), StripMap (SM) and SC. In summary, the two spotlight modes (HS and SL) are very similar and provide the highest geometrical resolution (up to 1 m resolution). The difference between these two modes is the scene coverage and the spatial resolution. In SL, the azimuth scene coverage increases (azimuth range: 10 km × 10 km) and the geometric azimuth resolution reduces (2 m); in HS mode, a high spatial resolution (1 m) is possible, but the observed area diminishes (azimuth range: 5 km × 10 km). The SM and SC modes are designed for imaging large areas. The SM mode acquires data by scanning the ground with a continuous sequence of pulses, allowing an acquisition length ranging from 30 km up to 1,650 km. The SC mode is similar to the SM mode, but with a much larger swath (100 km).

Since 21st June 2010, TSX has been in tandem formation with a second X-band satellite launched by the DLR named TDX. TDX is almost the exact replica of TSX and was built for interferometric applications, with the primary goal being to derive a global high-resolution digital elevation model (DEM) within a period of three years. In addition to this primary objective, the twin formation of TSX and TDX provides experimental interferometric modes such as the bistatic, alternating bistatic and monostatic modes. These experimental modes may be used with different techniques allowed by TSX/TDX, such as Along-Track Interferometry (ATI), Cross-Track Interferometry (CTI), polarimetric SAR interferometry (PollnSAR) and digital beamforming (DBF) (Huber and Krieger 2009).

The CSK mission introduced by the Italian Ministry of Research and Ministry of Defence, which was conducted by the Italian Space Agency (ASI), was designed to provide support for military and civilian activities. CSK consists of a constellation of four satellites launched between 8th June 2007 and 5th November 2010, and orbits a sun-synchronous polar orbit with a nominal altitude of 619 km. The full constellation has been operational since the launch of the fourth satellite. Each satellite is composed of an X-band radar sensor with a centre frequency of 9.6 GHz and a maximum bandwidth of 400 MHz. The sensors provide three acquisition modes, namely SL, SM Himage, SM pingpong and SC wide/huge modes. The SM modes consist of a trade-off between SL and SC modes, which allow either high spatial resolution or large swath coverage. Therefore, the SM Himage mode can provide data in single polarisation with a frame of 40 km × 40 km and with both ground range and azimuth spatial resolution of 3 m each. One of the main characteristics of CSK lies in its interferometric capabilities. The ASI proposes two interferometric modes, namely the Tandem-like and the Tandem interferometry. The Tandem-like mode provides multitemporal acquisitions, while the Tandem interferometry mode allows specific interferometric configurations for producing digital elevation models. Considering there is only one CSK satellite, the revisit time in a routine configuration (no urgencies) is approximately 16 days. Taking into account the tandem-like mode of three satellites, and by phasing one of the satellite's orbital position at 67.5° with one of the two other satellites, inteferometric data with a 1 day repeat-pass can be provided. With the full constellation and the same phasing orbital configuration, two consecutive acquisitions with a revisit time of 1 day can be ordered.

1.2.3 ENVILAND2

The work performed in this thesis has been part of the ENVILAND2 project. ENVILAND2 was the result of collaboration between several institutions, which aimed at developing robust and (semi-) automated methods for the generation of landcover products such as land use classification, change detection, urban mapping and biomass mapping (see Ackermann et al. 2012a, b). To achieve this objective, one focal point was the synergetic use of radar and multispectral remote sensing data (Fig. 1.2).

Among the different partners of ENVILAND2, the Jena-Optronik GmbH company managed and generally coordinated the project, while the Universities of Bonn (Zentrum für Fernerkundung Landoberfläche der an der Universität Bonn—ZBS), Ilmenau (Zentrum für Bild-und Signalverarbeitung eV Ilmenau—ZFL) and Jena (Friedrich-Schiller-Universität Jena—FSU) developed the approaches and algorithms. Furthermore, part of the funding of ENVILAND2 was contributed by the DLR.

ENVILAND2 constituted a continuation of ENVILAND1, which demonstrated the feasibility of producing landcover products based on the combination of radar and optical remote sensing information. ENVILAND1 began in October 2004 and ended in December 2007, while ENVILAND2 was in operation between October 2008 and September 2011 with a special extension to March 2012.

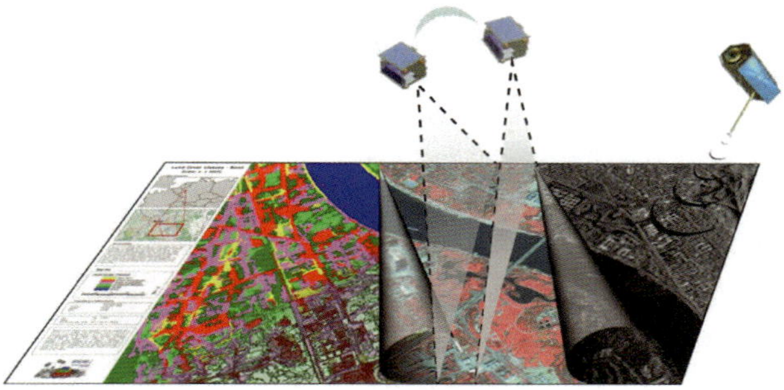

Fig. 1.2 ENVILAND2 logo (Ackermann et al. 2012b)

Because the DLR was a co-founder of ENVILAND2, the project involved the primary use of recent German remote sensing systems such as TSX and RapidEye (RE). In addition to these satellites, other types of instruments such as ALOS PALSAR, ENVISAT ASAR, SPOT5 and Kompsat2 were considered to extend the research field of the synergy between radar and optical systems. The final work of ENVILAND2 involved implementing new developed or adapted algorithms into software which aimed to provide, within the framework of the future GMES European programme, some important tools useful for commissioning and exploiting the upcoming sentinel satellites.

1.3 Structure of the Thesis

This thesis is composed of six principal chapters. The first chapter introduced the general context and background of the thesis. Based on these introductory elements, the second chapter presents a detailed review of the literature to highlight the important findings using SAR remote sensing for temperate forests. This review will serve as a foundation for discussing the open issues and scopes of the thesis. Chapter 3 describes the theoretical background related to SAR and forestry. This chapter provides some definitions and enables the understanding of the different techniques and concepts treated in this work. To achieve the objectives defined in Chap. 2, a specific test site was selected and earth observation (EO) data were acquired. Chapter 4 provides a description of the test site and also includes the material and methods used in this work. The focal point of this thesis is Chap. 5, which presents and discusses the results of this work. Finally, the last chapter concludes the thesis by summarising the main findings and proposing some perspectives for future studies.

References

Ackermann N, Becker F, Berger C, Bindel M, Eberle J, Elbertzhagen I, Franke K-H, Hecheltjen A, Koch T, Kubertschak T, Menz G, Riedel T, Schmullius C, Schwarz M, Thonfeld F, Weise K, Wolf B (2012a) ENVILAND 2 - Von multisensoralen Satellitenbildern zu operationellen Produkten. In: Borg E, Daedelow H, Johnson R (Hrsg) Proceedings of the 4th RESA workshop, Neustrelitz, Germany: RapidEye Science Archive (RESA): Vom Algorithmus zum Produkten, pp 123–149

Ackermann N, Becker F, Berger C, Bindel M, Eberle J, Elbertzhagen I, Franke K-H, Hecheltjen A, Koch T, Kubertschak T, Menz G, Riedel T, Schmullius C, Schwarz M, Thonfeld F, Weise K, Wolf B (2012b) ENVILAND2: synergistic usage of optical and SAR data

ASI (2007) Cosmo-SkyMed SAR products handbook. Technical report, Italian Space Agency (ASI)

Eineder M, Fritz T, Mittermayer J, Roth A, Börner E, Breit H (2009) TerraSAR-X ground segment basic product specification document. Technical report, DLR, Munich, Germany

Fritz T, Duque S, Bräutigram B, Bello LB (2012) TanDEM-X experimental product description. Technical report, DLR, Munich, Germany

Global forest resources assessment (2010) main report. Technical report, FAO, Rome, Italy 2010

Houghton RA (2007) Balancing the global carbon budget. Annu Rev Earth Planet Sci 35(1):313–347

Huber S, Krieger G (2009) The tandem-x mission: overview and interferometric performance. In: Radar conference, 2009. EuRAD 2009. European, pp 25–28

JAXA (2012) PALSAR User's Guide. Technical report, Japan Space Systems

Nabuurs GJ, Maser O, Andrasko K (2007) In climate change 2007: mitigation. Contribution of Working Group III to the fourth assessment report of the intergovernmental panel on climate change. Technical report, IPCC, Cambridge, UK, New York, NY

Pan Y, Birdsey RA, Fang J, Houghton R, Kauppi PE, Kurz WA, Phillips OL, Shvidenko A, Lewis SL, Canadell JG, Ciais P, Jackson RB, Pacala SW, McGuire AD, Piao S, Rautiainen A, Sitch S, Hayes D (2011) A large and persistent carbon sink in the world's forests. Science (New York, N.Y.) 333(6045):93–988

Pravettoni R (2009) Temperate forests. http://www.grida.no/graphicslib/detail/temperate-forests_1753

Reich PB, Frelich L (2002) Temperate deciduous forests. In: Munn T (ed) The earth system: biological and ecological dimensions of global environmental change, vol 2. Wiley, Chechester, pp 565–569

Solomon S, Qin D, Manning M, Chen Z, Marquis M, Averyt KB, Tignor M, Miller HL (2007) IPCC, 2007: climate change 2007: the physical science basis. Contribution of Working Group I to the fourth assessment report of the intergovernmental panel on climate change. Technical report, IPCC, Cambrigde, UK, New York, USA

Chapter 2
Literature Review

Over the past 10 years, the number of research activities in the field of radar remote sensing has increased considerably. To identify the main findings regarding SAR and forestry, a comprehensive review of the literature was conducted focusing on different SAR techniques and fusion of SAR information. Then, on the basis of the literature findings, the current issues in this regard were identified and the scope of this thesis was defined.

This chapter presents the literature review for this thesis and its objectives. The first section focuses on the remote sensing of temperate forests using SAR technology. This section particularly reviews the SAR backscatter intensity and SAR interferometry (InSAR), SAR polarimetry (PolSAR), polarimetric SAR interferometry (PolInSAR) and SAR tomography techniques. The second section deals with the methods applied for the fusion of SAR information from temperate forests. In this respect, the various approaches to combining SAR information are reviewed. The main objective of the present thesis is the estimation of *GSV* in temperate forests using SAR satellite imagery. The literature review focuses on temperate forests. However, some reports on boreal or tropical forests are also discussed to a lesser extent because these studies were also relevant to the assessment of SAR sensitivity in temperate forests. The reports on landcover or forest mapping are not always mentioned, and the fusion section reviews studies which focus on SAR information. After describing the literature findings, the last section of this chapter discusses the current issues identified from this literature review and introduces the objectives of this thesis.

2.1 SAR Remote Sensing of Forests

2.1.1 SAR Intensity

Radar remote sensing of forest biomass has been a domain of research since the early days of civilian imaging radar in the 1960s and the launch of the first SAR

N. Ackermann, *Growing Stock Volume Estimation in Temperate Forested Areas Using a Fusion Approach with SAR Satellites Imagery*,
Springer Theses, DOI 10.1007/978-3-319-13138-2_2

satellite Seasat in the 1970s (Leckie and Ranson 1998). The research activities in radar systems for forestry applications are attributed to the fact that radar microwaves are unaffected by weather conditions (Ulaby 1981; Imhoff et al. 1986; van der Sanden 1997) and solar illumination and are sensitive to the structure of the canopies (Imhoff et al. 1986). These capabilities together with the widespread use of and continuous surveys provided by spaceborne platforms make radar systems one of the primary domains of research in remote sensing for the estimation of forest biomass over a wide range of regions (Bergen et al. 1997).

2.1.1.1 SAR Parameters

Initially, investigations of radar imagery focused on gaining an understanding of the physical interactions between radar microwaves and forest canopies (Henderson and Lewis 1998). In particular, the microwave backscatter at different radar frequencies was considered. The examinations of conifer or broad-leaved forests mainly led to the same conclusions, i.e., there was a significant correlation between forest biomass and backscatter ($R^2 = 0.80 - 0.90$) at low radar frequencies (L-band, P-band) and poor correlation ($R^2 = 0.1 - 0.30$) with high-frequency systems (X-band, C-band) (Le Toan et al. 1992; Beaudoin et al. 1994; Rauste et al. 1994; Kasischke et al. 1995). Some researchers such as Imhoff (1995b) or Le Toan et al. (1992) analysed the relevant scattering mechanisms occurring at different frequencies and pointed out the limitations of different active systems. They showed that shorter wavelengths such as X-band and C-band are primarily scattered in forests by small components, typically foliage, twigs and branches of upper canopies. On the other hand, they indicated that the longer wavelengths (L-band and P-band) penetrate the canopy and are scattered by large elements such as branches and trunks, which constitute a major part of biomass. The analysis of P-band data presented a good correlation ($R^2 = 0.90$) between radar backscatter intensity and the main forest properties, including trunk biomass, age, basal area, diameter at breast height (*DBH*) and height (Le Toan et al. 1992; Mougin et al. 1999; Sandberg et al. 2011). Despite the improved correlation between low radar frequency and forest biomass, several studies showed that low biomass levels are also well correlated to short radar microwaves. This observation was highlighted in Wang et al. (1998), da Yanasse et al. (1997) among other studies, both of which concentrated on the correlation of C-band data to forest biomass and demonstrated the potential of this frequency band in the retrieval of biomass in early successional stages. With the penetration of short wavelengths being limited to the upper part of forest canopies, Hoekman and Varekamp (2001) also showed the potential of these wavelengths in the estimation of other forest biophysical parameters such as the forest canopy cover. In a tropical forest, it was possible to use the C-band to identify deforested areas as well as primary and secondary forests. Also, with respect to canopy cover, Natale et al. (2012) recently proposed a prospective study for future S-band missions, which involve the comparison of S-band and X-band data over a temperate forest. The researchers highlighted the higher performance of the S-band compared with the X-band in monitoring the canopy cover. Because longer wavelengths provide a better correlation between forest biomass and radar microwaves, some researchers have examined very high frequency (VHF) radar systems. For example, experiments employing the CARABAS VHF system have shown promising results (Fransson et al. 2000; Israelsson et al. 1997; Walter 1997). However, according to Goriachkin and Klovsky (1999), the exploitation of longer wavelengths such as P-band and VHF from space introduces new major technical difficulties (i.e. interference with other microwave sources, absorption of signals in the ionosphere), which limit the capabilities of a potential P-band or VHF future spaceborne sensor.

While investigating radar frequencies, researchers also noted the potential of using different microwave polarisations. For example, Le Toan et al. (1992), Dobson et al. (1992), Sader and Wu (1987) and Hussin et al. (1991) found that cross-polarisations (e.g. HV or VH polarisation) are more sensitive to forest canopies than like-polarisations (e.g. HH or VV polarisation). To support these observations, Beaudoin et al. (1994) developed a theoretical model for P-band backscatter. This study showed that cross-polarisation ($R^2 = 0.90$) is better correlated to forest biomass than like-polarisation ($R^2 = 0.75 - 0.85$). Moreover, the study demonstrated that HV backscatter is weakly sensitive to terrain conditions (soil moisture, roughness or local slopes) and HH polarisation is mostly returned as multiple trunk–ground interactions. Other researchers confirmed high direct surface scattering in HH or VV polarisation, especially under wet weather conditions (Ranson and Sun 2000; Balzter et al. 2002; Santoro et al. 2006). By comparing like-polarisations, many researchers have noticed dissimilarities between HH and VV. It was generally observed that HH polarisation presents higher backscatter than VV polarisation. This difference was related to several phenomena such as dihedrals (Dobson et al. 1992; Watanabe et al. 2006) or canopy absorption (van Zyl 1993; Santoro et al. 2009).

Because SAR remote sensing systems operate in the side-looking direction, scientists investigated the influence of varying look angles on backscatter intensity. With the help of simulations, researchers showed that backscatter generally decreases with increasing incidence angle (Sun et al. 1991; Westman and Paris 1987; Engheta and Elachi 1982). This trend could be verified using airborne and spaceborne SAR experiments (Magagi and Bernier 2002; Moghaddam and Saatchi 1993; Alasalmi et al. 1998; Rauste 1990; Ardila et al. 2010), which particularly highlight the importance of forest structure (i.e. density, canopy height, and branching structure) (Beaudoin et al. 1994; Imhoff et al. 2000) and ground layer characteristics (i.e. surface roughness and moisture) (Ardila et al. 2010; Westman and Paris 1987). In forested areas, it was especially noted that ground interactions were higher at steep incidence ($\theta = 20° - 30°$) angles, while canopy interactions were greater at shallow incidence angles ($\theta = 40° - 50°$) (Magagi and Bernier 2002; Rauste 1990; Sun et al. 1991; Westman and Paris 1987). On the basis of the considered frequency and polarisation, this observation suggested an increase in trunk–ground dihedrals with steep incidence angles (Rauste 1990; Sun et al. 1991; Westman and Paris 1987; Moghaddam and Saatchi 1993) and better sensitivity of SAR to biomass with the use of large incidence angles (Beaudoin et al. 1994).

From an analysis of the relationship between SAR backscatter intensity and biomass, it was found that the intensity generally increased with increasing biomass

Table 2.1 Saturation levels reported in literature for SAR backscatter intensity

Biome	Frequency	Saturation limit ($m^3 h^{-1}$)	Researchers(s)
Temperate	C	10–160	Imhoff (1995b), Rauste et al. (1994), Ranson and Guoqing (1994) and Wang et al. (1994)
	L	65–160	Dobson et al. (1992), He et al. (2012), Imhoff (1995b) and Rauste et al. (1994)
	P	>160–700	Dobson et al. (1992), Imhoff (1995b), Rauste et al. (1994) and Ranson and Guoqing (1994)
	VHF	>900	Melon et al. (2001)
Boreal	C	64	Fransson and Israelsson (1999)
	L	143	Fransson and Israelsson (1999)
	VHF	>550–625	Fransson et al. (2000), Imhoff (1995a) and Smith and Ulander (2000)
Tropical	C	30	Imhoff (1995a)
	L	65–160	de Araujo et al. (1999), Imhoff (1995a), Luckman (1997) and Rignot et al. (1997)
	P	>160	Imhoff (1995a)

When necessary, the density factor of 1.6 was considered to convert t ha^{-1} to m^3 ha^{-1}

until it reached a saturation level (Le Toan et al. 1992; Kasischke et al. 1995; Imhoff 1995b). The saturation of the SAR signal constitutes an important issue in the field of forest biomass estimation using radar data (Kasischke et al. 1997; Imhoff 1995a). Many previous studies attempted to define the saturation level and understand the different parameters contributing to the same (see Table 2.1). The values were reported either in t ha^{-1} for dry biomass or m^3 ha^{-1} for stem volume (see Chap. 3 for the unit conversion).

As a rule, it was found that the saturation level increased at lower frequencies (e.g. L- and P-band) (Fransson and Israelsson 1999; Imhoff 1995b). Experimental analyses using the airborne NASA/JPL AIRSAR system over temperate and tropical forests indicated a saturation at 20, 40 and 100 t ha^{-1} for C-, L- and P-bands, respectively (Imhoff 1995b). Other studies agreed on the limit of 10–160 m^3 ha^{-1} in C-band, 65–60 m^3 ha^{-1} in L-band and 160–700 m^3 ha^{-1} at P-band frequencies (see Table 2.1). In P-band, cases with no saturation were also reported (Rauste et al. 1994; Ranson and Guoqing 1994). The relative rapid saturation obtained at high frequency (small wavelength) was attributed to the large amount of foliage, twigs and small branches at the top of canopies, which considerably attenuated the signal. On the other hand, the saturation observed in L-band was primarily related to the smaller number of scatterers (large branches and trunk), which attenuated the EM microwaves to a lesser extent when compared with high-frequency systems (Le Toan et al. 1992). Consequently, by comparing the frequencies, it was finally noted that VHF sensors (Fransson et al. 2000; Melon et al. 2001; Smith and Ulander 2000) were the SAR systems which were affected to a lesser extent by the saturation effect.

In addition to the SAR frequency, the saturation point was investigated with various polarisations and incidence angles. For example, Watanabe et al. (2006) found that cross-polarised backscatter delays the saturation point to higher biomass level compared to like-polarisations, and Lu (2006) observed that a shallow incidence angle increased the saturation level, especially in low-frequency systems. It was shown that radar sensor configurations determined the biomass saturation level. However, as the forest structure influences the scattering mechanisms occurring in forests, numerous studies have also shown that the saturation level is site-dependent (Lu 2006).

The analysis of forest texture has been extensively investigated using optical remote sensing systems (Woodcock and Strahler 1987; Sarker and Nichol 2011; Kayitakire et al. 2006). With the development of high-resolution SAR sensors over the past decade, estimation of forest parameters using textures has attracted considerable research attention (Fukuda 2008; Wang et al. 2006). The most popular textural information refers to the gray level co-occurrence matrix (GLCM) defined by Haralick during the 1970s (Haralick et al. 1973). The publications dealing with the GLCM and forests mainly focused on tropical forests (Luckman et al. 1994, 1997; Oliver 2000), and, in particular, two studies demonstrated the potential of the Haralick parameters in improving the estimation of biomass in the tropics (Kuplich et al. 2005; Sarker et al. 2012). GLCM textures were also distinguished in boreal and temperate biomes for land classifications (Ulaby et al. 1986; Kurosu et al. 1999; Kurvonen and Hallikainen 1999) and retrieval of forest parameters (Weishampel et al. 1994; Champion et al. 2008) such as biomass (Champion et al. 2011; Kurvonen and Hallikainen 1999). One important issue in the examination of texture is the different scales of forest spatial patterns (Barros Filho and Sobreira 2008). To deal with the multiscalar textural property of a forest, researchers considered the concept of lacunarity ('gapiness' in Latin). This approach was initially described by Mandelbrot in 1983 and was subsequently further developed by Allain and Cloitre using the gliding box algorithm (Allain and Cloitre 1991; Plotnick et al. 1996). Lacunarity was investigated by a few studies which estimated forest parameters. However, only one of them was directly related to SAR imagery and temperate forests (Sun and Ranson 1998). The remaining studies mostly dealt with optical sensors and tropical forests (Malhi and Román-Cuesta 2008; Peralta and Mather 2000; Weishampel et al. 2001) or SAR sensors and landscape analysis (Plotnick et al. 1993; Hoechstetter et al. 2011; Henebry and Kux 1995; Kux and Henebry 1994; Mcintyre and Wiens 2000). In addition to the GLCM and lacunarity texture approaches, other methods were developed for investigating spatial patterns. A non-exhaustive list can be found in Kandaswamy et al. (2005) or Sarker and Nichol (2011).

2.1.1.2 Forest Properties

Realizing forest biomass estimation from SAR remote sensing systems requires thorough knowledge of radar principles and a very deep understanding of the forests and their environment. In this framework, many researchers sought to comprehend forest structural properties and their relationship with SAR backscatter (Beaudoin et al. 1994; Imhoff 1995a; Mougin et al. 1999; Kasischke et al. 1997; Ferrazzoli and Guerriero 1995). Considering forest structure, researchers often distinguished between three different classes, namely horizontal, vertical and branching structures (Spies 1998; Ferris and Humphrey 1999). The horizontal structure mostly referred to the basal area or stem density, the vertical structure concerned forest height or understory and the branching structure was denoted by the size shape or orientation of the branches and leaves at the canopy level (Kasischke and Christensen 1990; Wang et al. 1993; Mcdonald and Ulaby 1993).

First, with reference to the branching structure, to understand the complex relationship between radar backscatter and structural parameters of canopies, theoretical and semi-empirical models were developed. For example, Yueh et al. (1992) introduced a branching model which describes the backscatter of a soybean plant with its internal structure and the resulting clustering effects. This model was further used for modelling forest backscatter by combining it with the radiative transfer (RT) theory (Beaudoin et al. 1994; Hsu et al. 1994). In another example, Ferrazzoli and Guerriero (1995) focused on a physical model which relates geometrical tree properties such as the size of branches and their orientation with the backscattering coefficient. In this case, (1) deciduous leaves were represented by discs, (2) coniferous needles by thin cylinders and (3) branches and trunks by large cylinders. In general, the modelling of canopy structures confirmed the assumptions that tree morphological parameters have a significant impact on radar response (Mougin et al. 1993; Neumann et al. 2012).

As for the horizontal structure, researchers reported a significant contribution of the basal area and forest density parameters to the backscatter intensity (Brolly and Woodhouse 2012; Woodhouse 2006). For example, Dobson et al. (1995) observed that for equal biomass levels, there were clear variations between the backscatter responses of large, sparse forests and dense, young forests. More recently, Brolly and Woodhouse (2012) represented the forest as a collection of cylinders, which was therefore named ‘matchsticks’. Using the matchstick modelling approach, the researchers demonstrated that backscatter is not only sensitive to the quantity of biomass but also to the density and distribution of trees. On the basis of these findings, Woodhouse (2006) showed that backscatter intensity should increase until it reached an apparent saturation point, and then decrease with increasing quantities of biomass after the saturation level. These results, which were based on macroecology considerations, should provide further explanations regarding the unusual observations conducted by Rauste et al. (1994) in a conifer forest in Germany or by Ranson and Guoqing (1994) in a Northern mixed forest in Maine. They also highlight the differences between natural and managed forests and the importance of considering thinning and other forest-management activities in the implementation of a biomass-estimation methodology (Champion et al. 1998; Kuplich et al. 2000; Le Toan et al. 1992).

Among the vertical structural parameters of a temperate forest, forest height has been the subject of numerous studies. The height attribute was investigated using mainly InSAR or more advanced SAR techniques (discussed later in this section), although it potentially affects forest backscatter. At low frequencies, the vegetation beneath the forest canopy can be a source of considerable variability. However, this contribution has been discussed by few studies (Chauhan et al. 1991; Pulliainen et al. 1994; Silva and Dias 1996; Wang et al. 1998). For example, Wang et al. (1998) examined the backscatter variations induced by changes in forest floor properties. These studies reported the non-negligible contribution of the forest understory at the L-band frequency irrespective of the use of HH or VV polarisation. They also noted a decline in forest floor backscatter with increasing stem volume. At C-band frequencies, the same researchers showed that the forest floor contribution was significant for only steep incidence angles. A few studies have discussed the influence of forest understories on scattering mechanisms at the ground level. In particular for L-band data, it has been assumed that trunk–ground and crown–ground double bounces are significantly reduced owing to the attenuation of the canopy and roughness of the forest understory (Pulliainen et al. 1999; Karam et al. 1992; Santoro et al. 2006).

2.1.1.3 Forest Environmental Conditions

The environmental conditions in a forest affect radar backscatter and determine the growth properties of trees. In this context, several studies were conducted to highlight the effects of changing forest weather conditions on SAR backscatter intensity (Way et al. 1990; Rignot et al. 1994; Pulliainen et al. 1996). Bergen et al. (1997) focused on the influence of rainy precipitations on forest backscatter intensity. They showed that the amplitude of the signal generally increased with precipitation, especially at C-band frequencies, when compared with the case at L-band frequencies. Other researchers concentrated on soil humidity in forests. Wang et al. (1994) reported that at C-band frequencies, there was an increase in backscatter with increasing soil humidity. Simulations conducted to interpret these observations also showed that the increase in soil humidity for a biomass quantity extending between $40\,\mathrm{t\,ha^{-1}}$ and $100\,\mathrm{t\,ha^{-1}}$ was mainly presented as surface scattering. However, for a biomass quantity of greater than $130\,\mathrm{t\,ha^{-1}}$, the effect of soil humidity was found to be insignificant, and volume scattering from canopies was the main contribution to the total backscatter. These variations in soil and vegetation humidity generally affected the backscatter dynamic range and reduced its sensitivity to the biomass quantity (Salas et al. 2002; Cartus et al. 2012). By considering winter radar measurements, some researchers examined the impact of snow on forest SAR backscatter. Dry snow cover was found to have an insignificant effect on C-band and L-band backscatter (Alasalmi et al. 1998). However, with the melting of snow, the variations in the dielectric properties of the snow led to an increase in the backscatter intensity (Santoro 2003; Arslan et al. 2006) and variations in the SAR signal at both C-band (Koskinen et al. 2010; Santoro et al. 2011) and L-band (Santoro et al. 2006) frequencies. The consideration of snow cover over forested areas at X-band frequencies is restricted to only a few studies. The first experiment was conducted using the HUTSCAT scatterometer over a boreal forest (Hallikainen et al. 1997). This study showed that ground backscatter contribution was dominant for dry snow, while tree canopy backscatter was the most significant factor under wet-snow or snow-free conditions.

The environmental conditions in both temperate and boreal forests change considerably within seasons. During summer, water is present in the form of liquid; however, during winter, the temperatures at high latitudes decrease leading to frozen conditions, which produce observations different from the aforementioned ones. Thiel et al. (2009) recently reported contrasting results between winter and summer intensities at L-band frequencies. They noted a decrease in backscatter during winter and showed that the environmental conditions were much more stable during winter (Ni et al. 2011) when compared with those during summer. The decrease in backscatter during winter was found to be related to the dielectric properties of the forest floor and forest canopy (Rignot et al. 1994; Way et al. 1990). Santoro et al. (2011) investigated the sensitivity of stem volume to radar backscatter at C-band frequencies. They pointed out that under frozen conditions, the attenuation of microwaves in canopies decreased and the sensitivity of SAR backscatter to biomass quantity improved because the C-band microwaves were mostly sensitive to the trees' main components (i.e. trunk and branches in the canopy). The results reported by Santoro et al. (2006) showed that unfrozen conditions were most suitable for estimating biomass quantity at L-band frequencies. This was explained by the greater sensitivity of L-band microwaves to forest features (e.g. branches and stems) with an unfrozen canopy.

In addition, while examining the temperature conditions in forested areas, researchers often refer to a diurnal freezing/thaw cycle. According to Kimball et al. (2004), this cycle appears in autumn or during spring, namely when the air temperatures alternate between frozen and unfrozen conditions during night and day time, respectively. Under frozen conditions, the dielectric constant of trees decreases; also, sapflow ceases and the water content of the trees reduces. However, when the air temperature during day time again rises above the freezing point, the dielectric constant increases, and with the reactivation of sapflow, the water content of trees also increases (Way et al. 1990; Kidd and Scipal 2003). As radar microwaves are sensitive to water, the diurnally changing water content and dielectric constant of trees influence the radar backscatter intensity signal and may have an impact on the estimation of biomass quantity. The effects of diurnal variations from the trunk and canopy water status on radar backscatter have been examined by several studies (Way et al. 1990; Kidd and Scipal 2003; Bartsch et al. 2006; Kimball et al. 2004; Schmullius 1997; Rignot and Way 1991). Owing to the daily repetition of the freeze/thaw effect, most of the researchers considered scatterometer evaluations instead of SAR measurements. Scatterometers can provide high-resolution temporal data when compared with those provided by SAR sensors. The results of these studies showed a consistent relationship between backscatter intensity and freeze/thaw diurnal cycles. However, it is noted that none of these studies presented the impact of the freeze-thaw cycles with varying water content statuses on the estimation of biomass quantity.

Topography accounts for significant variations in forest radar backscatter (Beaudoin et al. 1995; Holecz et al. 1995; Tanase et al. 2011; Park et al. 2012). These variations are due to different phenomena such as modification of the ground illuminated area, change in the incidence angle and modification of dominant scattering mechanisms. To deal with these topographic issues, studies have proposed various approaches, which range from simple cosine corrections (Hinse et al. 1988; Bayer et al. 1991; Leclerc et al. 2001; Rees and Steel 2001; Soja et al. 2010; Zhou et al. 2011) or images ratio to more rigorous methods (Ranson and Saatchi 1995; Ranson et al. 2001; Wever and Bodechtel 1998) based on the digital elevation model (DEM) (Small et al. 1997; Loew and Mauser 2007). The ground illumination area has the greatest impact on the backscatter intensity (Holecz et al. 1995). To correct this effect, Ulander (1996) suggested a correction based on the projection angle ψ evaluated from the DEM. Small et al. (1997) introduced the possibility of integrating the SAR scattering areas under DEM slope 'facets'. He further developed this approach in 2004 (Small and Meier 2004) and 2009 (Small et al. 2009) and summarised it recently in 2011 (Small 2011). This approach was tested and compared with the projection-angle approach by Frey et al. (2013) and was reported to be very effective, particularly in layover areas. The relationship between topographic variations and forest canopies remain under investigation. The potential effects of an undulated terrain on the nature of scattering mechanisms were reported in the 1990s (van Zyl 1993; Israelsson and Askne 1993; Amar et al. 1993). For instance, van Zyl (1993) showed that ground–trunk double reflections are a major contribution at P-band frequencies. However, in the presence of a slope gradient, the nature of the dominant scattering changed because of the new angle formed between the vertical extended trunks and the surface local slope. More recently, Castel et al. (2001b) suggested a semi-empirical correction for L-band or C-band frequencies, which takes the changing path length of the microwaves through the canopy into account. The results showed that the correction for the canopy optical path length improves the normalisation of the backscatter in hilly areas. SAR backscatter may be affected in a different way by the abovementioned topographic effects. In fact, the topography determines several abiotic factors such as solar radiation, water availability or wind direction, which directly influence the growth conditions of a forest (Luckman 1998; Williamson 1975; Kellogg and Arber 1981; Telewski 1995; Watt et al. 2004) and possibly lead to systematic trends in the backscatter intensity (Luckman 1998). To date, the relationship between forest backscatter and physiological effects related to the topography of a forest have been barely discussed. In most cases, these effects have been neglected by assuming that the forested areas are homogeneous.

2.1.1.4 Modelling Techniques

The modelling of biomass and the forest parameters related with SAR remote sensing systems has been the subject of numerous studies over the past few decades. These studies were based on different approaches (Saatchi et al. 2007; Martinez et al. 2000; Arslan et al. 2000; Castel et al. 2002; Svoray and Shoshany 2002).

The most common approach is empirical regression, which was mainly used to investigate the relationship between the SAR information (i.e. backscatter intensity) and the ground-measured biophysical parameters (Le Toan et al. 1992). For the most part, the studies which presented regression analyses dealt with nonlinear

Table 2.2 Summary of the main models and techniques for the backscatter

Approach	Model	Researchers(s)
Empirical regression	Linear	Dobson et al. (1992) and Hussin et al. (1991)
	Multiple linear	Harrell et al. (1997), Luckman (1997) and Ranson and Sun (1997)
	Non-linear	Balzter et al. (2003b), Magnusson et al. (2007), Morel et al. (2011) and Tsolmon et al. (2002)
Theoretical modeling	MIMICS (RT)	Mcdonald et al. (1990) and Ulaby et al. (1990)
	Santa Barbara (RT)	Wang et al. (1993, 1994)
	AMAP (RT)	Castel et al. (2001a)
	MIT/CESBIO (RT)	Hsu (1996)
	WBE (RT)	Brolly and Woodhouse (2013), West et al. (1999) and Woodhouse (2006)
	Branching (RT)	Yueh et al. (1992)
	Full wave (DBA)	Angot et al. (2002), Bellez et al. (2009), Israelsson et al. (2000), Oh and Sarabandi (2002), Nguyen et al. (2006) and Ziade et al. (2008)
Semi-empirical regression	WCM (RT)	Askne et al. (1995, 2003), Attema and Ulaby (1978), Cartus et al. (2012), Fransson and Israelsson (1999), Kurvonen and Hallikainen (1999), Martinez et al. (2000), Pulliainen et al. (1994), Richards (1990) and Santoro et al. (2003a, 2004, 2010)
Non-parametric	ANN	Benediktsson and Sveinsson (1997), Del Frate and Solimini (2004), Kimes et al. (1997) and Wang and Dong (1997)
	k-NNN	Guo et al. (2011) and Holopainen et al. (2009)

models as the biomass quantity was found to be non-linearly related to backscatter (Balzter et al. 2003b; Magnusson et al. 2007; Tsolmon et al. 2002; Morel et al. 2011). However, linear trends were also reported between forest parameters and SAR backscatter, motivating the use of simple (Dobson et al. 1992; Hussin et al. 1991) and multiple (Harrell et al. 1997; Luckman 1997; Ranson and Sun 1997) linear regressions (Table 2.2).

Another approach relied on the physics of scattering processes. This approach was essentially considered for describing and understanding forest backscatter. Many scattering models have been cited in previous studies and were essentially classified as RT and distorted born approximation (DBA) approaches (Le Toan et al. 2002). The RT approach is based on energy conservation and relies only on incoherent modelling techniques. Of the RT models, either the Michigan Microwave Canopy Scattering Model (MIMICS) (Ulaby et al. 1990; Mcdonald et al. 1990), Santa Barbara microwave model (Wang et al. 1993, 1994) or the model proposed by Karam et al. (1992) may be cited. Each of these models has specific characteristics such as continued multilayers (Ulaby et al. 1990), discontinued multilayers (Sun et al. 1991; Mcdonald and Ulaby 1993) or second-order scattering. Although the aforementioned models were shown to validate the empirical observations (Moghaddam and Saatchi 1993; Imhoff 1995a; Romshoo and Shimada 2001), they only briefly discussed the structural properties of trees' canopies. In this respect, Hsu (1996) contributed with the MIT/CESBIO model structural effects by considering the branching model Yueh et al. (1992) and Castel et al. (2001a) introduced the AMAP model and Woodhouse (2006) and Brolly and Woodhouse (2013) presented a modelling approach based on the coupling of the RT theory and a macroecological model (West et al. 1999). The DBA approach was less popular than the RT approach. However, contrary to the RT models, it has the advantage of dependence on electromagnetic waves, thus allowing the coherent modelling of vegetation parameters (Le Toan et al. 2002). The radar modelling of red pine proposed by Lang et al. (1994) and the modelling assessment of the coherent effects for forest canopies given by Saatchi and McDonald (1997) provide two examples of the DBA models. Owing to the great concern regarding the retrieval of biomass quantity by low-frequency systems (Le Toan et al. 2012), and the limited ability of RT models to account for multiple reflections, there has been renewed interest in improving DBA models (Nguyen et al. 2006; Oh and Sarabandi 2002; Bellez et al. 2009; Ziade et al. 2008; Israelsson et al. 2000; Angot et al. 2002). For example, Bellez et al. (2009) applied a coherent 'full wave' simulation of forested areas and successfully modelled the main forest scattering mechanisms, including multiple reflections, using the method of moments (MoM).

Empirical models can be easily inversed, but they are also very site-dependent, while physical models can provide accurate estimations but often require a large number of unknown parameters (Martinez et al. 2000; Castel et al. 2002). To combine the benefits of both these modelling approaches, a third approach applying semi-empirical models has been the subject of discussions. The most common semi-empirical model is the water cloud model (WCM), which was developed by Attema and Ulaby (Attema and Ulaby 1978). This model is based on the RT theory and describes a forest as a single homogenous layer comprising cloud droplets which attenuate the radar signal (Martinez et al. 2000; Richards 1990; Santoro et al. 2003a, 2004; Fransson and Israelsson 1999; Kurvonen and Hallikainen 1999; Askne et al. 2003). The main strength of the WCM is its simplicity, which allows a straightforward inversion of the model and prediction of forest biomass quantity (Woodhouse 2006). However, the WCM also presents some limits. For example, it does not consider forest structures and high-order scattering such as dihedrals (Cartus et al. 2012). In this regard, some studies have presented extended versions of the WCM. For instance, Askne et al. (1995) modified the WCM by introducing a variable which accounts for vertical and horizontal discontinuities (gaps). Pulliainen et al. (1994) replaced the area-fill factor by *GSV*, which is more common in the field of forestry. Taking into consideration the updated versions of the WCM, some recent studies such as Santoro et al. (2010) and Cartus et al. (2012) successfully applied the WCM to large-scale forest biomass estimations. Besides the WCM model, some researchers proposed semi-empirical models which considered specific forest conditions. For example, Arslan et al. (2000) proposed a backscattering model which incorporates forest stem volume and snow water equivalent, while Kurvonen and Hallikainen (1999) implemented a semi-empirical model based on stem volume and soil moisture.

Finally, non-parametric machine learning algorithms was applied by studies for the estimation of a forest's biophysical parameters. This approach employing algorithms such as artificial neural networks (ANNs) and k-nearest neighbour (k-NN) have been widely applied to landcover classifications and biomass estimations based on optical data (Franco-Lopez et al. 2001; McRoberts et al. 2007; Baffetta et al. 2009; Tomppo 2004; Tomppo et al. 2008, 2009). The non-parametric processing techniques present a major advantage when compared to the aforementioned approaches; they generally require no prior knowledge about the data distribution (Mas and Flores 2008; McRoberts et al. 2007). Despite this advantage and the numerous investigations undertaken with optical sensors, there are few studies employing such techniques for radar remote sensing in the field of forestry. Several researchers have tested the retrieval of forest parameters using ANNs (Del Frate and Solimini 2004; Kimes et al. 1997; Wang and Dong 1997; Benediktsson and Sveinsson 1997), while others have focused on the k-NN algorithm using combined SAR and optical (Holmström and Fransson 2003) or light detection and ranging (LiDAR) data (Tian et al. 2012). To date, only two studies based on SAR data (Holopainen et al. 2009; Guo et al. 2011) have involved the examination of k-NN for the assessment of forest parameters.

2.1.1.5 Estimation Accuracy

Taking into account the capability of SAR systems in retrieving biomass, several studies have shown the accuracy of the models. To do so, researchers generally applied the root-mean-square error (*RMSE*) or the relative *RMSE* (*rRMSE*) for comparing different test sites. The achieved accuracy depended primarily on the SAR system configuration, the forest properties environmental conditions and the modelling technique. An accuracy of approximately 20 % was considered to be reasonable for envisaging the application of SAR systems to forestry. The estimated *RMSE* differed between the studies. Using the Japanese Earth resources satellite (JERS-1) L-band data, Santoro (2003) found an *rRMSE* of 30–50 % and 40–60 % with and without multitemporal retrieval in a boreal forest, respectively. The variations were attributed to weather conditions. In this respect, Santoro showed that an accuracy of 20–25 % can be achieved (Santoro et al. 2003a, 2006) by considering dry and unfrozen conditions as well as a multitemporal approach. Similar results were obtained by Sandberg et al. (2011) and Magnusson et al. (2007) with ALOS PALSAR data. The best accuracy obtained with L-band backscatter was reported by Cartus et al. (2012), who recently showed that an *RMSE* of 12.9 t ha^{-1} can be obtained with PALSAR data. At P-band and VHF-band frequencies, the ranges of errors were 20–30 % (40–60 t ha^{-1}) (Soja et al. 2010, 2013; Sandberg et al. 2011) and 15–20 % (50–65 m^3 ha^{-1}) (Fransson et al. 2000; Smith and Ulander 2000; Folkesson et al. 2008), respectively. The higher relative accuracies given by low-frequency sensors when compared with high-frequency sensors were mostly explained by their subsequent saturation in high biomass ranges (Castel et al. 2002; Saatchi et al. 2007).

2.1.2 SAR Interferometry

The potential of InSAR in retrieving forest parameters was introduced in the mid-1990s. Mainly, two different topics were discussed, namely sensitivity of forest variables to interferometric coherence and derivation of interferometric forest height.

2.1.2.1 InSAR Coherence

The first investigations referring to interferometric coherence were given by Zebker and Villasenor, who observed temporal decorrelations over forested areas with an L-band 17-day repeat-pass Seasat sensor (Zebker and Villasenor 1992). C-band phase correlation properties were subsequently examined a few years later using the the first European remote sensing satellite (ERS-1) 35 days repeat-pass data. In particular, Askne et al. (1997) highlighted the main decorrelation mechanisms, and Floury et al. (1996) and Beaudoin et al. (1996) showed the relationship between coherence and forest biomass. The latter two studies clearly depicted a decrease in coherence with an increase in stand age and bole volume, respectively. This trend was also presented in different subsequent studies (Smith et al. 1998; Santoro et al. 1999). Luckman (1997) and Luckman et al. (2000) were among the first to compare ERS-1/2 tandem data (Smith et al. 1998; Koskinen et al. 2001; Santoro et al. 2002) and L-band 44-day repeat-pass JERS-1 (Eriksson et al. 2002, 2003; Cartus et al. 2005), respectively, using forest biomass. These studies depicted a monotonic decrease in coherence with increasing vegetation cover and a stronger correlation with coherence than with backscatter intensity. Subsequent studies on ERS-1/2 tandem and JERS-1 coherence confirmed the observed trends. Among these, some studies examined the saturation level obtained with coherence (see Table 2.3). It was generally observed that with C-band 1-day repeat-pass data, the coherence in some cases reached a saturation between 100 and 400 $m^3\,ha^{-1}$, while no saturation was observed in other cases. For example, Castel et al. (2000) reported no saturation in a coniferous forest in Southern France up to 400 $m^3\,ha^{-1}$, while Santoro et al. (2002) reported a saturation at 350 $m^3\,ha^{-1}$ for a hemiboreal forest located in Sweden. With regard to L-band systems, a limited number of studies were performed. Eriksson et al. (2003) and Luckman et al. (2000) reported saturation levels at 100–130 $m^3\,ha^{-1}$ and 150 $m^3\,ha^{-1}$, respectively, for JERS-1 44 days repeat-pass system. More recently, Thiel and Schmullius (2012) reported a saturation at around 100 $m^3\,ha^{-1}$ for ALOS PALSAR 46-day repeat-pass system.

The wide range of saturation levels reported in literature was explained by the sensitivity of coherence to weather conditions and to the perpendicular baseline (Smith et al. 1996; Pulliainen et al. 2003; Koskinen et al. 2001; Askne et al. 2003). In terms

Table 2.3 Saturation levels reported in literature for SAR interferometric coherence

Biome	InSAR	Saturation limit ($m^3 h^{-1}$)	Researchers(s)
Temperate	ERS-1/2	>100–400	Cartus et al. (2005) and Castel et al. (2000)
Boreal	ERS-1/2	>100–400	Askne and Santoro (2007), Fransson (2001), Santoro et al. (2002), Smith et al. (1998) and Wagner et al. (2000, 2003)
	JERS-1	>100–130	Eriksson et al. (2003)
	PALSAR	100	Thiel and Schmullius (2012)
Tropical	ERS-1/2	150	Luckman et al. (2000)
	JERS-1	150	Luckman et al. (2000)

When necessary, a density factor of 1.6 was considered to convert $t\,ha^{-1}$ to $m^3\,ha^{-1}$

of spatial baseline configurations, large perpendicular baselines were usually recommended (Floury et al. 1996; Skinner et al. 2002). For example, Askne et al. (2003) and Santoro et al. (2008) suggested a normal baseline in the range 100–250 m to improve the *GSV* retrieval accuracy with ERS-1/2. Although the recent SAR spaceborne platforms allow the use of repeat pass and complete polarisation systems, a limited number of studies have discussed forest coherence signatures at different polarisations. Wegmuller et al. (1996) briefly reported differences in coherence between like- and cross-polarisations. He noted that with airborne SIR-C L-band data, coherence values for HH and VV were very close, with a slightly higher coherence for HH. However, comparing HH and HV polarisations, the researcher noted an obviously lower coherence for cross-polarisation compared to co-polarisation. More recently, Tanase et al. (2010) compared ALOS PALSAR HV and HH channels for different burn severity ranges. The coherence for HV polarisation was also found to be generally lower than that for HH polarisation. These results were very recently confirmed by Simard et al. (2012). The dependence of interferometric coherence on incidence angle has been briefly discussed by Shimada et al. (2010). The study did not report any significant variation in coherence with incidence angle. However, another study reported that an incidence angle of greater than 45° should be chosen because at this angle, spatial decorrelations due to layover and shadow over topographic areas are limited (Bamler and Hartl 1998).

2.1.2.2 Forest Properties

As for radar backscatter, it has been shown that a forest's horizontal structure can complicate the retrieval of biomass. As for interferometric coherence, the horizontal structure plays a significant role. It was reported that the forest structure may influence the coherence to a greater extent than the forest type or composition (Santoro et al. 2005). Although a few studies have noted the importance of the horizontal structure (Askne et al. 1997; Santoro et al. 2005, 2007) and included it in their modelling approaches (Askne et al. 2003; Drezet and Quegan 2006; Santoro et al. 2007), the assessment of its effect on coherence is in its preliminary stages (De Zan et al. 2013).

2.1.2.3 Forest Environmental Conditions

With respect to weather conditions, it was demonstrated that in boreal forests and for ERS-1/2 tandem data, winter interferometric acquisitions were more stable than summer acquisitions (Askne and Santoro 2007; Eriksson et al. 2002; Manninen et al. 2000; Santoro et al. 2002, 2005, 2007; Thiel et al. 2009; Koskinen et al. 2001). As for L-band data, studies investigating JERS-1 interferometric coherence reported similar results, namely more stable conditions with winter acquisitions. As an example, for JERS-1 winter coherences, Eriksson et al. (2002) and Askne et al. (2003) showed a large dynamic range and small standard deviations, which indicated promising results for the estimation of *GSV*. Among the different weather parameters affecting coherence, wind was found to be significant (Dammert et al. 1995; Smith et al. 1996; Beaudoin et al. 1996; Askne et al. 1997, 2003); soil and canopy moisture (Drezet and Quegan 2006; Luo et al. 2000) in the presence of rain (Santoro et al. 2002; Pulliainen et al. 2003; Wagner et al. 2003; Simard et al. 2012) or snow (Askne and Santoro 2005; Pulliainen et al. 2003) and temperature (Askne et al. 2003; Pulliainen et al. 2003), which modulates the forest dielectric properties, were also found to affect coherence. On the basis of these parameters, Santoro et al. (2007) defined stable conditions as acquisitions with 'no precipitation, no freeze/thaw, with temperatures constantly being at least a few degrees below zero and the presence of snow cover' and 'optimal conditions' as the additional presence of a moderate breeze. Forest understories have been examined by a few studies such as Drezet and Quegan (2006) or Neeff et al. (2005) and was modelled by Neumann et al. (2010). Although forest understories contribute to interferometric phase decorrelation, this parameter remains under investigation. Finally, the atmospheric conditions usually change between two radar acquisitions. The presence of heavy precipitations during one of the acquisitions was found to decrease interferometric coherence (Li et al. 2007). This observation was shown to be particularly significant at high frequencies (i.e. X-band and C-band). Owing to the temporal decorrelation induced by varying weather conditions, several researchers (Smith et al. 1996; Eriksson et al. 2002, 2003, 2008) suggested the use of small temporal baselines or low-frequency systems. In addition to the normal baseline and weather conditions, topography was found to affect the interferometric coherence and *GSV* retrieval (Wegmuller and Werner 1995; Cartus et al. 2005; Tanase et al. 2010, 2011). Two main effects were distinguished, namely spatial decorrelation due to the non-overlapping fraction of the range spectral band and volume decorrelation due to the varying path length of the microwaves in canopies within different terrain slopes and aspects (Castel et al. 2000; Lee and Liu 2001). The first issue could be solved using common-band filtering (Gatelli et al. 1994; Cartus et al. 2005, 2008; Santoro et al. 2007), while the second issue has not yet been discussed in literature.

2.1.2.4 Modelling Techniques

For evaluating backscatter intensity as well, modelling of coherence was mainly classified as one of the three different approaches, namely empirical, semi-empirical and theoretical approaches. Table 2.4 summarises the principal publications referring to these approaches.

Empirical regressions were applied to coherence in several studies. As an example, Koskinen et al. (2001) and Smith et al. (1998) suggested the use of linear relationships to describe ERS tandem coherence. Fransson evaluated ERS coherence using multiple linear regressions (Fransson 2001), Eriksson described L-band coherence with an exponential relation (Santoro et al. 2003b; Eriksson et al. 2003, 2005) and Wagner implemented the SAR Imaging for Boreal Ecology and Radar Interferometry Applications (SIBERIA) algorithm using ERS tandem coherence and JERS backscatter data (Wagner et al. 2003, 2000). Although empirical models had been successfully applied in these studies, they were shown to be limited when interferometric pairs were acquired under various conditions, namely with changing weather conditions and a different perpendicular baseline. To deal with multi-seasonal and multi-baseline datasets, semi-empirical models needed to be considered. In this respect, the retrieval of the stem volume using SAR interferometric coherence has attracted considerable attention in literature, particularly by a small community of researchers who developed the interferometric water cloud model (IWCM) [see Askne et al. (1997, 2003) and Santoro et al. (2002) for the complete description]. The IWCM is a semi-empirical model which was introduced by Askne et al. (1995). This two-layer model is an extension of the WCM and defines forestvolume decorrelation as the sum of ground and vegetation contributions. This model has been extensively applied to C-band ERS-1/2 tandem data in different regions of the world such as Siberia (Santoro et al. 2005, 2007; Cartus et al. 2008, 2011), Scandinavia (Askne et al. 2003;

Table 2.4 Summary of the main models and techniques for the interferometric coherence

Approach	Model	Researchers(s)
Empirical regression	Linear	Koskinen et al. (2001) and Smith et al. (1998)
	Multiple linear	Fransson (2001)
	Non-linear	Santoro et al. (2003b) and Eriksson et al. (2003, 2005)
	SIBERIA algorithm	Drezet and Quegan (2006) and Wagner et al. (2000, 2003)
Theoretical modelling	COSMO	Thirion et al. (2006)
Semi-empirical regression	IWCM (RT)	Askne et al. (1995, 1997, 2003), Askne and Santoro (2005, 2007, 2009), Cartus et al. (2005, 2008, 2011) and Santoro et al. (1999, 2000, 2003a, 2005, 2006, 2007)
	HUT (RT)	Engdahl et al. (2004), Koskinen et al. (2001) and Pulliainen et al. (2003)

Askne and Santoro 2005, 2007, 2009; Santoro et al. 1999, 2000, 2003a, 2005) and North China (Santoro et al. 2006; Cartus et al. 2008). The model was also tested in Europe over the Thuringian Forest in order to evaluate the influence of topography on *GSV* estimation (Cartus et al. 2005). With respect to physical modelling, scattering models for vegetation are well documented in literature. However, few studies have provided a theoretical description of the relationship between interferometric phase and coherence with forest canopies and underlining soils. Askne et al. (1997) and Treuhaft et al. (1996) highlighted the relationship between interferometric phase and scatterings from vegetation canopies and ground. Sarabandi and Lin (2000) presented a coherent scattering model based on Monte Carlo simulations, which predicted the interferometric response of forests. Liu et al. (2008) combined a tree growth model with a scattering model based on the RT2 theory to simulate the interferometric coherence and to validate the main assumptions of the IWCM model. With reference to this study, it was found that there was a good agreement between the RT2 and IWCM models and confirmed that C-band data conformed to the assumptions of the IWCM model. Finally, more recently, Thirion et al. (2006) presented the COSMO coherent model, which was aimed at simulating complex radar images with a view to provide a tool for interferometric as well as polarimetric applications.

2.1.2.5 Estimation Accuracy

The interferometric coherence accuracies reported in the literature were mainly discussed for ERS-1/2 tandem acquisitions. The studies showed that coherence generally provided an *rRMSE* of 20–30 % (25–60 $m^3\,ha^{-1}$), which is significantly better than using the intensity information on average (Fransson 2001; Askne and Santoro 2007; Santoro et al. 2005, 2007; Cartus et al. 2008; Smith et al. 1998). In extreme cases, the *rRMSE* could increase beyond 48 % (Pulliainen et al. 2003) or even 60 % (Askne and Santoro 2007) and could also fall below 5 % (10 $m^3\,ha^{-1}$) (Askne et al. 2003). These results highlight the instability of the coherence due to weather conditions and the influence of the SAR acquisition configuration (perpendicular baseline) and forest properties. The L-band studies conducted by Eriksson showed that an *rRMSE* of 35–39 % could be achieved using a JERS-1 44 repeat-pass system.

2.1.2.6 InSAR Forest Height

In addition to using coherence to retrieve forest variables, estimation of tree height on the basis of interferometric phase has attracted considerable attention. The concept of extracting forest height using interferometric techniques was introduced by Hagberg et al. (1995), Wegmuller and Werner (1995), Rodriguez and Michel (1995), Ulander et al. (1995) and Dammert et al. (1995). These studies have proven that InSAR is a useful tool for estimating canopy height and forest biomass through allometric relations. Different approaches were proposed in literature for the retrieval of forest height (Balzter 2001; Balzter et al. 2007). The most common approach was the evaluation of the difference between InSAR elevation and a DEM. Kellndorfer et al. (2004) evaluated the feasibility of deriving vegetation canopy height from the Shuttle Radar Topography Mission (SRTM) C-band, namely the DEM derived in 2000 at C-band from the SRTM and National Elevation Dataset (NED), which is a reference surface available in the U.S. On the basis of preliminary investigations from Brown and Sarabandi (2003), Kellndorfer showed that it is possible to estimate the forest height with an *RMSE* of 1.1 and 4.5 m, respectively, for two different test sites. More recent studies have proposed additional methods by using SRTM C-band data (Brown et al. 2010; Sexton et al. 2009; Kenyi et al. 2009) and have investigated the feasibility of using SRTM X-band elevation data (Weydahl et al. 2007; Solberg et al. 2010; Walker et al. 2007) to derive forest height. Owing to variation in the position of the scattering phase center for different frequencies, the success of height retrieval was closely related to the frequency systems. The investigations showed that high frequencies were more suitable for estimating forest height because the scattering phase center was mostly located in the upper part of the canopies (Yong et al. 2003; Balzter et al. 2003a, 2007). However, low-frequency systems also appeared to have the potential to retrieve vegetation parameters (Lei et al. 2012). In addition to SAR frequency, some studies pointed out the influence of temporal decorrelation using repeat-pass systems. To limit the inaccuracies induced by changing weather conditions, it was recommended that single-pass interferometric systems be considered (Balzter et al. 2003a). With the availability of polarimetric interferometric SAR sensors, new methods based on full polarimetry were considered for the estimation of forest height. The literature review referring to these advanced techniques are discussed later in this chapter.

2.1.3 SAR Polarimetry

While the potential of polarimetry was demonstrated in the 1950s by studies on light scattering (Fabelinskii 1957), attention was directed toward radar polarimetric techniques in the early 1980s with the deployment of the first radar polarimetric sensor.

2.1.3.1 PolSAR Parameters

The first investigations of radar polarimetric data over vegetation cover were performed by Evans, 1986. The study highlighted the capability of full polarimetry to map forest cover by retrieving the density and structure of canopies. With the increasing number of sensors, which enable full polarimetric acquisitions (e.g. AIRSAR, PiSAR, E-SAR, SIR-C C/X, ALOS PALSAR, and RADARSAT-2), the number of publications in this regard exponentially increased in the subsequent years. These studies examined forest parameters using diverse techniques and polarimetric parameters. Some researchers directly compared forest biophysical parameters with the amplitude signal of linear polarisations (Herold et al. 2001; Balzter et al. 2002; Shimada et al. 2009). Other researchers explored polarimetric ratios (Mougin et al. 1999; Proisy et al. 2000), polarimetric phase difference (Ulaby and El-rayes 1987; Proisy et al. 2000; Kwok et al. 1993; Shimada et al. 2009; Ranson and Guoqing 1994; Thiel et al. 2007) and polarimetric coherence (Proisy et al. 2000). The outcomes mainly showed that polarimetric coherence decreased constantly with increasing vegetation density. However, the correlation of polarimetric coherence with biomass remained lower than the correlation shown by backscattering intensity with biomass in HV polarisation.

2.1.3.2 Modelling Techniques

There have been many expectations regarding the use of PolSAR in studies aimed at enhancing the knowledge of the scattering mechanisms occurring in forested areas and other landcovers. Various polarimetric decomposition techniques have been developed in this regard. Most of these can be divided into two categories. The first one is referred to as coherent decomposition, e.g. Pauli, Krogager (SHD), Cameron and Hyunen decompositions, while the second one is referred to as incoherent decomposition, e.g. eigenvector-based, Freeman–Durden and Yamaguchi decompositions. The eigenvector-based decomposition was introduced by Cloude (1985) and was performed on the popular Cloude–Pottier Entropy/Alpha unsupervised classification scheme (Cloude and Pottier 1997; Jong-Sen et al. 2004). Another renowned incoherent decomposition is the Freeman–Durden model (FDD) (Freeman and Durden 1998). In forested areas, this model distinguishes between three main scattering mechanisms, namely volume scattering from canopies, double bounces between ground and trunks and surface scattering from the underlying ground surface. Canopy volume scattering was modelled as a cloud of uniformed distributed dipoles with a fixed volume component in the coherency matrix. This provided potentially unrealistic negative values for the surface and double bounce scattering components (Sato et al. 2012; Cui et al. 2012). The FDD model has been widely used, and improvements have been proposed for the same by several studies (Freeman 2007; Yamaguchi et al. 2005, 2006; Arii et al. 2011; Sato et al. 2012; Cui et al. 2012). One significant contribution made by Yamaguchi et al. (2005) is the addition of a helix component to the FDD model (Yamaguchi et al. 2005). This fourth scattering component deals with targets which do not meet the reflection symmetry assumption of the FDD model, such as forests or man-made structures. Yamaguchi et al. (2005) also improved the model for forests by adding a vertically orientated volume scattering (Yamaguchi et al. 2005). Another important contribution to the FDD model was provided by Arii et al. (2011) and van Zyl et al. (2011) and more recently by Sato et al. (2012) and Cui et al. (2012), who proposed different solutions to the negative-power issue.

2.1.3.3 Forest Environmental Conditions

A topographic correction which is specifically dedicated to polarimetric datasets has been developed over the past decade. By performing topographic measurements from polarimetric SAR data (Schuler and De Grandi 1996; Schuler et al. 1998), Schuler highlighted that azimuth slopes need to be corrected and proposed a compensation method based on the variations in the EM wave orientation angle (Schuler et al. 1999). This method was further investigated in subsequent studies (Lee et al. 1999, 2000a, b, 2004) and was introduced as a de-orientation concept for the different target-decomposition algorithms (Xu and Jin 2005; An et al. 2010; Yamaguchi et al. 2011; Sugimoto et al. 2012; Lee and Ainsworth 2011). In addition to azimuth-slope corrections, a few studies focused on the forest radar signal over a topographic terrain. For example, Park et al. (2012) underlined the scattering mechanisms, and Villard et al. (2012) proposed a new backscattering coefficient for angular corrections of topography-induced scattering variations.

2.1.4 Polarimetric SAR Interferometry

While researchers have been actively developing separate algorithms for polarimetry and interferometry, over the past decade, there has been growing interest in coherently combining these two techniques to create SAR systems which are able to retrieve the three-dimensional structure of forest canopies. The joint use of polarimetry and interferometry is referred to as polarimetric interferometric SAR (PolInSAR). These instruments are particularly attractive because they provide an increased number of independent observables and, therefore, limit the complexity of the theoretical polarimetric models (Neumann et al. 2010; Praks et al. 2012a). The application of PolInSAR was first reported by Cloude and Papathanassiou (1997) and Cloude and Papathanassiou (1998), who suggested a coherent optimisation approach for separating the phase centers of different scattering mechanisms. These studies were further considered by Papathanassiou and Cloude (2001), who derived the formulation of a generalised complex interferometric coherence.

PolInSAR has become an important application for ecology and forestry as it can provide reliable estimates of forest height and biomass (Mette et al. 2002; Mette 2007; Hajnsek et al. 2009). Studies on PolInSAR focused on one parameter, namely forest height, but other components such as canopy extinction and ground-to-volume scattering ratio have also attracted considerable attention. The relationship between polarimetry and interferometry in the estimation of canopy heights is based on the concept that phase differences could be corrected using coherent wave scattering models (Cloude and Papathanassiou 2003). The best known PolInSAR model is the random volume over ground (RVoG) model, which is widely used in the scientific community because of the good tradeoff between its physical description and model complexity (Cloude and Papathanassiou 2003). RVoG was introduced by Treuhaft and Cloude (1999) and was further extended to full polarimetry by Papathanassion et al. (2000). This two-layer model was also described by Papathanassiou and Cloude (2001), who proposed an inversion algorithm and loci geometrical projection for plotting the complex coherence on an Argand diagram. The inversion of the RVoG model was developed for PolInSAR data (Cloude and Papathanassiou 2003). However, single-pol interferometric data were also considered under certain conditions, such as use of an external DEM, fixing of a forest extinction coefficient value and removal of the ground scattering contribution, which can be neglected at high SAR frequencies (Praks et al. 2007a, b, 2012a; Hajnsek et al. 2009; Garestier et al. 2008).

Several studies have demonstrated PolInSAR height estimation over temperate and boreal forests (Papathanassiou and Cloude 2001; Kugler et al. 2006; Mette 2007; Woodhouse et al. 2003; Garestier et al. 2008; Praks et al. 2007b). For example, Papathanassiou and Cloude (2001) used an L-band 10-min repeat-pass airborne system to show that it is possible to retrieve the forest height with a standard deviation of approximately 2.5 m. The RVoG model was assessed at different frequencies. Although a few publications reported P-band investigations (Garestier et al. 2008, 2009; Lee et al. 2009), most of the analyses concerned X-, C- and L-band frequencies (Kugler et al. 2006, 2007; Praks et al. 2007a; Mette and Papathanassiou 2004; Hajnsek et al. 2009). These studies showed that the accuracy of the inversed PolInSAR height depends on several factors. Among these factors, (1) forest height, (2) density and coefficient of extinction, (3) soil moisture, (4) terrain topography and (5) SAR frequency and temporal baseline have been discussed in literature (Hajnsek et al. 2009; Praks et al. 2012a; Lavalle 2009; Le Toan et al. 2012; Kugler et al. 2006). In repeat-pass PolInSAR systems, the temporal baseline consisted of one of the most limiting components (Lee et al. 2010; Neumann et al. 2010, 2012). The change in weather conditions or vegetation properties between two PolSAR acquisitions induces temporal decorrelation; thus, the forest height is overestimated and the phase deviation is increased, which affects the accuracy of height retrieval (Lee et al. 2009; Li and Guo 2012). To circumvent this issue, some studies have discussed the use of multi-baseline acquisitions (Lee et al. 2010; Neumann et al. 2010, 2012) or a combination of RVoG with a temporal decorrelation model (Papathanassiou and Cloude 2003; Li and Guo 2012; Lavalle et al. 2012). Although these methods improved the accuracy of the height estimates, they did not completely remove the temporal effect. In this context, the use of single-pass PolInSAR systems, such as the recently launched spaceborne TSX/TDX or future TanDEM-L systems, should be very promising (Kugler et al. 2010; Torano Caicoya et al. 2012; Praks 2012).

2.1.5 SAR Tomography

An extended approach to PolInSAR involves polarimetric SAR tomography. Introduced by Reigber et al. (2000) during the early millennium, this approach relies on the coherent combination of multi-baseline InSAR acquisitions and allows the localisation of scattering contributions along the vertical direction of the targets (Tebaldini and Rocca 2012). Tomography considerably extends the capabilities of SAR for the extraction of forest vertical structure information (Reigber 2001) and estimation of forest parameters such as biomass (Dinh et al. 2012). The studies related to SAR tomography are very recent and are focused on ameliorating the processing algorithms (Frey et al. 2007; Cloude 2006, 2007; Lombardini 2005; Tebaldini and Rocca 2009; Tebaldini et al. 2010; Zhang et al. 2012). However, the past three years have also seen a growing number of publications on the analysis of tomographic SAR data in forested areas (Frey and Meier 2010; Tebaldini and Rocca 2009, 2012; Tebaldini et al. 2010). For example, Tebaldini et al. (2010) and Tebaldini and Rocca (2012) compared the scattering contributions for P-band and L-band SAR tomography in a boreal forest. These studies reported a relatively uniform scattering distribution in L-band and a significant ground-level contribution in P-band in co-polar channels as well as HV polarisation. Frey and Meier (2010) reported similar results by investigating P-band and L-band datasets in a temperate forest. Dinh et al. (2012) evaluated the potential of tomographic SAR signals to retrieve biomass at a P-band frequency in a tropical forest and obtained the best sensitivity for a vegetation layer located above height of 30 ± 10 m and biomass quantities in the range 250–450 t ha^{-1} at a specific test site. These studies and future studies regarding tomography will be essential for the planned TanDEM-L, DESDynI and BIOMASS spaceborne missions.

2.2 Fusion of SAR Information

With the number of SAR techniques and datasets having increased over the past few years, there is a need to improve the estimation of forest parameters such as biomass by merging different sources of SAR information. The combination of SAR information in multiple ways was conceived. First, with reference to the different SAR techniques, a combination of two sensors for evaluating interferometric phase as well as the integration of polarimetry and interferometry in PolInSAR could be considered as fusion techniques. Therefore, publications describing InSAR, PolSAR, PolInSAR or Tomography approaches were seen in some cases as fusion approaches (Lavalle 2009). Then, researchers showed a variety of possible combinations by integrating different frequencies, polarisations, incidence angles or temporal acquisitions in a single fusion approach by using different SAR acquisition parameters. Finally, the approaches were combined in different ways such as using simple ratios, Bayesian rules, neural networks, multiple linear regressions or physical models.

The combination of various SAR acquisition parameters and the use of different fusion approaches provide the ability of improving the retrieval of forest variables. While the number of reports on SAR techniques is relatively large, the number of studies dealing with the fusion potential of SAR parameters remains limited. One of the most cited fusion methods involved combining multi-temporal datasets. In this regard, the researchers proposed diverse approaches such as merging (Goodenough et al. 2005), weighting (Santoro et al. 2003a, 2008, 2011; Askne et al. 2003; Askne and Santoro 2007) and linear combination (Quegan et al. 2000; Quegan 2001;

Bruzzone et al. 2004; Gineste 1999) of multi-temporal data. A combination of different SAR frequencies has been proven to increase saturation levels and reduce uncertainties in biomass estimates. For example, Englhart et al. (2011) recently demonstrated that combining X-band and L-band SAR data over a tropical forest using multi-regression models would increase the saturation level from approximately 100–300 t ha^{-1} and reduce the *RMSE* from approximately 110–79 t ha^{-1}. The combination of different polarisations has been presented in a few studies, and it has been generally shown that the use of a polarisation ratio such as HH/HV improved the estimation of forest biophysical parameters (Wu 1987; Sarker et al. 2012; Mougin et al. 1999). No study has supposedly demonstrated the potential integration of multi-angle or multi-pass direction SAR data. One major issue for retrieving large scale biomass maps lies in the development of algorithms which are independent of ground inventory data. To solve this problem, some researchers have combined SAR and optical data (Santoro et al. 2010, 2011; Cartus et al. 2011, 2012). However, to date, although some studies have proposed the combination of backscatter intensity and coherence (Wagner et al. 2003, 2000), no study has reported the possibility of deriving *GSV* from only SAR information without considering reference data.

2.3 Open Issues and Scope of the Thesis

With respect to the literature review performed in Sects. 2.1 and 2.2, four different topics were identified as important issues in forest biomass estimation from remote sensing systems in temperate forests. The first issue is topography. Although it is known that topography affects the SAR signal and corrections for sloping terrains have been proposed, the effects of steep slopes in forests on the returned SAR signals remain under investigation (Castel et al. 2001a; Luckman 1998; Cartus et al. 2005, 2008; Castel et al. 2000). The second issue is the forest's horizontal structure, which directly refers to the results reported by Woodhouse (2006) and Brolly and Woodhouse (2012). The researchers introduced a new approach for the uncommon trend observed in several studies, namely a decrease in backscatter intensity with increasing forest biomass quantity (Rauste et al. 1994; Ranson and Guoqing 1994; Dobson et al. 1995; Woodhouse 2006). The third issue is the *GSV* range. Most studies examining the relationship between *GSV* and SAR data only considered forests with GSV values up to 400 m^3 ha^{-1}. However, temperate forested areas may show GSV values greater than 700 m^3 ha^{-1}. With the aim of estimating forest biomass wordwide, it is necessary to cover the different ranges of GSV and investigate the related effects on remote sensing systems. After reviewing several sensors used in forestry applications, Wulder et al. (2004) mentioned that 'the best sensor is often more than one sensor'. In this regard, the last issue highlights the significant potential of combining SAR information and the small number of studies published in this field. On the basis of the four aforementioned issues, the following objectives were defined for this thesis:

- Highlight the scattering and decorrelation mechanisms occurring in temperate forests with a topographic terrain;
- Examine potential effects related to forest horizontal structure and high *GSV* values;
- Determine the optimal SAR acquisition parameters for the estimation of forest *GSV*;
- Estimate *GSV* from spaceborne remote sensing sensors using algorithms presented in literature;
- Investigate and develop an integrated approach to deriving a *GSV* map from the fusion of SAR information.

The above objectives were formulated to provide answers to the open issues and to present novel scientific knowledge to the radar community. To limit the framework of this thesis, the investigations focused on X-band and L-band spaceborne data (see the SAR missions presented in Sect. 1.2). In addition, although PolSAR and PolInSAR techniques may be explored, greater attention will be paid on backscatter intensity and interferometric coherence data. Finally, as mentioned in Sect. 1.2, the examinations performed in this thesis were incorporated into the ENVILAND2 project and consequently also involved investigations of optical remote sensing. Owing to time constraints, only findings obtained from radar data are presented in this thesis.

References

Alasalmi H, Praks J, Arslan A, Koskinen J, Hallikainen M (1998) Investigation of snow and forest properties by using airborne SAR data. In: Second international workshop on retrieval of bio- and geo-physical parameters from SAR data for land applications, ESA-ESTEC, Nordwijk, Netherlands

Allain C, Cloitre M (1991) Characterizing the lacunarity of random and deterministic fractal sets. Phys Rev A 44(6):3552–3558

Amar F, Fung AK, De Grandi G, Lavalle C, Sieber A (1993) Backscattering from forest canopies over slanted terrain. In: Proceedings of IEEE international geoscience and remote sensing symposium, IGARSS '93, vol 2, pp 576–579

An W, Cui Y, Yang J (2010) Three-component model-based decomposition for polarimetric SAR data. IEEE Trans Geosci Remote Sens 48(6):2732–2739

Angot L, Roussel H, Tabbara W (2002) A full wave three dimensional analysis of forest remote sensing using VHF electromagnetic wave. Prog Electromagnet Res 38:311–331

Ardila JP, Tolpekin V, Bijker W (2010) Angular backscatter variation in L-band ALOS scanSAR images of tropical forest areas. IEEE Trans Geosci Remote Sens Lett 7(4):821–825

Arii M, van Zyl JJ, Kim Y (2011) Adaptive model-based decomposition of polarimetric SAR covariance matrices. IEEE Trans Geosci Remote Sens 49(3):1104–1113

Arslan AN, Koskinen J, Pulliainen J, Hallikainen M (2000) A semi empirical backscattering model of forest canopy covered by snow using SAR data. In: Proceedings of IEEE 2000 international geoscience and remote sensing symposium. Taking the pulse of the planet: the role of remote sensing in managing the environment, IGARSS 2000, Cat. No.00CH37120, vol 5, issue 4, pp 1904–1906

Arslan AN, Pulliainen J, Hallikainen M (2006) Observations of l-and c-band backscatter and a semi-empirical backscattering model approach from a forest-snow-ground system. Prog Electromagnet Res 56(56):263–281

Askne J, Santoro M (2005) Multitemporal repeat pass SAR interferometry of boreal forests. IEEE Trans Geosci Remote Sens 43(6):1219–1228
Askne J, Santoro M (2007) Selection of forest stands for stem volume retrieval from stable ERS tandem InSAR observations. IEEE Trans Geosci Remote Sens 4(1):46–50
Askne J, Santoro M (2009) Automatic model-based estimation of boreal forest stem volume from repeat pass C-band InSAR coherence. IEEE Trans Geosci Remote Sens 47(2):513–516
Askne J, Dammert P, Fransson J, Israelsson H, Ulander LMH (1995) Retrieval of forest parameters using intensity and repeat-pass interferometric SAR information. Retrieval of bio- and geophysical parameters from SAR data for land applications, Symposium held in Toulouse, France, 10–13 October 1995, (ACTES), pp 119–129
Askne J, Dammert PBG, Uhlander LMH, Smith G (1997) C-band repeat-pass interferometric SAR observations of the forest. IEEE Trans Geosci Remote Sens 35(1):25–35
Askne J, Santoro M, Smith G, Fransson JES (2003) Multitemporal repeat-pass SAR interferometry of boreal forests. IEEE Trans Geosci Remote Sens 41(7):1540–1550
Attema EPW, Ulaby FT (1978) Vegetation modeled as a water cloud. Radio Sci 13(2):357–364
Baffetta F, Fattorini L, Franceschi S, Corona P (2009) Design-based approach to k-nearest neighbours technique for coupling field and remotely sensed data in forest surveys. Remote Sens Environ 113(3):463–475
Balzter H (2001) Forest mapping and monitoring with interferometric synthetic aperture radar (InSAR). Prog Phys Geogr 25(2):159–177
Balzter H, Baker JR, Hallikainen MT, Tomppo E (2002) Retrieval of timber volume and snow water equivalent over a Finnish boreal forest from airborne polarimetric synthetic aperture radar. Int J Remote Sens 23(16):3185–3208
Balzter H, Cox R, Rowland C, Saich P, Wood M, Ripton A (2003a) Forest canopy height mapping from dual-wavelength sar interferometry. In: Proceedings of the workshop on POLinSAR—applications of SAR polarimetry and polarimetric interferometry, ESA SP-529, p 51.1
Balzter H, Skinner L, Luckman AJ, Brooke R (2003b) Estimation of tree growth in a conifer plantation over 19 years from multi-satellite L-band SAR. Remote Sens Environ 84:184–191
Balzter H, Rowland CS, Saich P (2007) Forest canopy height and carbon estimation at Monks Wood National Nature Reserve, UK, using dual-wavelength SAR interferometry. Remote Sens Environ 108:224–239
Bamler R, Hartl P (1998) Synthetic aperture radar interferometry. Inverse Prob 14(4):R1–R54
Barros Filho MN and Sobreira FJA (2008) Accuracy of lacunarity algorithms in texture classification of high spatial resolution images from urban areas. In: Jun C, Jie J, Förstner W (eds) XXIst ISPRS congress. The international archives of the photogrammetry, remote sensing and spatial information science, vol XXXVII. International society for photogrammetry and remote sensing, Beijing
Bartsch A, Kidd R, Wagner W, Bartalis Z (2006) Temporal and spatial variability of the beginning and end of daily spring freeze/thaw cycles derived from scatterometer data. Remote Sens Environ 106(3):360–374
Bayer T, Winter R, Schreier G (1991) Terrain influences in SAR backscatter and attempts to their correction. IEEE Trans Geosci Remote Sens 29(3):451–462
Beaudoin A, Le Toan T, Goze S, Nezry E, Lopes A, Mougin E, Hsu CC, Han HC, Kong JA, Shin RT (1994) Retrieval of forest biomass from SAR data. Int J Remote Sens 15(14):2777–2796
Beaudoin A, Stussi N, Troufleau D, Desbois N, Piet L, Deshayes M (1995) On the use of ERS-1 SAR data over hilly terrain: necessity of radiometric corrections for thematic applications. In: 1995 international geoscience and remote sensing symposium, IGARSS '95. Quantitative remote sensing for science and applications, vol 3, Montpellier, France, pp 2179–2182
Beaudoin A, Castel T, Rabaute T (1996) Forest monitoring over hilly terrain using ERS INSAR data. In: Guyenne T-D, Danesy D (eds) Proceedings of the fringe 96 workshop, Zurich, Switzerland, p 105

Bellez S, Dahon C, Roussel H (2009) Analysis of the main scattering mechanisms in forested areas: an integral representation approach for monostatic radar configurations. IEEE Trans Geosci Remote Sens 47(12):4153–4166

Benediktsson JA, Sveinsson JR (1997) Feature extraction for multisource data classification with artificial neural networks. Int J Remote Sens 18(4):727–740

Bergen KM, Dobson MC, Pierce LE, Ulaby FT (1997) Effects of within-season dielectric variations on terrain classification using SIR-C, X-SAR. In: Proceedings of 1997 IEEE international geoscience and remote sensing IGARSS'97 symposium, remote sensing—a scientific vision for sustainable development, vol 2, pp 1072–1074

Brolly M, Woodhouse IH (2012) A Matchstick model of microwave backscatter from a forest. Ecol Model 237–238:74–87

Brolly M, Woodhouse IH (2013) Vertical backscatter profile of forests predicted by a macroecological plant model. Int J Remote Sens 34(4):1026–1040

Brown CG, Sarabandi K (2003) Estimation of red pine tree height using shuttle radar topography mission and ancillary data. In: Proceedings of 2003 IEEE international geoscience and remote sensing symposium, IGARSS 2003, IEEE Cat. No.03CH37477, vol 4, pp 2850–2852

Brown CG, Sarabandi K, Pierce LE (2010) Model-based estimation of forest canopy height in red and Austrian pine stands using shuttle radar topography mission and ancillary data: a proof-of-concept study. IEEE Trans Geosci Remote Sens 48(3):1105–1118

Bruzzone L, Marconcini M, Wegmuller U, Wiesmann A (2004) An advanced system for the automatic classification of multitemporal SAR images. IEEE Trans Geosci Remote Sens 42(6):1321–1334

Cartus O, Santoro M, Kellndorfer J (2012) Mapping forest aboveground biomass in the northeastern United States with ALOS PALSAR dual-polarization L-band. Remote Sens Environ 124:466–478

Cartus O, Santoro M, Schmullius C (2005) Feasibility of forest parameter retrieval in mountainous areas using ERS and JERS SAR interferometry. In: Proceedings of 9th international symposium on physical measurements and signatures in remote sensing, Beijing, China, Friedrich-Schiller University Jena, pp 530–532

Cartus O, Santoro M, Schmullius C, Li Z (2011) Large area forest stem volume mapping in the boreal zone using synergy of ERS-1/2 tandem coherence and MODIS vegetation continuous fields. Remote Sens Environ 115(3):931–943

Cartus O, Santoro M, Schmullius C, Yong P, Zengyuan L, Erxue C (2008) Creation of large area forest biomass maps for northeast China using ERS-1/2 Tandem coherence. In: Proceedings of dragon 1 programme final results 2004–2007, Beijing, China, ESA SP–655

Castel T, Martinez J-M, Beaudoin A, Wegmüller U, Strozzi T (2000) ERS INSAR data for remote sensing hilly forested areas. Remote Sens Environ 73:73–86

Castel T, Guerra F, Caraglio Y, Houllier F (2002) Retrieval biomass of a large Venezuelan pine plantation using JERS-1 SAR data. Analysis of forest structure impact on radar signature. Remote Sens Environ 79:30–41

Castel T, Beaudoin A, Floury N, Le Toan T, Caraglio Y, Barczi J-F (2001a) Deriving forest canopy parameters for backscatter models using the AMAP architectural plant model. IEEE Trans Geosci Remote Sens 39(3):571–583

Castel T, Beaudoin A, Stach N, Stussi N (2001b) Sensitivity of space-borne SAR data to forest parameters over sloping terrain. Theory and experiment. Int J Remote Sens 22(12):2351–2376

Champion I, Dubois-Fernandez P, Dupuis X (2011) Retrieving forest biomass from the texture of SAR images. In: EARSeL eProceedings, Prague, Czech Republic, pp 102–109

Champion I, Dubois-Fernandez P, Guyon D, Cottrel M (2008) Radar image texture as a function of forest stand age. Int J Remote Sens 29(6):1795–1800

Champion I, Guyon D, Riom J, Le Toan T, Beaudoin A (1998) Effect of forest thinning on the radar backscattering coefficient at L-band. Int J Remote Sens 19(11):2233–2238

Chauhan NS, Lang RH, Ranson KJ (1991) Radar modeling of a boreal forest. IEEE Trans Geosci Remote Sens 29(4):627–638

Cloude SR (1985) Target decomposition theorems in radar scattering. Electr Lett 21(1):22

Cloude SR (2006) Polarization coherence tomography. Radio Sci 41(4):RS4017
Cloude SR (2007) Dual-baseline coherence tomography. IEEE Geosci Remote Sens Lett 4(1):127–131
Cloude SR, Papathanassiou KP (1997) Polarimetric optimisation in radar interferometry. Electron Lett 33(13):1176–1178
Cloude SR, Pottier E (1997) An entropy based classification scheme for land applications of polarimetric SAR. IEEE Trans Geosci Remote Sens 35(1):68–78
Cloude SR, Papathanassiou KP (1998) Polarimetric SAR interferometry. IEEE Trans Geosci Remote Sens 36(5):1551–1565
Cloude SR, Papathanassiou KP (2003) Three-stage inversion process for polarimetric SAR interferometry. IEE Proc Radar Sonar Navig 150(3):125
Cui Y, Yamaguchi Y, Yang J, Park S-E, Kobayashi H, Singh G (2012) Three-component power decomposition for polarimetric SAR data based on adaptive volume scatter modeling. Remote Sens 4(12):1559–1572
Dammert PB, Ulander LM, Askne J (1995) SAR interferometry for detecting forest stands and tree heights. Proc SPIE 2584(1):384–390
da Yanasse CCF, Sant'Anna S, Frery AC, Rennó CD, Soares JV, Luckman AJ (1997) Exploratory study of the relationship between tropical forest regeneration stages and SIR-C L and C data. Remote Sens Environ 59(2):180–190
de Araujo SL, dos Santos RJ, da Costa Freitas C, Abrahim Magalhaes Xaud H (1999) The use of microwave and optical data for estimating aerial biomass of the savanna and forest formations at Roraima State, Brazil. In: IEEE 1999 international geoscience and remote sensing symposium, IGARSS'99, Cat. No.99CH36293, vol 5, pp 2762–2764
De Zan F, Krieger G, Lopez-Dekker P (2013) On some spectral properties of TanDEM-X interferograms over forested areas. IEEE Geosci Remote Sens Lett 10(1):71–75
Del Frate F, Solimini D (2004) On neural network algorithms for retrieving forest biomass from SAR data. IEEE Trans Geosci Remote Sens 42(1):24–34
Dinh HTM, Rocca F, Tebaldini S, Mariotti D'Alessandro M, LeToan T, Villard L (2012) Relating tropical forest biomass to P-band SAR tomography. In: 2012 IEEE international geoscience and remote sensing symposium, July 2012, pp 7589–7592
Dobson MC, Ulaby FT, LeToan T, Beaudoin A, Kasischke ES, Christensen N (1992) Dependence of radar backscatter on coniferous forest biomass. IEEE Trans Geosci Remote Sens 30(2):412–415
Dobson MC, Ulaby FT, Pierce LE, Sharik TL, Bergen KM, Kellndorfer J, Kendra JR, Li E, Lin YC, Nashashibi A, Sarabandi K, Siqueira P (1995) Estimation of forest biophysical characteristics in northern Michigan with SIR-C/X-SAR. IEEE Trans Geosci Remote Sens 33(4):877–895
Drezet PML, Quegan S (2006) Environmental effects on the interferometric repeat-pass coherence of forests. IEEE Trans Geosci Remote Sens 44(4):825–837
Engdahl ME, Pulliainen JT, Hallikainen MT (2004) Boreal forest coherence-based measures of interferometric pair suitability for operational stem volume retrieval. IEEE Geosci Remote Sens Lett 1(3):228–231
Engheta N, Elachi C (1982) Radar scattering from a diffuse vegetation layer over a smooth surface. IEEE Trans Geosci Remote Sens 20(2):212–216
Englhart S, Keuck V, Siegert F (2011) Aboveground biomass retrieval in tropical forests—the potential of combined X- and L-band SAR data use. Remote Sens Environ 115(5):1260–1271
Eriksson LEB, Askne J, Santoro M, Schmullius C, Wiesmann A (2005) Stem volume retrieval with spaceborne L-band repeat-pass coherence:multi-temporal combination for boreal forest. In: Proceedings of 2005 IEEE international geoscience and remote sensing symposium, IGARSS '05, vol 5, Seoul, Korea, pp 3591–3594
Eriksson LEB, Schmullius C, Riedel T, Wiesmann A (2002) Multi-temporal JERS coherence for observation of Siberian forest. In: Proceedings of IEEE international geoscience and remote sensing symposium, Toronto, Canada, pp 2896–2898

Eriksson LEB, Santoro M, Wiesmann A, Schmullius C (2003) Multitemporal JERS repeat-pass coherence for growing-stock volume estimation of Siberian forest. IEEE Trans Geosci Remote Sens 41(7):1561–1570

Eriksson LEB, Santoro M, Fransson JES (2008) Temporal decorrelation for forested areas observed in spaceborne L-band SAR interferometry. In: 2008 IEEE international geoscience and remote sensing symposium, IGARSS 2008, vol 1, pp 283–285

Fabelinskii IL (1957) Theory of light scattering in liquids and solids. Adv Phys Sci (USSR) 63:474

Ferrazzoli P, Guerriero L (1995) Radar sensitivity to tree geometry and woody volume: a model analysis. IEEE Trans Geosci Remote Sens 33(2):360–371

Ferris R, Humphrey JW (1999) A review of potential biodiversity indicators for application in British forests. Forestry 72(4):313–328

Floury N, Le Toan T, Souyris JC (1996) Relating forest parameters to interferometric data. In: 1996 international geoscience and remote sensing symposium, IGARSS '96, vol 2, pp 975–977

Folkesson K, Smith-Jonforsen G, Ulander LMH (2008) Validating backscatter models for CARABAS SAR images of coniferous forests. Can J Remote Sens 34(5):480–495

Franco-Lopez H, Ek AR, Bauer ME (2001) Estimation and mapping of forest stand density, volume, and cover type using the k-nearest neighbors method. Remote Sens Environ 77(3):251–274

Fransson JES (2001) Stem volume estimation in boreal forests using ERS-1/2 coherence and SPOT XS optical data. Int J Remote Sens 22(14):2777–2791

Fransson JES, Israelsson H (1999) Estimation of stem volume in boreal forests using ERS-1 C- and JERS-1 L-band SAR data. Int J Remote Sens 20(1):123–137

Fransson JES, Walter F, Ulander LMH (2000) Estimation of forest parameters using CARABAS-II VHF SAR data. IEEE Trans Geosci Remote Sens 38(2):720–727

Freeman A (2007) Fitting a two-component scattering model to polarimetric SAR data from forests. IEEE Trans Geosci Remote Sens 45(8):2583–2592

Freeman A, Durden S (1998) A three-component scattering model for polarimetric SAR data. IEEE Trans Geosci Remote Sens 36(3):963–973

Frey O, Meier E (2010) Analyzing tomographic SAR data of a forest with respect to frequency, polarization, and focusing technique. In: 2010 IEEE international geoscience and remote sensing symposium, vol 49, issue 10, July 2010, pp 150–153

Frey O, Morsdorf F, Meier E (2007) Tomographic processing of multi-baseline P-band SAR data for imaging of a forested area. In: 2007 IEEE international geoscience and remote sensing symposium, pp 156–159

Frey O, Santoro M, Werner CL, Wegmuller U (2013) DEM-based SAR pixel-area estimation for enhanced geocoding refinement and radiometric normalization. IEEE Geosci Remote Sens Lett 10(1):48–52

Fukuda S (2008) Forest spatial structure enhancing non-Gaussian texture in airborne L-band polsar images. In: 2008 IEEE international geoscience and remote sensing symposium, IGARSS 2008, pp II-637–II-640

Garestier F, Dubois-Fernandez PC, Papathanassiou KP (2008) Pine forest height inversion using single-pass X-band PolInSAR data. IEEE Trans Geosci Remote Sens 46(1):59–68

Garestier F, Dubois-Fernandez PC, Guyon D, Le Toan T (2009) Forest biophysical parameter estimation using L- and P-band polarimetric SAR data. IEEE Trans Geosci Remote Sens 47(10):3379–3388

Gatelli F, Guamieri AM, Parizzi F, Pasquali P, Prati C, Rocca F (1994) The wavenumber shift in SAR inferometry. IEEE Trans Geosci Remote Sens 32(4):855–865

Gineste P (1999) A simple, efficient filter for multitemporal SAR images. Int J Remote Sens 20(13):2565–2576

Goodenough DG, Dyk A, Carey S (2005) Multitemporal evaluation with ASAR of boreal forests. In: Proceedings of 2005 IEEE international geoscience and remote sensing symposium, IGARSS '05, vol 3, Seoul, Korea, pp 1662–1665

Goriachkin OV, Klovsky DD (1999) Some problems of realization spaceborne SAR in P, UHF, VHF bands. In: IEEE 1999 international geoscience and remote sensing symposium, IGARSS'99, vol 2, June 1999, pp 1271–1273

Guo Y, Li Z, Chen E, Zhang X (2011) Fast and automatic forest volume estimation based on K nearest neighbor and SAR. In: Li J (ed) International symposium on lidar and radar mapping 2011: technologies and applications, June 2011, pp 82861D–82861D-6

Hagberg JO, Uhlander LMH, Askne J, Ulander LMH (1995) Repeat-pass SAR interferometry over forested terrain. IEEE Trans Geosci Remote Sens 33(2):331–340

Hajnsek I, Kugler F, Lee S-K, Papathanassiou KP (2009) Tropical-forest-parameter estimation by means of Pol-InSAR: the INDREX-II campaign. IEEE Trans Geosci Remote Sens 47(2):481–493

Hallikainen M, Makynen M, Pullainen J, Vanska T (1997) Radar backscatter from boreal forest in winter. In: Proceedings of 1997 IEEE international geoscience and remote sensing symposium. Remote sensing—a scientific vision for sustainable development, IGARSS'97, vol 2, pp 803–805

Haralick RM, Shanmugam K, Dinstein I (1973) Textural features for image classification. IEEE Trans Syst Man Cybern 3(6):610–621

Harrell PA, Kasischke ES, Bourgeau-Chavez LL, Haney EM, Christensen NL (1997) Evaluation of approaches to estimating aboveground biomass in southern pine forests using SIR-C data. Remote Sens Environ 59(2):223–233

He Q, Cao C, Chen E, Sun G, Ling F, Pang Y, Zhang H, Ni W, Xu M, Li Z, Li X (2012) Forest stand biomass estimation using ALOS PALSAR data based on LiDAR-derived prior knowledge in the Qilian mountain, western China. Int J Remote Sens 33(3):710–729

Henderson FM, Lewis AJ (1998) Principles and applications of imaging radar, 3rd edn. Wiley, New York

Henebry GM, Kux HJH (1995) Lacunarity as a texture measure for SAR imagery. Int J Remote Sens 16(3):565–571

Herold M, Pathe C, Schmullius C (2001) The effect of free vegetation water on the multi-frequency and polarimetric radar backscatter—first results from the TerraDew 2000 campaign. In: Proceedings of IEEE international geoscience and remote sensing symposium, Sydney, NSW, Australia, pp 2445–2447

Hinse M, Gwyn QHJ, Bonn F (1988) Radiometric correction of C-band imagery for topographic effects in regions of moderate relief. IEEE Trans Geosci Remote Sens 26(2):122–132

Hoechstetter S, Walz U, Thinh NX (2011) Adapting lacunarity techniques for gradient-based analyses of landscape surfaces. Ecol Complex 8(3):229–238

Hoekman DH, Varekamp C (2001) Observation of tropical rain forest trees by airborne high-resolution interferometric radar. IEEE Trans Geosci Remote Sens 39(3):584–594

Holecz F, Wegmuller U, Rignot E, Wang Y (1995) Observed radar backscatter from forested areas with terrain variations. In: 1995 international geoscience and remote sensing symposium, IGARSS '95. Quantitative remote sensing for science and applications, vol 1, Firenze, Italy, pp 613–615

Holmström H, Fransson JES (2003) Optical and radar data in kNN-estimation of forest variables. Soc Am For 49(3):409–418

Holopainen M, Haapanen R, Karjalainen M, Vastaranta M, Yu X, Tuominen S (2009) Combination of low-pulse ALS data and TerraSar-X radar images in the estimation of plot-level forest variables. In: ISPRS workshop laserscanning'09, vol 38, pp 135–140

Hsu CC (1996) Theoretical models for microwave remote sensing of forests and vegetation. Massachusetts Institute of Technology, Massachusetts, p 191

Hsu CC, Han HC, Shin RT, Kong JA, Beaudoin A, Le Toan T (1994) Radiative transfer theory for polarimetric remote sensing of pine forest at P band. Int J Remote Sens 15(14):2943–2954

Hussin YA, Reich RM, Hoffer RM (1991) Estimating splash pine biomass using radar backscatter. IEEE Trans Geosci Remote Sens 29(3):427–431

Imhoff ML, Johnson P, Holford W, Hyer J, May L, Lawrence W, Harcombe P (2000) BioSARTM: an inexpensive airborne VHF multiband SAR system for vegetation biomass measurement. IEEE Trans Geosci Remote Sens 38(3):1458–1462

Imhoff ML (1995a) A theoretical analysis of the effect of forest structure on synthetic aperture radar backscatter and the remote sensing of biomass. IEEE Trans Geosci Remote Sens 33(2):341–352
Imhoff ML (1995b) Radar backscatter and biomass saturation: ramifications for global biomass inventory. IEEE Trans Geosci Remote Sens 33(2):511–518
Imhoff ML, Story M, Vermillion C, Khan F, Polcyn F (1986) Forest canopy characterization and vegetation penetration assessment with space-borne radar. IEEE Trans Geosci Remote Sens 24(4):535–542
Israelsson H, Askne J (1993) The effect of leaning trunks on forest radar backscattering. In: Proceedings of IEEE international geoscience and remote sensing symposium, IGARSS '93, pp 57–59
Israelsson H, Uhlander LMH, Askne J, Fransson JES, Frölind P-O, Gustavsson A, Hellsten H (1997) Retrieval of forest stem volume using VHF SAR. IEEE Trans Geosci Remote Sens 35(1):36–40
Israelsson H, Ulander LMH, Martin T, Askne JIH (2000) A coherent scattering model to determine forest backscattering in the VHF-band. IEEE Trans Geosci Remote Sens 38(1):238–248
Jong-Sen L, Grunes MR, Pottier E, Ferro-Famil L (2004) Unsupervised terrain classification preserving polarimetric scattering characteristics. IEEE Trans Geosci Remote Sens 42(4):722–731
Kandaswamy U, Adjeroh DA, Lee MC (2005) Efficient texture analysis of SAR imagery. IEEE Trans Geosci Remote Sens 43(9):2075–2083
Karam MA, Fung AK, Lang RH, Chauhan NS (1992) A microwave scattering model for layered vegetation. IEEE Trans Geosci Remote Sens 30(4):767–784
Kasischke ES, Christensen NL (1990) Connecting forest ecosystem and microwave backscatter models. Int J Remote Sens 11(7):1277–1298
Kasischke ES, Christensen NL Jr, Bourgeau-Chavez LL (1995) Correlating radar backscatter with components of biomass in loblolly pine forests. IEEE Trans Geosci Remote Sens 33(3):643–659
Kasischke ES, Melack JM, Dobson MC (1997) The use of imaging radars for ecological applications—a review. Remote Sens Environ 59(2):141–156
Kayitakire F, Hamel C, Defourny P (2006) Retrieving forest structure variables based on image texture analysis and IKONOS-2 imagery. Remote Sens Environ 102(3–4):390–401
Kellndorfer J, Walker W, Pierce L, Dobson C, Fites JA, Hunsaker C, Vona J, Clutter M (2004) Vegetation height estimation from shuttle radar topography mission and national elevation datasets. Remote Sens Environ 93(3):339–358
Kellogg RM, Arber FJ (1981) Stem eccentricity in coastal western hemlock. Can J For Res 11(3):714–718
Kenyi LW, Dubayah R, Hofton M, Schardt M (2009) Comparative analysis of SRTM-NED vegetation canopy height to LIDAR-derived vegetation canopy metrics. Int J Remote Sens 30(11):2797–2811
Kidd R, Scipal K (2003) A diurnal difference indicator for freeze-thaw monitoring from Ku band scatterometer applied within the Siberia II project. In: Proceedings of IEEE international geoscience and remote sensing symposium, IGARSS '04, vol 3, pp 1671–1674
Kimball JS, McDonald KC, Running SW, Frolking SE (2004) Satellite radar remote sensing of seasonal growing seasons for boreal and subalpine evergreen forests. Remote Sens Environ 90(2):243–258
Kimes DS, Ranson KJ, Sun G (1997) Inversion of a forest backscatter model using neural networks. Int J Remote Sens 18(10):2181–2199
Koskinen JT, Palliainen JT, Hyyppa JM, Engdahl ME, Hallikainen MT (2001) The seasonal behavior of interferometric coherence in boreal forest. IEEE Trans Geosci Remote Sens 39(4):820–829
Koskinen JT, Pulliainen JT, Luojus KP, Takala M (2010) Monitoring of snow-cover properties during the spring melting period in forested areas. IEEE Trans Geosci Remote Sens 48(1):50–58
Kugler F, Koudogbo FN, Papathanassiou KP (2006) Frequency effects in Pol-InSAR forest Height estimation. In: European conference on synthetic aperture radar (EUSAR), pp 1–4
Kugler F, Papathanassiou K, Hajnsek I, Coscia A (2007) Potential of forest height estimation using X band by means of two different inversion scenarios. In: 2007 IEEE international geoscience and remote sensing symposium, pp 1132–1135

Kugler F, Sauer S, Lee S-K, Papathanassiou K, Hajnsek I (2010) Potential of TanDEM-X for forest parameter estimation X-band Inversion test sites and results. In: 2010 8th european conference on synthetic aperture radar (EUSAR), pp 178–181

Kuplich TM, Salvatori V, Curran PJ (2000) JERS-1/SAR backscatter and its relationship with biomass of regenerating forests. Int J Remote Sens 21(12):2513–2518

Kuplich TM, Curran PJ, Atkinson PM (2005) Relating SAR image texture to the biomass of regenerating tropical forests. Int J Remote Sens 26(21):4829–4854

Kurosu T, Uratsuka S, Maeno H, Kozu T (1999) Texture statistics for classification of land use with multitemporal JERS-1 SAR single-look imagery. IEEE Trans Geosci Remote Sens 37(1):227–235

Kurvonen L, Hallikainen MT (1999) Textural information of multitemporal ERS-1 and JERS-1 SAR images with applications to land and forest type classification in boreal zone. IEEE Trans Geosci Remote Sens 37(2):680–689

Kux HJH, Henebry GM (1994) Multi-scale texture in SAR imagery: landscape dynamics of the Pantanal, Brazil. In: Proceedings of 1994 IEEE international geoscience and remote sensing symposium, IGARSS '94, vol 2, pp 1069–1071

Kwok R, Way J, Rignot E, Freeman A, Holt J (1993) Polarization signatures of frozen and thawed forests of varying biomass. In: 10th annual international symposium on geoscience and remote sensing, vol 32, issue 2, pp 337–340

Lang RH, Chauhan NS, Ranson KJ, Kilic O (1994) Modeling P-band SAR returns from a red pine stand. Remote Sens Environ 47(2):132–141

Lavalle M (2009) The synergy of SAR polarimetry and interferometry for forest parameters retrieval. In: European radar conference, EuRAD 2009, pp 160–163

Lavalle M, Simard M, Hensley S (2012) A temporal decorrelation model for polarimetric radar interferometers. IEEE Trans Geosci Remote Sens 50(7):2880–2888

Le Toan T, Beaudoin A, Riom J, Gyon D (1992) Relating forest biomass to SAR data. IEEE Trans Geosci Remote Sens 30(2):403–411

Le Toan T, Picard G, Martinez J-M, Melon P, Davidson M (2002) On the relationships between radar measurements and forest structure and biomass. In: Proceedings of the third international symposium on retrieval of bio- and geophysical parameters from SAR data for land applications, Noordwijk, Netherlands. ESA Publications Division, pp 3–12

Le Toan T, Ulander L, Papathanassiou K, Rocca F, Quegan S, Davidson M, Scipal K (2012) The BIOMASS mission retrieval algorithms: results from recent campaigns. In: 2012 IEEE international geoscience and remote sensing symposium, July 2012, pp 5546–5549

Leckie DG, Ranson KJ (1998) Forestry applications using imaging radar, 3rd edn. Wiley, New York

Leclerc G, Beaulieu N, Bonn F (2001) A simple method to account for topography in the radiometric correction of radar imagery. Int J Remote Sens 22(17):3553–3570

Lee J-S, Ainsworth TL (2011) The effect of orientation angle compensation on coherency matrix and polarimetric target decompositions. IEEE Trans Geosci Remote Sens 49(1):53–64

Lee J-S, Schuler DL, Ainsworth TL, Boerner WM (1999) POLSAR data compensation for terrain azimuth slope variation. In: IEEE 1999 international geoscience and remote sensing symposium, IGARSS'99, Cat. No.99CH36293, vol 5, pp 2437–2439

Lee J, Schuler DL, Ainsworth TL (2000a) Polarimetric SAR data compensation for terrain azimuth slope variation. IEEE Trans Geosci Remote Sens 38(5):2153–2163

Lee J-S, Krogager E, Schuler DL, Ainsworth TL, Boerner WM (2000b) On the estimation of polarization orientation angles induced from azimuth slopes using polarimetric SAR data. In: Proceedings of IEEE 2000 international geoscience and remote sensing symposium taking the pulse of the planet: the role of remote sensing in managing the environment, IGARSS 2000, Cat. No.00CH37120, vol 3, issue 202, pp 1310–1312

Lee J-S, Schuler DL, Ainsworth TL (2004) A review of polarization orientation estimation from polarimetric SAR data. In: Proceedings of the workshop on POLinSAR—applications of SAR polarimetry and polarimetric interferometry, Frascati, Italy, CD-ROM, p 3.1

Lee S-K, Kugler F, Papathanassiou K, Hajnsek I (2009) Polarimetric SAR interferometry for forest application at P-band: potentials and challenges. In: 2009 IEEE international geoscience and remote sensing symposium, pp IV-13-IV-16

Lee S-K, Kugler F, Hajnsek I, Papathanassiou K (2010) Multi-baseline Pol-InSAR forest height estimation in the presence of temporal decorrleation Pol-InSAR inversion and temporal decorrelation. In: 2010 8th european conference synthetic aperture radar (EUSAR), vol i, pp 829–832

Lee H, Liu JG (2001) Analysis of topographic decorrelation in SAR interferometry using ratio coherence imagery. IEEE Trans Geosci Remote Sens 39(2):223–232

Lei Y, Siqueira P, Clewley D, Lucas R (2012) Observation of vegetation vertical structure and disturbance using L-band InSAR over the Injune region in Australia. In: 2012 IEEE international geoscience and remote sensing symposium, July 2012, pp 1637–1640

Li ZW, Ding XL, Huang C, Zou ZR, Chen YL (2007) Atmospheric effects on repeat-pass InSAR measurements over Shanghai region. J Atmos Solar-Terr Phys 69(12):1344–1356

Li Z, Guo M (2012) A new three-stage inversion procedure of forest height with the improved temporal decorrelation RVoG model. In: 2012 IEEE international geoscience and remote sensing symposium, vol 2, issue 1, July 2012, pp 5141–5144

Liu D, Du Y, Sun G, Yan W, Wu B (2008) Analysis of InSAR sensitivity to forest structure based on radar scattering model. Prog Electromagnet Res 84:149–171

Loew A, Mauser W (2007) Generation of geometrically and radiometrically terrain correctedSAR image products. Remote Sens Environ 106(3):337–349

Lombardini F (2005) Differential tomography: a new framework for SAR interferometry. In: Proceedings of 2003 IEEE international geoscience and remote sensing symposium, IGARSS 2003, IEEE Cat. No.03CH37477, vol 2, issue 1, pp 1206–1208

Lu D (2006) The potential and challenge of remote sensing-based biomass estimation. Int J Remote Sens 27(7):1297–1328

Luckman A (1997) A study of the relationship between radar backscatter and regenerating tropical forest biomass for spaceborne SAR instruments. Remote Sens Environ 60(1):1–13

Luckman AJ (1998) The effects of topography on mechanisms of radar backscatter from coniferous forest and upland pasture. IEEE Trans Geosci Remote Sens 36(5):1830–1834

Luckman A, Groom G, Baker J (1994) Forest age discrimination from texture measures of SAR imagery. In: Proceedings 1994 IEEE international geoscience and remote sensing, IGARSS '94, vol 1, pp 104–107

Luckman AJ, Frery AC, Yanasse CCF, Groom GB (1997) Texture in airborne SAR imagery of tropical forest and its relationship to forest regeneration stage. Int J Remote Sens 18(6):1333–1349

Luckman AJ, Baker J, Wegmüller U (2000) Repeat-pass interferometric coherence measurements of disturbed tropical forest from JERS and ERS satellites. Remote Sens Environ 73(3):350–360

Luo X, Askne J, Smith G, Dammert P (2000) Coherence characteristics of radar signals from rough soil—abstract. J Electromagnet Waves Appl 14(11):1555–1557

Magagi R, Bernier M (2002) Quantitative analysis of RADARSAT SAR data over a sparse forest canopy. IEEE Trans Geosci Remote Sens 40(6):1301–1313

Magnusson M, Fransson JES, Eriksson LEB, Sandberg G, Smith-Jonforsen G, Ulander LMH (2007) Estimation of forest stem volume using ALOS PALSAR satellite images. In: 2007 IEEE international geoscience and remote sensing symposium, pp 4343–4346

Malhi Y, Román-Cuesta RM (2008) Analysis of lacunarity and scales of spatial homogeneity in IKONOS images of Amazonian tropical forest canopies. Remote Sens Environ 112(5):2074–2087

Manninen T, Parmes E, Häme T, Sephton A, Bach H, Borgeaud M (2000) ERS coherence and SLC images in forest characterisation. In: Proceedings of ERS-ENVISAT-symposium 2000, vol 5, Gothenburg, Sweden, ESA

Martinez J-M, Floury N, Le Toan T, Beaudoin A, Hallikainen MT, Mäkynen M (2000) Measurements and modeling of vertical backscatter distribution in forest canopy. IEEE Trans Geosci Remote Sens 38(2):710–719

Mas JF, Flores JJ (2008) The application of artificial neural networks to the analysis of remotely sensed data. Int J Remote Sens 29(3):617–663
Mcdonald KC, Dobson MC, Ulaby FT (1990) Using mimics to model L-band multiangle and multitemporal backscatter from a walnut orchard. IEEE Trans Geosci Remote Sens 28(4):477–491
Mcdonald KC, Ulaby FT (1993) Radiative transfer modelling of discontinuous tree canopies at microwave frequencies. Int J Remote Sens 14(11):2097–2128
Mcintyre NE, Wiens JA (2000) A novel use of the lacunarity index to discern landscape function. J Wildl Manag 15:313–321
McRoberts RE, Tomppo EO, Finley AO, Heikkinen J (2007) Estimating areal means and variances of forest attributes using the k-nearest neighbors technique and satellite imagery. Remote Sens Environ 111(4):466–480
Melon P, Martinez JM, Le Toan T, Ulander LMH, Beaudoin A (2001) On the retrieving of forest stem volume from VHF SAR data: observation and modeling. IEEE Trans Geosci Remote Sens 39(11):2364–2372
Mette T (2007) Forest Biomass Estimation from Polarimetric SAR Interferometry. Technische Universität München, Doktorarbeit
Mette T, Papathanassiou K (2004) Biomass estimation from polarimetric SAR interferometry over heterogeneous forest terrain. In: Proceedings of IEEE international geoscience and remote sensing symposium, IGARSS '04, vol 1, issue C, pp 511–514
Mette T, Papathanassiou KP, Hajnsek I, Zimmermann R (2002) Forest biomass estimation using polarimetric SAR interferometry. In: Proceedings of IEEE international geoscience and remote sensing symposium, Toronto, Canada, pp 817–819
Moghaddam M, Saatchi S (1993) Analysis of scattering mechanisms in SAR imagery over boreal forest: results from BOREAS '93. IEEE Trans Geosci Remote Sens 33(5):1290–1296
Morel ACS, Saatchi SS, Malhi Y, Berry NJ, Banin L, Burslem D, Nilus R, Ong RC (2011) Estimating aboveground biomass in forest and oil palm plantation in Sabah, Malaysian Borneo using ALOS PALSAR data. For Ecol Manag 262(9):1786–1798
Mougin E, Lopes A, Karam MA, Fung AK (1993) Effect of tree structure on X-band microwave signature of conifers. IEEE Trans Geosci Remote Sens 31(3):655–667
Mougin E, Proisy C, Marty G, Fromard F, Puig H, Betoulle JL, Rudant JP (1999) Multifrequency and multipolarization radar backscattering from mangrove forests. IEEE Trans Geosci Remote Sens 37(1):94–102
Natale A, Guida R, Bird R, Whittaker P, Hall D, Cohen M (2012) Validation of S-band data performance for future spaceborne SAR missions. In: 9th European conference on synthetic aperture radar, EUSAR, pp 75–78
Neeff T, Dutra LV, dos Santos JR, da Costa Freitas C, Araujo LS (2005) Tropical forest measurement by interferometric height modeling and P-band radar backscatter. For Sci 51(6):585–594
Neumann M, Ferro-Famil L, Reigber A (2010) Estimation of forest structure, ground, and canopy layer characteristics from multibaseline polarimetric interferometric SAR data. IEEE Trans Geosci Remote Sens 48(3):1086–1104
Neumann M, Saatchi SS, Ulander LMH, Fransson JES (2012) Assessing performance of L- and P-band polarimetric interferometric SAR data in estimating boreal forest above-ground biomass. IEEE Trans Geosci Remote Sens 50(3):714–726
Nguyen H, Roussel H, Tabbara W (2006) A coherent model of forest scattering and SAR imaging in the VHF and UHF-band. IEEE Trans Geosci Remote Sens 44(4):838–848
Ni W, Guo Z, Zhang Z, Sun G (2011) The annual behavior of backscattering and coherence of PALSAR data. In: 2011 IEEE international geoscience and remote sensing symposium, July 2011, pp 2472–2475
Oh Y, Sarabandi K (2002) Full-wave analysis of microwave scattering from short vegetation: an investigation on the effect of multiple scattering. IEEE Trans Geosci Remote Sens 40(11):2522–2526

Oliver CJ (2000) Rain forest classification based on SAR texture. IEEE Trans Geosci Remote Sens 38(2):1095–1104
Papathanassiou KP, Cloude SR (2001) Single-baseline polarimetric SAR interferometry. IEEE Trans Geosci Remote Sens 39(11):2352–2363
Papathanassiou KP, Cloude SR (2003) The effect of temporal decorrelation on the inversion of forest parameters from Pol-InSAR data. In: Proceedings of IGARSS
Papathanassion KP, Cloude SR, Reiber A, Boerner WM (200) Multi-baseline polarimetric SAR interferometry for vegetation parameters estimation. In: Proceedings of IEEE 2000 international geoscience and remote sensing symposium. Taking the pulse of the planet: the role of remote sensing in managing the environment, IGARSS 2000, Cat. No.00CH37120, vol 6, issue 44, pp 2762–2764
Park SE, Moon WM, Pottier E (2012) Assessment of scattering mechanism of polarimetric SAR signal from mountainous forest areas. ieeexplore.ieee.org 50(11):4711–4719
Peralta P, Mather P (2000) An analysis of deforestation patterns in the extractive reserves of Acre, Amazonia from satellite imagery: a landscape ecological approach. Int J Remote Sens 21(13–14):2555–2570
Plotnick RE, Gardner RH, Hargrove WW, Prestegaard K, Perlmutter M (1996) Lacunarity analysis : a general technique for the analysis of spatial patterns. Phys Rev E 53(5):5461–5468
Plotnick RE, Gardner RH, O'Neill RV (1993) Lacunarity indices as measures of landscape texture. Landscape Ecol 8(3):201–211
Praks J (2012) Radar polarimetry and interferometry for remote sensing of boreal forest. PhD thesis, School of Electrical Engineering, Department of Radio Science and Engineering, Space Technology
Praks J, Antropov O, Hallikainen MT (2012a) LIDAR-Aided SAR interferometry studies in Boreal forest: scattering phase center and extinction coefficient at X- and L-band. IEEE Trans Geosci Remote Sens 50:3831–3843
Praks J, Hallikainen M, Kugler F, Papathanassiou KP (2007a) X-band extinction in boreal forest: estimation by using E-SAR POLInSAR and HUTSCAT. In: 2007 IEEE international geoscience and remote sensing symposium, pp 1128–1131
Praks J, Kugler F, Papathanassiou KP, Hajnsek I, Hallikainen M (2007b) Height estimation of Boreal forest: interferometric model-based inversion at L- and X-band versus HUTSCAT profiling scatterometer. IEEE Geosci Remote Sens Lett 4(3):466–470
Proisy C, Mougin E, Fromard F, Karam MA (2000) Interpretation of polarimetric radar signatures of mangrove forests. Remote Sens Environ 71(1):56–66
Pulliainen JT, Heiska K, Hyyppa J, Hallikainen MT (1994) Backscattering properties of boreal forests at the C- and X-bands. IEEE Trans Geosci Remote Sens 32(5):1041–1050
Pulliainen JT, Mikhela PJ, Hallikainen MT, Ikonen J-P (1996) Seasonal dynamics of C-band backscatter of boreal forests with applications to biomass and soil moisture estimation. IEEE Trans Geosci Remote Sens 34(3):758–770
Pulliainen JT, Kurvonen L, Hallikainen MT (1999) Multitemporal behavior of L- and C-band SAR observations of boreal forests. IEEE Trans Geosci Remote Sens 37(2):927–937
Pulliainen JT, Engdahl ME, Hallikainen MT (2003) Feasibility of multi-temporal interferometric SAR data for stand-level estimation of boreal forest stem volume. Remote Sens Environ 85:397–409
Quegan S (2001) Filtering of multichannel SAR images. IEEE Trans Geosci Remote Sens 39(11):2373–2379
Quegan S, Le Toan T, Yu JJ, Ribbes F, Floury N (2000) Multitemporal ERS SAR analysis applied to forest mapping. IEEE Trans Geosci Remote Sens 38(2):741–753
Ranson KJ, Guoqing S (1994) Mapping biomass of a northern forest using multifrequency SAR data. IEEE Trans Geosci Remote Sens 32(2):388–396
Ranson KJ, Saatchi S (1995) Boreal forest ecosystem characterization with SIR-C/XSAR. IEEE Trans Geosci Remote Sens 33(4):867–876

Ranson KJ, Sun G (1997) Effect of environmental temperatures on SAR forest biomass estimates. In: Proceedings of 1997 IEEE international geoscience and remote sensing symposium, IGARSS'97. Remote sensing—a scientific vision for sustainable development, vol 4, pp 1722–1724

Ranson KJ, Sun G (2000) Effects of environmental conditions on boreal forest classification and biomass estimates with SAR. IEEE Trans Geosci Remote Sens 38(3):1242–1252

Ranson KJ, Sun G, Kharuk VI, Kovacs K (2001) Characterization of forests in western Sayani mountains, Siberia from SIR-C SAR data. Remote Sens Environ 75:188–200

Rauste Y (1990) Incidence-angle dependence in forested and non-forested areas in Seasat SAR data. Int J Remote Sens 11(7):1267–1276

Rauste Y, Häme T, Pulliainen JT, Heiska K, Hallikainen MT (1994) Radar-based forest biomass estimation. Int J Remote Sens 15(14):2797–2808

Rees WG, Steel AM (2001) Simplified radar mapping equations for terrain correction of space-borne SAR images. Int J Remote Sens 22(18):3643–3649

Reigber A (2001) Polarimetric sar tomography Techniques, PhD thesis, University of Stuttgart

Reigber A, Papathanassiou KP, Cloude SR, Moreira A (2000) SAR tomography and interferometry for the remote sensing of forested terrain. EUSAR 2000 Special Issue, Frequenz, Zeitschrift für Telekommunikation (Journal of Telecommunications), 55:119–122, March/April 2001

Richards JA (1990) Radar backscatter modelling of forests: a review of current trends. Int J Remote Sens 11(7):1299–1312

Rignot E, Salas WA, Skole DL (1997) Mapping deforestation and secondary growth in Rondonia, Brazil, using imaging radar and thematic mapper data. Remote Sens Environ 59(2):167–179

Rignot E, Way JB (1991) Monitoring freeze-thaw along north-south Alaskan transects using ERS-1 SAR. In: Proceedings of IEEE international geoscience and remote sensing symposium, IGARSS '93, pp 1453–1455

Rignot E, Williams C, McDonald K, Way JB, Zimmermann R, Viereck L (1994) Monitoring of freeze/thaw transitions in Taiga forests using ERS-1 SAR. In: Proceedings of 1994 IEEE international geoscience and remote sensing symposium, IGARSS '94, vol 1, Nov 1993, p 225

Rodriguez E, Michel T (1995) Estimation of penetration of forest canopies by interferometric SAR measurements. In: Electromagnetics research symposium (PIERS95), Seattle, WA, USA

Romshoo SA, Shimada M (2001) Employing SAR for biomass retrieval from tropical forests of southeast Asia. In: Proceedings of 22nd Asian conference on remote sensing, Singapore, pp 627–632

Saatchi SS, McDonald KC (1997) Coherent effects in microwave backscattering models for forest canopies. IEEE Trans Geosci Remote Sens 35(4):1032–1044

Saatchi SS, Halligan K, Despain DG, Crabtree RL (2007) Estimation of forest fuel load from radar remote sensing. IEEE Trans Geosci Remote Sens 45(6):1726–1740

Sader SA, Wu S-T (1987) Multipolarization SAR data for surface feature delineation and forest vegetation characterization. IEEE Trans Geosci Remote Sens 25(1):67–76

Salas WA, Ducey MJ, Rignot E, Skole D (2002) Assessment of JERS-1 SAR for monitoring secondary vegetation in Amazonia: I. Spatial and temporal variability in backscatter across a chrono-sequence of secondary vegetation stands in Rondonia. Int J Remote Sens 23(7):1357–1379

Sandberg G, Ulander LMH, Fransson JES, Holmgren J, Le Toan T (2011) L- and P-band backscatter intensity for biomass retrieval in hemiboreal forest. Remote Sens Environ 115(11):2874–2886

Santoro M (2003) Estimation of biophysical parameters in boreal forests from ERS and JERS SAR interferometry. Doctor of philosophy, Friedlich-Schiller University Jena

Santoro M, Askne J, Dammert PBG, Fransson JES, Smith G (1999) Retrieval of biomass in boreal forest from multi-temporal ERS-1/2 interferometry. In: Fringe 99: second international workshop on ERS SAR interferometry

Santoro M, Askne J, Smith G, Dammert PBG, Fransson JES (2000) Boreal forest monitoring with ERS coherence. In: Proceedings of the ESA envisat symposium 2000, Gothenburg, Sweden, ESA

Santoro M, Askne J, Smith G, Fransson JES (2002) Stem volume retrieval in boreal forests from ERS-1/2 interferometry. Remote Sens Environ 81:19–35

Santoro M, Askne J, Eriksson L, Schmullius C, Wiesmann A, Fransson J (2003a) Seasonal dynamics and stem volume retrieval in boreal forests using JERS-1 backscatter. In: Owe M, D'Urso G, Toulios L (eds) Proceedings of IEEE international geoscience and remote sensing symposium, 2004, IGARSS '04, vol 1, Agia Pelagia, Crete, Greece, March 2003, pp 231–242

Santoro M, Eriksson LEB, Schmullius C, Wiesmann A (2003b) The potential of ALOS single polarization InSAR for estimation of growing stock volume in boreal forest. In: Proceedings of IEEE international geoscience and remote sensing symposium, Toulouse, France, pp 1939–1941

Santoro M, Schmullius C, Askne J (2004) Evaluation of JERS-1 L-band SAR backscatter for stem volume retrieval in boreal forest. In: Proceedings of IEEE international geoscience and remote sensing symposium, IGARSS '04, vol 1, issue C, pp 515–518

Santoro M, Shvidenko A, McCallum I, Askne J, Schmullius C (2005) Analysis of large area ERS-1/2 coherence signatures of boreal forests and implications for retrieval of stem volume. In: Proceedings of 9th international symposium on physical measurements and signatures in remote sensing, vol M, Beijing, China, pp 14–16

Santoro M, Eriksson LEB, Askne J, Schmullius C (2006) Assessment of stand-wise stem volume retrieval in boreal forest from JERS-1 L-band SAR backscatter. Int J Remote Sens 27(16):3425–3454

Santoro M, Askne J, Werner CL, Wegmüller U (2007) Observations, modeling, and applications of ERS-ENVISAT coherence over land surfaces. IEEE Trans Geosci Remote Sens 45(8):2600–2611

Santoro M, Askne J, Beer C, Cartus O, Schmullius C, Wegmuller U, Wiesmann A (2008) Automatic model inversion of multi-temporal C-band coherence and backscatter measurements for forest stem volume retrieval. In: 2008 IEEE international geoscience and remote sensing symposium, IGARSS 2008, pp 124–127

Santoro M, Fransson JES, Eriksson LEB, Magnusson M, Ulander LMH, Olsson H (2009) Signatures of ALOS PALSAR L-band backscatter in Swedish forest. IEEE Trans Geosci Remote Sens 47(12):4001–4019

Santoro M, Schmullius C, Beer C, Cartus O, Eriksson L, Leiterer R, Reiche J, Thiel CA, Thiel CH (2010) Forest mapping of the northern hemisphere with spaceborne rada. In: ESA living planet symposium, Bergen, Norway, June 2010

Santoro M, Beer C, Cartus O, Schmullius C, Shvidenko A, McCallum I, Wegmüller U, Wiesmann A (2011) Retrieval of growing stock volume in boreal forest using hyper-temporal series of Envisat ASAR ScanSAR backscatter measurements. Remote Sens Environ 115(2):490–507

Sarabandi K, Lin YC (2000) Simulation of interferometric SAR response for characterizing the scattering phase center statistics of forest canopies. IEEE Trans Geosci Remote Sens 38(1):115–125

Sarker LR, Nichol JE (2011) Improved forest biomass estimates using ALOS AVNIR-2 texture indices. Remote Sens Environ 115(4):968–977

Sarker LR, Nichol J, Ahmad B, Busu I, Rahman AA (2012) Potential of texture measurements of two-date dual polarization PALSAR data for the improvement of forest biomass estimation. ISPRS J Photogramm Remote Sens 69:146–166

Sato A, Yamaguchi Y, Singh G (2012) Four-component scattering power decomposition with extended volume scattering model. IEEE Geosci Remote Sens Lett 9(2):166–170

Schmullius C (1997) Monitoring Siberian forests and agriculture with the ERS-1 Windscatterometer. IEEE Trans Geosci Remote Sens 35(5):1363–1366

Schuler DL, De Grandi G (1996) Measurement of topography using polarimetric SAR images. IEEE Trans Geosci Remote Sens 34(5):1266–1277

Schuler DL, Ainsworth TL, Lee J-S, De Grandi G (1998) Topographic mapping using polarimetric SAR data. Int J Remote Sens 19(1):141–160

Schuler DL, Lee J-S, Ainsworth TL (1999) Compensation of terrain azimuthal slope effects in geophysical parameter studies using polarimetric SAR data. Remote Sens Environ 69(2):139–155

Sexton JO, Bax T, Siqueira P, Swenson JJ, Hensley S (2009) A comparison of lidar, radar, and field measurements of canopy height in pine and hardwood forests of southeastern north America. For Ecol Manag 257(3):1136–1147

Shimada M, Isoguchi O, Tadono T, Isono K (2009) PALSAR radiometric and geometric calibration. IEEE Trans Geosci Remote Sens 47(12):3915–3932

Shimada M, Ohki M, Noguchi H (2010) Incidence angle dependence of PALSAR repeat pass interferometry Masanobu Shimada, Japan aerospace exploration agency, earth observation research center, Japan Masato Ohki, Japan aerospace exploration agency, earth observation research center, Japan Hideyu. In: 2010 8th European conference on synthetic aperture radar (EUSAR), pp 449–452

Silva TAM, Dias JMB (1996) The effect of forest understory on synthetic aperture radar backscatter. In: Proceedings of 1997 IEEE international geoscience and remote sensing symposium, IGARSS'97. Remote sensing—a scientific vision for sustainable development, vol 2, July 1995, pp 773–777

Simard M, Hensley S, Lavalle M, Dubayah R, Pinto N, Hofton M (2012) An empirical assessment of temporal decorrelation using the uninhabited aerial vehicle synthetic aperture radar over forested landscapes. Remote Sens 4(12):975–986

Skinner L, Luckman AJ, Balzter H (2002) Estimating forest stand height using SAR interferometry: a case study at thetford forest using spacebome and airborne interferometric systems. In: Proceedings of the ForestSAT symposium, Edinburgh, Scotland, CD-ROM

Small D (2011) Flattening gamma: radiometric terrain correction for SAR imagery. IEEE Trans Geosci Remote Sens 49(8):3081–3093

Small D, Holecz F, Meier E, Nüesch D, Barmettler A (1997) Geometric and radiometric calibration of RADARSAT images. In: Proceedings of the geomatics in the era of radarsat (GER'97) symposium, Ottawa, Canada, pp 24–30,

Small D, Meier E (2004) Robust radiometric terrain correction for SAR image comparisons. In: Proceedings of IEEE international geoscience and remote sensing symposium, IGARSS '04, vol 3, pp 1730–1733

Small D, Miranda N, Meier E (2009) A revised radiometric normalisation standard for SAR. In: 2009 IEEE international geoscience and remote sensing symposium, pp IV-566–IV-569

Smith G, Dammert PB, Askne J (1996) Decorrelation mechanisms in C-band SAR interferometry over boreal forest. In: Franceschetti G, Oliver CJ, Rubertone FS, Tajbakhsh S (eds) Proceedings of SPIE, microwave sensing and synthetic aperture radar, Dec 1996, pp 300–310

Smith G, Dammert PBG, Santoro M, Fransson JES, Askne J, Smith G (1998) Biomass retrieval in boreal forest using ERS and JERS SAR. In: Proceedings of the 2nd international workshop on retrieval of bio- and geophysical parameters from SAR data for land applications, volume ESTEC, Noordwijk, Netherlands. ESA, pp 293–300

Smith G, Ulander LMH (2000) A model relating VHF-band backscatter to stem volume of coniferous boreal forest. IEEE Trans Geosci Remote Sens 38(2):728–740

Soja MJ, Sandberg G, Ulander LMH (2010) Topographic correction for biomass retrieval from P-band SAR data in boreal forests. In: 2010 IEEE international geoscience and remote sensing symposium, July 2010, pp 4776–4779

Soja MJ, Sandberg G, Ulander LMH (2013) Regression-based retrieval of boreal forest biomass in sloping terrain using P-band SAR backscatter intensity data. IEEE Trans Geosci Remote Sens 51(5):2646–2665

Solberg S, Astrup R, Gobakken T, Næsset E, Weydahl DJ (2010) Estimating spruce and pine biomass with interferometric X-band SAR. Remote Sens Environ 114(10):2353–2360

Spies TA (1998) Forest structure: a key to the ecosystem. Northwest Sci 72(2):34–39

Sugimoto M, Ouchi K, Nakamura Y (2012) Four-component scattering power decomposition algorithm with rotation of covariance matrix using ALOS-PALSAR polarimetric data. Remote Sens 4(12):2199–2209

Sun G, Ranson KJ (1998) Radar modelling of forest spatial patterns. Int J Remote Sens 19(9):1769–1791

Sun G, Simonett DS, Strahler AH (1991) A radar backscatter model for discontinuous coniferous forests. IEEE Trans Geosci Remote Sens 29(4):639–650

Svoray T, Shoshany M (2002) SAR-based estimation of areal aboveground biomass (AAB) of herbaceous vegetation in the semi-arid zone: a modification of the water-cloud model. Int J Remote Sens 23(19):4089–4100

Tanase MA, Perez-Cabello F, de la Riva J, Santoro M (2010) TerraSAR-X data for burn severity evaluation in mediterranean forests on sloped terrain. IEEE Trans Geosci Remote Sens 48(2):917–929

Tanase M, de la Riva J, Santoro M, Pérez-Cabello F, Kasischke ES (2011) Sensitivity of SAR data to post-fire forest regrowth in Mediterranean and boreal forests. Remote Sens Environ 115(8):2075–2085

Tebaldini S, D'Alessandro MM, Rocca F (2010) SAR imaging of forest structure at longer wavelengths. In: 2010 IEEE radar conference, pp 811–815

Tebaldini S, Rocca F (2009) Polarimetric options for SAR tomography of forested areas. In: Proceedings of the fourth international workshop on science and applications of SAR polarimetry and polarimetric interferometry, PolInSAR 2009, p 45

Tebaldini S, Rocca F (2012) Multibaseline polarimetric SAR tomography of a Boreal forest at P- and L-bands. IEEE Trans Geosci Remote Sens 50(1):232–246

Telewski FW (1995) Wind-induced physiological and developmental responses in trees. In: Wind and trees, chapter 14. Cambridge University Press, Cambridge, pp 237–263

Thiel CH, Thiel CA, Reiche J, Leiterer R, Santoro M, Schmullius C (2007) Polarimetric PALSAR SAR data for forest cover mapping in Siberia. In: Proceedings CD of first joint PI symposium of ALOS data nodes for ALOS science program, Kyoto, Japan

Thiel CH, Thiel CA, Schmullius C (2009) Operational large-area forest monitoring in Siberia using ALOS PALSAR summer intensities and winter coherence. Earth 47(12):3993–4000

Thiel C, Schmullius C (2012) Effect of tree species on PALSAR INSAR coherence over Siberian forest at frozen and unfrozen conditions. In: 2012 IEEE international geoscience and remote sensing symposium, July 2012, pp 190–193

Thirion L, Colin E, Dahon C (2006) Capabilities of a forest coherent scattering model applied to radiometry, interferometry, and polarimetry at P- and L-band. IEEE Trans Geosci Remote Sens 44(4):849–862

Tian X, Su Z, Chen E, Li Z, van der Tol C, Guo J, He Q (2012) Estimation of forest above-ground biomass using multi-parameter remote sensing data over a cold and arid area. Int J Appl Earth Obs Geoinf 14(1):160–168

Tomppo E (2004) Using coarse scale forest variables as ancillary information and weighting of variables in k-NN estimation: a genetic algorithm approach. Remote Sens Environ 92(1):1–20

Tomppo EO, Gagliano C, De Natale F, Katila M, McRoberts RE (2009) Predicting categorical forest variables using an improved k-nearest neighbour estimator and landsat imagery. Remote Sens Environ 113(3):500–517

Tomppo E, Olsson HK, Ståhl G, Nilsson M, Hagner O, Katila M (2008) Combining national forest inventory field plots and remote sensing data for forest databases. Remote Sens Environ 112(5):1982–1999

Torano Caicoya A, Kugler F, Hajnsek I, Papathanassiou K (2012) Boreal forest biomass classification with TanDEM-X. In: 2012 IEEE international geoscience and remote sensing symposium, July 2012, pp 3439–3442

Treuhaft RN, Cloude SR (1999) The structure of oriented vegetation from polarimetric interferometry. IEEE Trans Geosci Remote Sens 37:2620

Treuhaft RN, Madsen SN, Moghaddam M, van Zyl JJ (1996) Vegetation characteristics and underlying topography from interferometric radar. Radio Sci 31(6):1449–1485

Tsolmon R, Tateishi R, Tetuko JS (2002) A method to estimate forest biomass and its application to monitor Mongolian Taiga using JERS-1 SAR data. Int J Remote Sens 23(22):4971–4978

Ulaby FT (1981) Microwave response of vegetation. Adv Space Res 1(10):55–70

Ulaby FT, El-rayes M (1987) Microwave dielectric spectrum of vegetation—part II: dual-dispersion model. IEEE Trans Geosci Remote Sens GE-25(5):550–557

Ulaby FT, Kouyate F, Brisco B, Williams TH (1986) Textural information in SAR images. IEEE Trans Geosci Remote Sens 24(2):235–245

Ulaby FT, Sarabandi K, Mcdonald K, Whitt M, Dobson MC (1990) Michigan microwave canopy scattering model. Int J Remote Sens 11(7):1223–1253

Ulander LMH (1996) Radiometric slope correction of synthetic-aperture radar images. IEEE Trans Geosci Remote Sens 34(5):1115–1122

Ulander LMH, Dammert PBG, Hagberg JO (1995) Measuring tree height using ERS-1 SAR interferometry. In: International geoscience and remote sensing symposium, IGARSS '95. Quantitative remote sensing for science and applications, vol 3, pp 2189–2193

van der Sanden JJ (1997) Radar remote sensing to support tropical forest management. PhD thesis, Wageningen

van Zyl JJ (1993) The effect of topography on radar scattering from vegetated areas. IEEE Trans Geosci Remote Sens 31(1):153–160

van Zyl JJ, Arii M, Kim Y (2011) Model-based decomposition of polarimetric SAR covariance matrices constrained for nonnegative eigenvalues. IEEE Trans Geosci Remote Sens 49(9):3452–3459

Villard L, Le Toan T, Lasne Y, Mermoz S (2012) Specific biomass indicator for tropical dense forests over hilly terrains derived from the P-band SAR coherency matrix. In: 2012 IEEE international geoscience and remote sensing symposium, vol 2, issue 1, July 2012, pp 5344–5347

Wagner W, Luckman AJ, Vietmeier J, Tansey K, Balzter H, Schmullius C, Davidson M, Gaveau DLA, Gluck M, Le Toan T, Shaun Q, Shvidenko A, Wiesmann A, Yu JJ (2003) Large-scale mapping of boreal forest in SIBERIA using ERS tandem coherence and JERS backscatter data. Remote Sens Environ 85(2):125–144

Wagner W, Vietmeier J, Schmullius C, Davidson M, Le Toan T, Luckman A, Tansey K, Balzter H, Gaveau D (2000) The use of coherence information from ERS tandem pairs for determining forest stock volume in SIBERIA. In: Proceedings of IEEE 2000 international geoscience and remote sensing symposium. Taking the pulse of the planet: the role of remote sensing in managing the environment, IGARSS 2000, Cat. No.00CH37120, vol 4, Honolulu, Hawaii, pp 1396–1398

Walker WS, Kellndorfer JM, LaPoint E, Hoppus ML, Westfall J (2007) An empirical InSAR-optical fusion approach to mapping vegetation canopy height. Remote Sens Environ 109:482–499

Walter F (1997) Extraction of forest tree volume from CARABAS SAR data. Scand J For Res 12(4):370–374

Wang Y, Kasischke ES, Melack JM, Davis FW, Christensen NL Jr (1994) The effects of changes in loblolly pine biomass and soil moisture on ERS-1 SAR backscatter. Remote Sens Environ 49:25–31

Wang Y, Day JL, Davis FW (1998) Sensitivity of modeled C- and L-band radar backscatter to ground surface parameters in loblolly pine forest. Remote Sens Environ 66:331–342

Wang Y, Day J, Sun G (1993) Santa Barbara microwave backscattering model for woodlands. Int J Remote Sens 14(8):1477–1493

Wang Y, Dong D (1997) Retrieving forest stand parameters from SAR backscatter data using a neural network trained by a canopy backscatter model. Int J Remote Sens 18(4):981–989

Wang H, Ouchi K, Watanabe M, Shimada M, Tadono T, Rosenqvist Å, Romshoo SA, Matsuoka M, Moriyama T, Uratsuka S (2006) In search of the statistical properties of high-resolution polarimetric SAR data for the measurements of forest biomass beyond the RCS saturation limits. Remote Sens Environ Lett 3(4):495–499

Watanabe M, Shimada M, Rosenqvist A, Tadono T, Matsuoka M, Romshoo SA, Ohta K, Furuta R, Nakamura K, Moriyama T (2006) Forest structure dependency of the relation between L-band sigma nought and biophysical parameters. IEEE Trans Geosci Remote Sens 44(11):3154–3165

Watt MS, Moore JR, McKinlay B (2004) The influence of wind on branch characteristics of Pinus radiata. Trees 19(1):58–65

Way J, Paris J, Kasischke ES, Slaughter C, Viereck L, Christensen NL, Dobson MC, Ulaby FT, Richards J, Milne A, Sieber A, Ahern FJ, Simonett D, Hoffer R, Imhoff ML, Weber J (1990) The effect of changing environmental conditions on microwave signatures of forest ecosystems: preliminary results of the (March 1988) Alaskan aircraft SAR experiment. Int J Remote Sens 11(7):1119–1144

Wegmuller U, Werner CL (1995) SAR interferometric signatures of forest. IEEE Trans Geosci Remote Sens 33(5):1153–1161

Wegmuller U, Strozzi T, Werner C (1996) Forest applications of ERS, JERS, and SIR-C SAR interferometry. In: Proceedings of 1997 IEEE international geoscience and remote sensing symposium. Remote sensing—a scientific vision for sustainable development, IGARSS'97, vol 2, pp 790–792

Weishampel JF, Godin JR, Henebry GM (2001) Pantropical dynamics of intact rain forest canopy texture. Glob Ecol Biogeogr 10(4):389–397

Weishampel JF, Sun G, Ranson KJ, LeJeune KD, Shugart HH (1994) Forest textural properties from simulated microwave backscatter: the influence of spatial resolution. Remote Sens Environ 47:120–131

West GB, Brown JH, Enquist BJ (1999) A general model for the structure and allometry of plant vascular systems. Lett Nat 400:664–667

Westman WE, Paris JF (1987) Detecting forest structure and biomass with C-band multipolarization radar: physical model and field tests. Remote Sens Environ 22(2):249–269

Wever T, Bodechtel J (1998) Different processing levels of SIR-C/X-SAR radar data for the correction of relief induced distortions in mountainous areas. Int J Remote Sens 19(2):349–357

Weydahl DJ, Sagstuen J, Dick B, Rønning H (2007) SRTM DEM accuracy assessment over vegetated areas in Norway. Int J Remote Sens 28(16):3513–3527

Williamson L (1975) Out-of-roundness in Douglas-fir stems. For Sci 21(4):365–370

Woodcock CE, Strahler AH (1987) The factor of scale in remote sensing. Remote Sens Environ 21(3):311–332

Woodhouse IH (2006) Predicting backscatter-biomass and height-biomass trends using a macroecology model. IEEE Trans Geosci Remote Sens 44(4):871–877

Woodhouse IH, Cloude S, Papathanassiou K, Hutchinson C (2003) Evaluating POLinSAR tree height and topography retrievals in Glen Affric. In: Lacoste H (ed) Proceedings of the workshop on POLinSAR—applications of SAR polarimetry and polarimetric interferometry (ESA SP-529), Frascati, Italy, p 21.1

Wulder MA, Skakun RS, Kurz WA, White JC (2004) Estimating time since forest harvest using segmented Landsat ETM+ imagery. Remote Sens Environ 93(1–2):179–187

Wu S (1987) Potential application of multipolarization SAR for pine-plantation biomass estimation. IEEE Trans Geosci Remote Sens GE-25(3):403–409

Xu F, Jin Y (2005) Deorientation theory of polarimetric scattering targets and application to terrain surface classification. IEEE Trans Geosci Remote Sens 43(10):2351–2364

Yamaguchi Y, Moriyama T, Ishido M, Yamada H (2005) Four-component scattering model for polarimetric SAR image decomposition. IEEE Trans Geosci Remote Sens 43(8):1699–1706

Yamaguchi Y, Sato A, Boerner W-F, Sato R, Yamada H (2011) Four-component scattering power decomposition with rotation of coherency matrix. IEEE Trans Geosci Remote Sens 49(6):2251–2258

Yamaguchi Y, Yajima Y, Yamada H (2006) A four-component decomposition of POLSAR images based on the coherency matrix. IEEE Geosci Remote Sens Lett 3(3):292–296

Yong P, Li Z, Sun G, Erxue C, Xuejian C (2003) Comparison of tree height estimations from C and L-band InSAR data. In: Proceedings of IGARSS 2003, 2003 IEEE international geoscience and remote sensing symposium, IEEE Cat. No. 03CH37477, vol 4, issue C, pp 2586–2588

Yueh SH, Kong JA, Jao JK, Shin RT, Le Toan T (1992) Branching model for vegetation. IEEE Trans Geosci Remote Sens 30(2):390–402

Zebker HA, Villasenor J (1992) Decorrelation in interferometric radar echoes. IEEE Trans Geosci Remote Sens 30(5):950–959

Zhang Q, Huang Y, Schwaebisch M, Mercer B, Wei M (2012) Forest height estimation using single-pass dual-baseline L-band PolInSAR data. In: 2012 IEEE international geoscience and remote sensing symposium, July 2012, pp 7055–7058

Zhou Z, Lehmann E, Wu X, Caccetta P, Mcneill S, Mitchell A, Milne A, Tapley I, Lowell K (2011) Terrain slope correction and precise registration of SAR data for forest mapping and monitoring. In: International symposium for remote sensing of the environment

Ziade Y, Roussel H, Lesturgie M, Tabbara W (2008) A coherent model of forest propagation—application to detection and localization of targets using the DORT method. IEEE Trans Antennas Propag 56(4):1048–1057

Chapter 3
Theory and Techniques

To comprehend the various forest parameters and SAR techniques considered in this thesis, the biophysical attributes of a forest and SAR theory have been reviewed in this chapter. The first section introduces the important parameters and phenologies related to a forest. In particular, biomass quantity, GSV and other related forest parameters, which are essential for this study, are defined, and the influence of temperature, topography and wind on forest growth is briefly described. The second section describes SAR remote sensing systems. More specifically, the SAR fundamentals together with the InSAR, PolSAR and PolInSAR techniques, which are applied in this thesis, have been reviewed from the perspective of forestry. The final section focuses on the fusion of SAR information. The analysis of the fusion methods employed in this thesis does not require any technical background. However, the definitions of fusion and its synonyms, namely synergy, integration or combination, may be ambiguous. Therefore, the section on fusion presents the definitions of fusion and terms which are closely related to it and briefly describes the different fusion levels. For conciseness, this chapter provides only the main theoretical background and formula required to understand this thesis. The reader is referred to references for further details.

3.1 Forestry Fundamentals

3.1.1 Forest Biophysical Parameters

On the basis of the domain of interest (e.g. ecology, bioenergy, forest-management), the term forest biomass encompasses several components of a forest and can therefore have different interpretations. For clarity, it is necessary to accurately define forest biomass and GSV and describe their characteristics.

Many researchers have defined biomass or forest biomass (Young 1980; Bonnor 1985; Levine 1996; Köhl et al. 2006). For instance, Köhl described biomass as 'the

N. Ackermann, *Growing Stock Volume Estimation in Temperate Forested Areas Using a Fusion Approach with SAR Satellites Imagery*,
Springer Theses, DOI 10.1007/978-3-319-13138-2_3

plant material being produced by or resulting from photosynthesis'. A general and accepted definition of biomass has also been reported in FAO (2005). According to this report, biomass is the total amount of 'organic material above and below the ground, both living and dead'. In forested areas, organic material naturally refers to trees and their understories, which are often represented by shrubs, saplings, epiphytes and other herbaceous plants. The biomass in understories is generally low in natural forests (less than 3–5 % of mature forest biomass), while it can represent more than 30 % of biomass of mature trees in secondary forests depending on the age of the forest and openness of canopies (Levine 1996). According to FAO (2005), above- and below-ground biomass can be distinguished. Above-ground forest biomass generally comprises leaves, twigs, branches, main bole and tree barks (van Laar and Akça 2007; Levine 1996), while below-ground biomass includes roots and the bottom part of stumps. Distinctions have also been drawn between living and dead organic matter. With a particular vertical extension, the entire living matter in forests is generally observed above the ground, while dead matter, which is composed of fine litter and coarse woody debris such as standing dead trees and dead wood, is mostly found above the ground on the forest floor and below the ground in an advanced state of decomposition. Plants and trees are hygroscopic, which implies that the biomass of living organic matter includes the weight of moisture. The water content can reach 80–90 % of the plants' weight and may vary according to several parameters, such as tree species composition, components of trees, tree size, location of growth and seasons as well as time of the day (Köhl et al. 2006). In this regard, the study of biomass generally specifies whether it is a dry or fresh biomass (West 2009).

Wood is principally composed of carbon (C), hydrogen (H) and oxygen (O). These elements combined form three major organic polymers, namely cellulose, hemicelluloses and lignin. Of the organic compounds which constitute wood cell walls, carbon is the dominant structural component. Researchers have reported that carbon constitutes 50 % of dry biomass (Köhl et al. 2006; Miller 1999).

To estimate the amount of carbon or to enable comparisons between different forests, it is required that biomass be measured over a unit area. In this regard, forest biomass expressed in kilograms (kg) or tonnes (t) of dry weight is defined by the forest biomass density, which is usually reported in tonnes per hectares ($\mathrm{t\,ha^{-1}}$) (Köhl et al. 2006). Biomass is evaluated using two different methods, destructive and non-destructive sampling. Destructive sampling involves estimation of the weight (mass) of vegetation. In practice, each part of a tree needs to be felled, collected and dried in order to determine the weight of each component and obtain the total dry weight of the assortments. This method cannot be used for forest monitoring as it requires logging of the trees under measurement. In the framework of realizing the implementation of forest inventories, methods based on dendrometric tree measurements have been introduced as an alternative to the destructive sampling approach. These techniques involve indirect measurement of biomass by determining the stem volume of a forest. Stem volume can be converted to biomass using the following relation:

$$B = GSV \cdot \rho , \tag{3.1}$$

where GSV is the growing stock volume and ρ the wood density. The wood density is expressed in $\mathrm{kg\,m^{-3}}$ or $\mathrm{g\,cm^{-3}}$, is species-specific and varies with growth conditions (i.e. seasons) and the measured parts of the tree. Over the past few decades, wood density has been widely studied, and estimates for most species can now be found in previous studies (Penman et al. 2003). The values for the different species were usually approximated to 0.5–0.6 $\mathrm{g\,cm^{-3}}$. However, wood densities of broadleaves were generally found to be higher than those of conifers (Penman et al. 2003; Tolunay 2011). GSV is a measure commonly used by foresters for forest-management. It is expressed in $\mathrm{m^3\,ha^{-1}}$ and is commonly referred to as bole volume, stem volume or growing stem volume. GSV can be defined as the tree stems for all living species per unit area, including the bark ignoring branches or stumps (FAO 2000; Santoro et al. 2011). According to FAO (2000), GSV refers to trees with a diameter at breast height (*DBH*) of greater than 10 cm. However, the fixed value of 10 cm may vary from country to country. Generally, for GSV estimation, foresters consider a tree as a cylinder with the basal area as the base and the height of the tree as the height of the cylinder. This approximation leads to the following allometric equation

$$GSV_{\mathrm{tree}} = h\pi \left(\frac{1}{2}DBH\right)^2 f_Z\,, \tag{3.2}$$

where h is the tree height, *DBH* is the stem diameter at breast height (1.3 m) and f_Z is a form factor accounting for the shapes of different tree species. The tree height is 'the distance between its top and base and is measured along a perpendicular dropped from the top' (van Laar and Akça 2007). *DBH* is measured over the bark at a fixed distance from the base. In general, the point of measurement is defined at a distance of 1.3 m above ground level, but this may vary with countries (Brack 2000). The form factor is expressed as the ratio of the actual stem volume to the volume of a reference cylinder and has been evaluated for several species in previous studies (Gray 1956; Zianis et al. 2005). To retrieve the GSV for an entire forest, it is common to divide the forest into patches, referred to as forest stands. A forest stand is defined as 'an aggregation of trees occupying a specific area and sufficiently uniform in composition, age, arrangement and condition so that this area is distinguishable from the forest in adjoining areas' (Northcote and Hartman 2004). The GSV for a forest stand can be evaluated as follows:

$$GSV_{\mathrm{stand}} = h_{\mathrm{g}} N_{\mathrm{trees}} \pi \left(\frac{1}{2}DBH\right)^2 f_Z\,, \tag{3.3}$$

$$= h_{\mathrm{g}} G f_Z\,, \tag{3.4}$$

with

$$G = N_{\mathrm{trees}} \pi \left(\frac{1}{2}DBH\right)^2 . \tag{3.5}$$

The variable h_g is the arithmetic mean height (m), G is the basal area ($m^2\,ha^{-1}$) and N is the number of trees per hectare. The stand-level height is expressed in a different manner depending on the condition of the stand. For even-aged stands, the mean height (h_g) would be chosen, while for inhomogeneous stands (owing to the effects of thinning and mortality), the maximum height (h_{100}) would be preferred (van Laar and Akça 2007; West 2009). Height measurement is time consuming and more complicated than DBH assessment. To improve the GSV determination process, foresters have developed species-specific dendrometric equations for relating the trees' DBH to their height. These equations can be fitted by least-square minimisation to select the measured samples of DBH-height pairs and further used to calculate h_g or h_{100} (see Sect. 4.2). The basal area of a stand is determined by the sum of all stems' (over the bark) cross-sectional areas, either at the breast height or at a specific height above the ground level. Owing to wind, slope steepness and other site-related factors, the stems of the trees often have an elliptical form instead of a circular shape (see Sect. 3.1). In this case, the quadratic mean of the stem diameter is more representative of the stand volume than the simple arithmetic mean and is, therefore, chosen for the determination of the basal area (Curtis and Marshall 2000; West 2009). The density parameter N_{trees} (trees ha^{-1}) can be considered as either an absolute or relative value. Absolute density refers to the number of trees in a specific area, while relative density is determined by comparing the absolute density with a pre-established standard. Hence, by enabling comparisons of forests at differing stages of development, relative density provides information about the growth conditions and competition in a stand and is preferred for silvicultural practices (planting, logging, thinning, etc.) (Bradley 1963). With the need for standards in the monitoring of forest stands, foresters have generalised the term relative density to imply relative stocking (Arner et al. 2001). By definition, a full-stocked stand corresponds to a theoretically optimal stem production under optimal growth conditions, while under- and overstocked stands represent poor productivity with lower and higher site-occupancy values, respectively, compared to a full-stocked stand (del Rio et al. 2001; Husch et al. 2003). Harvesting or thinning activities, which aim to control the forest stands' relative stocking, modify the forest structure to an extent. The term forest structure defines the arrangement of and relationship between tree components such as trunks, branches and leaves (Press 2013). It consists of attributes such as the type, size, shape and spatial distribution of the forest (Spies 1998; McElhinny et al. 2005; Pommerening 2002). The structure of natural forests is shaped by natural forces such as fire, wind and succession (Spies 1998), while that of managed forests is mainly defined by forestry activities. For economical purposes, foresters do not produce stem volumes in managed stands at the same rate as in natural stands. Therefore, the structure of a managed forest differs significantly from that of natural forests (Spies 1998).

3.1.2 Forest Phenologies

Temperate ecosystems are marked by four distinct seasons which introduce various annual forest phenologies. Phenology refers to the study of organism–environment relationships on the basis of the periodic environmental variations and the associated life cycle of the organisms (Liang and Schwartz 2009; Badeck et al. 2004). The most well-known and visible seasonal phenomenon is senescence, which is observed in autumn, and its occurrence is indicated by the leaf colour and leaf fall in many deciduous trees. Leaf fall is usually facilitated by wind. However, in the absence of strong winds, it is observed that the leaves remain on tree branches throughout winter until a new foliation develops in spring. This phenomenon is commonly referred to as marcescence (Abadia et al. 1996; Escudero and del Arco 1987). The summer season consists of a vegetative period with a peak in photosynthetic activity and tree development. In contrast, the winter season is the dormancy period with limited cellular activities (Lapointe 2001). The resting period is essentially driven by a reduction in the photoperiod and a decrease in temperature. It is particularly indicated by the cessation of sap flow and a decrease in water content. Sap flow and water transport in the stem are usually interrupted in autumn or early winter by embolism mechanisms. At this time, the temperature alternates above and below the freezing point, producing freeze-thaw cycles. The latter induces the formation of small gas bubbles in the stem transport system (vessels), which obstruct sap transmissions (Cochard and Granier 1999; Cruiziat et al. 2002; Cochard et al. 2000, 2001). Freeze-thaw cycles, which occur in autumn or early winter, are essential for preparing trees for the new vegetative period in spring. Indeed, the alternate warm and frozen conditions modify the internal pressure of trees and roots, facilitating the storage of nutrients in roots for winter and replenishment of the sap flow through trees after winter (Mayr et al. 2003). However, embolism of the vessels produced at the beginning of winter may prevent sap release in spring. In this case, if the tree cannot recover the damaged vessels, its growth season may be considerably shortened (Cochard et al. 2000, 2001). Owing to embolism mechanisms, conifers, which comprise thinner xylem conduct structures compared to broadleaves, are found at higher altitudes and latitudes (Lebourgeois 2006).

Similar to the case of temperature variations, wind may be related to several forest phenological aspects. One of these aspects is wind loading, which affects the morphology of trees. With a recurrent wind blowing in one direction, mechanical deformations appear on trunks. More specifically, the trunks tend to be elliptic, with their major axes being parallel to the prevailing winds (Hewson et al. 1978; Bannan and Bindra 1970; Telewski 1995). In topographic areas, similar effects can be found. Indeed, owing to both the orographic wind effect and the steep terrain, stems tend to be eccentric with the longest diameter oriented in the same direction as the slopes (Williamson 1975; Kellogg and Arber 1981). The consideration of wind and topography also determines the height of trees. Indeed, with a constant wind load on the canopies, trees located on hilltops tend to be more exposed to wind and are, therefore, found to be smaller than the trees located at the bottom of the slopes (Wade and Hewson 1979).

Topography involves not only the influence of wind but also several other climatic properties which are directly related to forest phenologies. One of these properties is the soil water content. Forests located at the top of hills or mountains are normally observed to have limited growth compared to forests situated at the bottom of slopes. This can be explained by the convex terrain formed at the hilltops, which is particularly subject to soil water-deficient conditions compared to the concave terrain down the hills or mountains resulting in water accumulation (Littell et al. 2008; Cornu 1998; FAO 2005). Solar radiation is another example of an environmental variable which varies with topography and influences the growth of forests. Typically, slopes exposed to the South present high illumination compared to those in the North, which mostly remain in the shade. Therefore, a forest's growth on south-facing slopes is more significant than that on north-facing slopes (Ellis et al. 2011; Bonan and Shugart 1989).

3.2 SAR Remote Sensing Systems

3.2.1 SAR Fundamentals

3.2.1.1 Electromagnetic Wave Theory

Fundamentally, remote sensing involves capturing and storing the energy of reflected electromagnetic radiation. This radiation can be represented by two perpendicular vectors, electric field ($\vec{E}$) and magnetic field ($\vec{H}$). The mathematical description of electromagnetic waves in space and time is given by Maxwell's equations.

$$\nabla \times \vec{E} = -\frac{\partial \vec{B}}{\partial t}, \tag{3.6}$$

$$\nabla \times \vec{H} = \vec{J} + \frac{\partial \vec{D}}{\partial t}, \tag{3.7}$$

$$\nabla \cdot \vec{B} = 0, \tag{3.8}$$

$$\nabla \cdot \vec{D} = \rho. \tag{3.9}$$

The above four equations describe the propagation of electromagnetic waves. The components of these equations are electric field $\vec{E}$, magnetic field $\vec{H}$, electric flux density $\vec{D}$, magnetic flux density $\vec{B}$, electric current density $\vec{J}$ and electric charge density ρ. In total, Maxwell's equations have 12 unknowns and constitute 8 equations. It can be assumed that the source terms ($\vec{J}$, ρ) are known, and the continuity law for current and charge densities can be followed. Hence, the following relation is obtained:

$$\nabla \cdot \vec{J} + \frac{\partial \rho}{\partial t} = 0. \tag{3.10}$$

Taking Relation (3.10) into consideration, there are only six independent Maxwell's equations for 12 variables. Thus, to obtain the complete system, six additional equations are required. They are given by the definition of the medium:

$$\vec{D} = \varepsilon \vec{E}\,, \tag{3.11}$$

$$\vec{B} = \mu \vec{H}\,. \tag{3.12}$$

Here ε and μ are the relative permittivity (dielectric constant) and permeability of the medium, respectively. Equations (3.11) and (3.12) complete the system, which is now solvable. At this stage, it is possible to obtain a simple electromagnetic wave in an environment which does not interrupt its propagation. Assuming that $\vec{J} = \rho = 0$ and the presence of a linear medium ($\vec{D} = \varepsilon \vec{E}$ and $\vec{B} = \mu \vec{H}$),

$$\vec{E}(\vec{r}, t) = \vec{E}_0 \cos(\vec{k} \cdot \vec{r} - \omega t + \varphi_0)\,. \tag{3.13}$$

Here $\vec{k} \cdot \vec{E} = 0$ and $k = \omega\sqrt{\mu\varepsilon}$. Relation (3.13) represents a monochromatic electromagnetic plane wave propagating in vacuum and mathematically describes the wave by a wave function which combines time t and space $\vec{r}$. $\vec{k}$ is the wave vector and its norm $\|\vec{k}\|$ is the wave number. This wave is defined by the wavelength $\lambda = \frac{2\pi}{k}$, the wave velocity $v = \frac{\omega}{k} = \frac{1}{\sqrt{\mu\varepsilon}}$, the period $T_\mathrm{p} = \frac{2\pi}{\omega} = \frac{1}{f}$ (with f being the carrier frequency) and the constant φ_0, which represents the initial phase. The equation of the EM wave in Relation (3.13) can also be described as a complex number, as shown in the following relation:

$$\vec{E}(\vec{r}, t) = A_0 e^{i(\vec{k}\cdot\vec{r} - \omega t + \varphi_0)} = A_0 e^{i\varphi} = A(\cos(\varphi) + i\sin(\varphi))\,. \tag{3.14}$$

The parameters $A_0 = E_0$ and $\varphi = kz - \omega t + \varphi_0$ describe the amplitude and phase of the signal, respectively. The intensity of the wave can be related to the amplitude as $I = A_0^2$. With reference to Eq. (3.14), electromagnetic radiation is characterised by five important quantities:

- intensity I;
- wavelength λ;
- polarisation P;
- phase φ;
- direction of propagation C.

These parameters carry information related to the source of radiation as well as the target under study. Therefore, the analysis of SAR signals in remote sensing mostly involves the evaluation of these five fundamental parameters, which are discussed further along the theoretical background provided in this section.

3.2.1.2 SAR Systems

Fundamentals

SAR systems are active sensors and are configured in lateral sight geometry; they are generally mounted on airborne or spaceborne platforms. For the same sensor operating from an aerial or spaceborne platform, important differences can be distinguished. First, a SAR sensor which records measurements from an aircraft provides a smaller swath and coverage area but has a higher spatial resolution compared to the same sensor records measurements from space. Second, in terms of incidence angle, the illumination is more uniform and precise for spaceborne platforms. Finally, a satellite is more stable than an aircraft; thus, for measurements recorded by a spaceborne SAR system, only minor corrections are required in trajectory and better position accuracy is provided.

The radio spectrum of SAR microwaves ranges between 2.5 mm and 10 m, corresponding to the carrier frequency range of 120 GHz–30 MHz, respectively ($f = c/\lambda$). The most common frequencies used in SAR remote sensing are X- (8–12 GHz), C- (4–8 GHz) and L- (1–2 GHz) bands. Some experiments also consider S-band (2–4 GHz), P-band (225–390 MHz) and VHF (30–300 MHz) sensors. The advantage of using microwave-spectrum regions is the smaller influence of atmospheric vapours, especially at low frequencies. However, electromagnetic waves at P- and VHF bands are affected by dispersed ionosphere particles (Snoeij et al. 2001). Therefore, use of such frequencies on spaceborne platforms is generally avoided.

The principles of SAR were originally derived from the real aperture radar (RAR) and radio detection and ranging (RADAR) instruments. These sensors consist of a transmitter, a receiver, an antenna and an electronic system to process and store data. As the platform moves, the transmitter generates short pulses with a certain pulse repetition frequency (*PRF*), and the receiver detects the returned signal. The surface illuminated by a single pulse describes a footprint which extends in two directions, namely range and azimuth. The former is orthogonal to the direction of motion of the aircraft (or spacecraft), while the latter is parallel to the platform flying direction.

Geometry—Spatial Resolutions

The principles of SAR are based on measurement of distance between the sensor and the features present in a scene. The range direction is commonly referred to as slant-range geometry. The slant range is projected onto the ground range on the basis of the incidence angle. The incidence angle is defined as the angle between the direction of the propagation of the waves and the normal to the surface (Woodhouse 2005). With the definition of the incidence angle, the depression angle may also be defined. This angle is given between an horizontal plane and the line of sight of the radar. The depression angle differs from the incidence angle in that it does not take into account for the effects of Earth curvature. On the basis of these characteristics, the ground range resolution can be calculated as follows (Massonnet and Souyris 2008):

$$\delta_{\mathrm{g}} \geq \frac{c\tau}{2\sin(\theta)}\,. \tag{3.15}$$

Here c is the speed of light ($c = 3 \times 10^8\ \mathrm{m\,s^{-1}}$), τ is the pulse duration (μs) and θ is the incidence angle (°). The factor 2 indicates the length of the pulse travelling to the target and back to the sensor. The pulse duration is approximately the reciprocal of the bandwidth. The sine of the incidence angle is used to project the slant-range resolution onto the ground-range resolution. Relation (3.15) shows that the ground-range resolution is proportional to the length of the pulses. Indeed, the shorter the pulse length, the finer the range resolution. Hence, to improve the range resolution, radar pulses should be as short as possible. However, a small pulse length implies less energy being returned to the sensor, thus limiting the detection of the signal. To overcome this problem, an electronic approach known as 'chirping' is used. As a result, the pulse length increases, thus improving the range resolution.

The azimuth resolution describes the ability of an imaging radar system to separate two closely spaced scatterers in the direction parallel to the motion vector of the sensor (Woodhouse 2005). In RAR systems, the azimuth resolution is technically low, particularly owing to the size of the antenna. To overcome this issue, the SAR technique, which considers the forward motion of the radar and synthesises the received pulse sequences, was developed. This technique, which takes the Doppler effect into consideration, synthetically increases the size of the antenna, thus improving the azimuth resolution. SAR azimuth resolution can be derived as follows (Massonnet and Souyris 2008):

$$\delta_{\mathrm{a}} \geq \frac{L}{2}\,. \tag{3.16}$$

As shown in Eq. (3.16), the SAR slant-range resolution is only related to antenna length L (m). This implies that theoretically, the spatial resolution of a SAR image in the azimuth direction at different altitudes would be the same. Moreover, in contrast to RAR systems, the slant-range distance does not influence the SAR azimuth resolution.

Signal Information—Phase and Amplitude

As described in Eq. (3.14), a SAR signal contains two important pieces of information, namely magnitude (also referred to amplitude) and phase. The phase can be geometrically expressed by the two-way path distance between the sensor and the target plus some perturbations elements. Its relationship can be expresses as follows:

$$\varphi = -\left(\frac{4\pi}{\lambda}\right)R + \varphi_{\mathrm{scatterer}} + \varphi_{\mathrm{delay}}\,. \tag{3.17}$$

Here R represents the slant-range distance, $\varphi_{\mathrm{scatter}}$ indicates the scattering effects introduced by the target and φ_{delay} refers to perturbations induced by the medium between the radar and the target, such as atmospheric or ionospheric effects. The magnitude of phase is indicated by the energy returned from a target. This quantity is provided by the fundamental radar equation (Massonnet and Souyris 2008)

$$P_{\mathrm{r}} = \frac{P_{\mathrm{t}} \cdot G_{\mathrm{t}} \cdot \sigma \cdot A_{\mathrm{r}}}{(4\pi)^2 \cdot R^4}. \tag{3.18}$$

The radar equation measures the proportion of transmitted energy scattered back from a target. P_{r} and P_{t} are the received and transmitted powers, respectively; R is the slant-range distance separating the target from the emitter, A_{r} is the effective area of the antenna, G_{t} is the gain of the antenna characterizing the antenna sensitivity pattern and σ is the radar cross-section. The parameters of the radar equation depend on the sensor configuration and imaging geometry, with the exception of σ, which also characterises the effects of the terrain on the radar signal.

The parameter σ is used to evaluate signals from point targets. However, in the context of radar EO, the targets are often distributed areas. In this case, the radar cross-section is replaced by a normalised parameter called backscatter coefficient (Massonnet and Souyris 2008)

$$\sigma_0 = \frac{\sigma}{A_\sigma}. \tag{3.19}$$

The radar backscatter coefficient determines the fraction of electromagnetic radiation reflected back to the radar for a specific cell resolution. It is a unitless quantity, but for ease of interpretation, it is usually expressed in decibels (dB). The backscatter coefficient considers several terrain characteristics such as geometry, surface roughness and moisture content, as well as radar system parameters such as wavelength, polarisation and incidence angle. The backscatter coefficient is usually in the range from -30 to 5 dB.

Speckle

The illumination of an area made of many scattering targets implies possible interference between the electromagnetic waves, which may be either constructive or destructive. Constructive interference results in high-intensity return signals, while destructive interference yields weak return signals. This intrinsic SAR effect is commonly called speckle and produces ‘salt and pepper’ noise, which degrades the image quality. To quantify noise, a theoretical noise value derived from the radar equation for a given SAR system has been defined. This value refers to the signal-to-noise ratio (*SNR*), which can be derived as follows (Kim 2009):

$$\frac{S}{N} = \frac{P_{\mathrm{t}} \cdot G_{\mathrm{t}} \cdot G_{\mathrm{r}} \cdot \lambda^3 \cdot \sigma_0 \cdot c \cdot \tau \cdot PRF}{(4\pi)^3 \cdot R^3 \cdot k \cdot T_0 \cdot NF \cdot B \cdot L_{\mathrm{tot}} \cdot 4 \cdot V_{\mathrm{s}} \cdot \sin(\theta - \alpha)} = \frac{\sigma_0}{NESZ}, \tag{3.20}$$

where

$$NESZ = \frac{(4\pi)^3 \cdot R^3 \cdot k \cdot T_0 \cdot NF \cdot W \cdot L_{tot} \cdot 4 \cdot V_s \cdot \sin(\theta - \alpha)}{P_t \cdot G_t \cdot G_r \cdot \lambda^3 \cdot c \cdot \tau \cdot PRF} . \quad (3.21)$$

As shown in Eq. (3.21), *SNR* is the ratio of signal power to noise power. The higher the *SNR*, the better the performance of a SAR system. An *SNR* value of 1 signifies that the signal is equal to the noise; in other words, the signal comprises only noise. By considering $SNR = 1$, the noise-equivalent sigma zero (*NESZ*) for a given system (Kim 2009) can be derived. According to Eq. (3.21), *NESZ* is evaluated in particular from the bandwidth W, the Boltzmann constant k, the equivalent noise temperature T_0, the total losses L_{tot}, the SAR platform flight speed V_s, the system noise figure *NF* and the antenna gain at the receiver G_r. *NESZ* indicates the sensitivity of a system to areas of low radar backscatter. The evaluation of *NESZ* in Relation (3.21) is valid for a monostatic system. In the case of a bistatic system, the reader is referred to Fiedler et al. (2002).

To reduce speckle, spatial, frequency and temporal filters can be used. One common way of reducing noise and improving the estimates of backscatter intensity is to incoherently evaluate the average intensity of a collection of L pixels. This process is called multi-looking, and it generates an L-look intensity image. Owing to the incoherent nature of the process, phase information is lost after multi-looking. Additionally, the spatial resolution of an L-look image is reduced by a factor L. For a homogenous area, the multi-looked-intensity image can be modelled using a gamma distribution (Oliver and Quegan 2004)

$$P_I(I) = \frac{1}{\Gamma(L)} (\frac{L}{\sigma})^L I^{L-1} e^{-\frac{LI}{\sigma}} . \quad (3.22)$$

Here Γ is the gamma function, L is the number of looks, I is the estimated intensity and σ is the backscatter coefficient. With an increasing number of looks, the multi-looked intensity distribution tends to normality. The performance of multi-looking and the filters used can be evaluated using equivalent number of looks (*ENL*). In this respect, two approaches can be conceived, namely theoretical (Massonnet and Souyris 2008; Marino 2012) and statistical (Oliver and Quegan 2004)

$$ENL_t = \frac{\rho_a}{\delta_a} \frac{\rho_r}{\delta_r} L_a L_r N_s , \quad (3.23)$$

$$ENL_s = \frac{mean^2}{variance} . \quad (3.24)$$

Here ρ_a and δ_a are the azimuth pixel spacing and spatial resolution, respectively, ρ_r and δ_r are the slant range pixel spacing and spatial resolution, respectively, L_a and L_r are the range and azimuth multi-looking factors, respectively, and N_s is the number of samples. The theoretical ENL ENL_t corresponds to the ideal case and represents the maximum number of independent looks for a given SAR configuration, considering the applied processing scheme (multi-looking, spatial filtering, etc.). The statistical ENL relies on the MoM and assumes homogenous areas as well as a normal distribution. In this case, the higher the variance, the lower the ENL_s and the noisier the data. A comparison of ENL_t and ENL_s generally leads to dissimilar quantities, which is explained by the large amount of noise in the data (see Sect. 4.3).

SAR Parameters

SAR images are described by three important components: frequency, incidence angle and polarisation. As introduced previously, the SAR carrier frequency defines the size of the wavelengths. To understand the relationship between the frequency and the target under study, it can be considered that the radar sees objects whose sizes are greater than or equal to the SAR wavelength. In this case, L-band (23.5 cm) and P-band (65 cm) systems are sensitive to the trunk and large branches of the trees, while X-band (3.1 cm) and C-band (5.6 cm) microwaves essentially scatter from the leaves, needles and small branches of the canopies. The penetration of high-frequency systems is limited to the top layer of the forest, while low-frequency systems are able to penetrate into the canopies and interact with the ground. The incidence angle, which was defined by Eq. (3.15), usually varies across the range direction of the radar image. While these variations may be significant in aerial data, they are not significant in spaceborne images. When dealing with hilly or mountainous areas, the local incidence angle is more appropriate than the actual incidence angle. Indeed, in contrast to the incidence angle which leads to the approximation (curvature of the Earth) that a terrain is a flat area, the local incidence angle considers the local surface normal, which includes information about the rugged terrain. The polarisation parameter is determined by the SAR systems' capability of transmitting and receiving microwaves in different polarisation planes. In general, four polarisations are distinguished: HH, VV, HV and VH. These linear polarisations provide information necessary for the discrimination of scattering mechanisms of targets. Their study, which referred to as PolSAR techniques, is considered to be a research domain. Further details on polarisations and their related techniques are given in Sect. 3.2.3.

Geometrical Distortions

The particular side-looking geometry of SAR systems and their slant-range distance measurements induce geometrical distortions. Three different effects are observed: foreshortening, layover and shadowing. The first two phenomena occur when the microwaves reach the summit of a high target (hill, mountain) before (layover) or after (foreshortening) touching its base. Both induce bright features and compressions in SAR images, especially at steep incidence angles. As for optical remote systems, the shadowing effect occurs when some surfaces cannot be illuminated. This particularly appears for the backslopes of topographic areas.

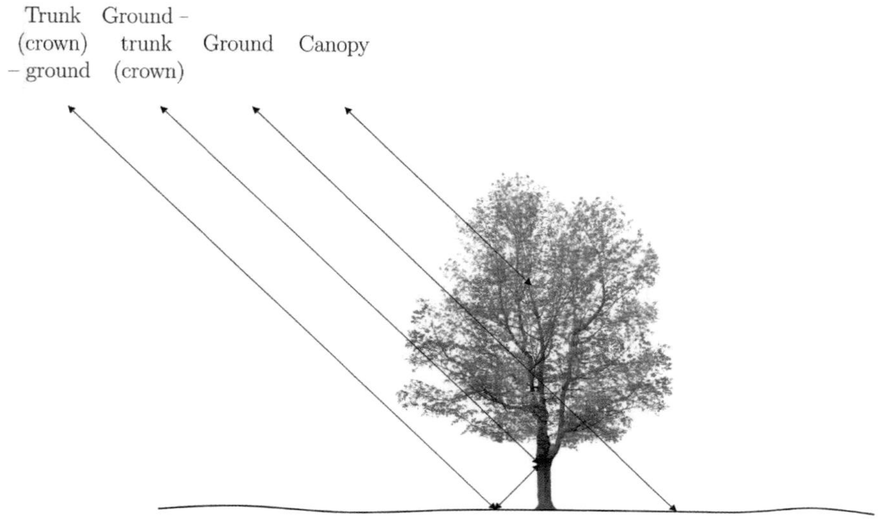

Fig. 3.1 Scattering pathways in forested areas

SAR Scattering from Forests

Forests may be schematically characterised by a ground surface with vertically extending complex objects represented by the stems and crowns of trees. Measuring radar backscatter in these areas requires the consideration of several scattering pathways (Fig. 3.1).

One first possible path is reaching the crown and scattering within this layer (1). Direct canopy (branches, trunk) contributions can be obtained from this path. A second potential path is given by the penetration of microwaves through the canopy, via either spaces resulting from canopy openness or direct penetration into tree components, which is enabled by the microwaves' penetration capabilities. In this case, if the microwaves interact with the ground and return directly to the sensor through the canopy (2), a direct ground contribution can be considered. If the microwaves are reflected twice [(on the ground and on a tree component (trunk, crown) (3) or vice-versa (4)], a double-reflection contribution can be considered. The scattering pathways of the various microwaves and the consequent contributions of the crown, ground or multiple ground-tree/tree-ground reflections are generally summarised by three important scattering mechanisms, namely surface, volume and double bounce. The relative strength and dominance of one of these scattering mechanisms depend on several factors, including the SAR configuration, forest properties and environmental weather conditions (see Chap. 2 in literature review).

Surface Scattering

Surface scattering in forests is dependent mainly on the properties of the ground beneath the canopy. The latter includes two important factors which affect the strength of the backscattering coefficient, namely the roughness and the dielectric constant. The roughness is characterised by the Raleigh criterion

$$h_{\text{smooth}} < \frac{\lambda}{8\cos(\theta_1)}. \tag{3.25}$$

Here the surface height variation h_{smooth} is associated with the radar wavelength λ and the local incidence angle θ_1. According to Relation (3.25), the smoothness of the surface is directly proportional to the wavelength and inversely proportional to local incidence angles. Thus, the surface smoothness or roughness is essentially related to the radar configuration. In terms of backscatter intensity, in the case of a smooth surface, the returned energy is focused with coherent specular reflections onto a narrow beam away from the sensor. In such a configuration, the strenght of the backscattered signal is relatively small. In the case of a rough surface, the redistributed energy is characterised by diffuse scattering, also known as incoherent scattering. In this case, the strength of the measured signal is important.

The relative dependence of roughness on wavelength given in the Rayleigh criterion follows the fundamentals of optical scattering. The theory of optical scattering developed by Lord Rayleigh and Gustav Mie states that scattering mechanisms occurring at a given incident wavelength are equivalent to those at another wavelength if the sizes of the target remain proportional. If the wavelengths are different, three scattering regions can be proposed. One is defined as Rayleigh scattering, another as optical scattering, and the last one as Mie scattering. On the assumption that there is a perfectly smooth spherical particle of radius a, the first scattering region was specified by a target at least ten times larger than the wavelength ($2\pi a \gg \lambda$). The second region was denoted by a target at least ten times smaller than the wavelength ($2\pi a \ll \lambda$). The Mie scattering region was between these two extremes ($0.1\lambda < 2\pi a < 10\lambda$).

Specular reflections which apply to smooth surfaces are described by Fresnel equations. Fresnel showed that the propagation of EM waves at the interface of two idealised stratified media involves two main components, namely reflection and transmission. His equations provide the magnitudes of the reflected and transmitted radiations as functions of the incidence angle and dielectric constant (relative permittivity) of the transmitting medium

$$R_{\text{HH}} = \frac{\mu\cos(\theta_1) - \sqrt{\mu\varepsilon - \sin(\theta_1)^2}}{\mu\cos(\theta_1) + \sqrt{\mu\varepsilon - \sin(\theta_1)^2}}, \tag{3.26}$$

$$R_{\text{VV}} = \frac{\varepsilon\cos(\theta_1) - \sqrt{\mu\varepsilon - \sin(\theta_1)^2}}{\varepsilon\cos(\theta_1) + \sqrt{\mu\varepsilon - \sin(\theta_1)^2}}. \tag{3.27}$$

Here R_{HH} and R_{VV} are the reflection coefficients for the HH and VV polarisations, respectively, ε is the dielectric constant of the transmitting medium, μ is the permeability of the medium, and θ_l is the local incidence angle of the incoming waves. The Fresnel coefficient is a complex quantity ($\varepsilon_r = \varepsilon' - i\varepsilon''$). The real and imaginary parts of the dielectric constant describe the relative permittivity and absorption of energy, respectively. For natural surfaces, the influence of magnetic fields on the scattering medium can be neglected because they are usually not considered to be ferromagnetic. In this case, μ can be made equal to one. The coefficients for HH and VV polarisations are separated because their reflections show varying incidence angles. The general tendency of both polarisations suggests that reflection coefficients increase with incidence angle. However, in the case of VV polarisation, the reflection first decreases until it reaches a minimum. This minimum corresponds to the angle for which the transmitted wave is completely absorbed by the dielectric medium. It is referred to as Brewster angle.

The surface reflections under rough terrain conditions can be described by Lambertian scattering. Such conditions produce an isotropic (in all directions) diffusion of the scattering waves. In terms of backscatter, three different regions of incidence angle is distinguished. The first region refers with shallow incidence angles to the specular zone. The second region at mid-range denotes the flat zone. The last region depicts with large incidence angles the grazing zone (Massonnet and Souyris 2008). Compared to smooth surfaces, backscattering decreases with increasing incidence angle. Also, the dependence of backscatter on incidence angle decreases in this case. Another important feature of surface roughness is the particular resonant scattering produced when the terrain roughness presents periodic patterns. In contrast to the case of specular reflection from smooth surfaces, which focuses the returned pulses on a single direction, this specific effect known as Bragg scattering coherently adds the EM waves in various directions.

Volume Scattering

The description of surface scattering showed that at the boundary of two media, a fraction of the incoming radiation is reflected and another fraction is transmitted through the medium. In the latter case, the radiation interacts with the bulk material. Assuming the medium to be a random layer of scatterers, these interactions are commonly referred to as volume diffusion or volume scattering. Volume scattering is the effect of multiple scatterings occurring within a layer formed by various elementary scatterers of different sizes, shapes, orientations and dielectric properties. In forested areas, these elements are essentially represented by foliage and twigs, as well as by small and large branches of a canopy.

Two particular effects can be introduced with volume scattering. The first one is referred to as signal extinction. When radar microwaves travel through a volume medium, a fraction of the radiation is scattered and absorbed each time a dielectric scatterer is encountered. This phenomenon attenuates the signal and decreases the backscattered intensity. Volume extinction may be indirectly related to the vertical extension (i.e. depth) of the volume and to the size of the radar wavelengths. The second effect which is closely related to volume scattering consists of depolarisation of the EM waves. This phenomenon involves rotation of the polarisation vector with curved scatterers such as discs or cylinders. The rotation is induced by the specific angle formed between the incident waves and the main axis of the oriented scatterers. If the angle extends between 0° and 90°, the EM wave is depolarised (Woodhouse 2005).

The discrete elements composing a volume each generally comprise a specific orientation. However, considering all elements as a whole, it can be assumed that they are randomly oriented and distributed. In this case, radiation is scattered in all directions, and volume scattering shows no or only weak dependence on the incidence angle. This effect is an important feature of volume scattering. Another important property is the effect of the dielectric constant. A forest canopy can be represented as an infinite horizontal layer comprising distributed water droplets in suspension, each of which expresses the dielectric discontinuities given by the various dielectric constants of the canopy components. The greater the number of droplets, the greater the absorption of the EM waves and the extinction of the signal. This concept may be useful for modelling a forest canopy and is the basic idea of the WCM (Attema and Ulaby 1978). With reference to the WCM, volume scattering can be formulated as follows (Mette 2007):

$$\sigma_{\mathrm{vol}} = N_{\mathrm{b}} \int_{0}^{h_{\mathrm{V}}} \sigma_{\mathrm{veg}}(z) \exp\left(\frac{-2\alpha_{\mathrm{ext}} z}{\cos(\theta)}\right) \mathrm{d}z \,. \tag{3.28}$$

Volume backscatter σ_{vol} is the sum of the individual scatterers along the forest layer height h_{V}, including the density of scatterers N_{b} (scatterers m^{-3}), the extinction coefficient α_{ext} and the radar cross-section of the vegetation σ_{veg}.

Double Bounce Scattering

Double bounces occur when two conducting surfaces are locally orthogonal. This specific configuration creates the so-called 'dihedral' corner reflector, which scatters the EM waves back toward the sensor with two 45° specular reflections. In forests, dihedrals are typically produced by trunks growing perpendicular to a flat terrain. However, this specific configuration is non-existent in sloping terrains, as the trees generally maintain vertical extensions, even with the inclination of the ground. The implication of double reflections is an increase in the energy of the signal returned to the transmitter, thus resulting in an increase in backscatter intensity. The effect of double bounce scattering also generally creates a phase shift between HH and VV polarisations of half the wavelength (π radians). This shift explains why the polarimetric phase difference between HH and VV polarisations is usually used for studying double reflections.

3.2.2 SAR Interferometry

The EM phase of a single SAR acquisition does not provide any useful information. However, with the application of InSAR, when two SAR acquisitions are obtained for the same area, this situation changes.

SAR interferometric techniques rely on the acquisition of two SAR images measured by repeat- or single-passes (see Fig. 3.2). Repeat-pass acquisitions refer to interferometric data acquired at two different epochs, while single-pass acquisitions describe a pair of images measured simultaneously. For interferometric acquisitions, both approaches may use one or two sensors having the same instruments and may be operated in equivalent configurations. The position of the SAR sensors may be slightly shifted between the first and second acquisitions; this introduces a distance known as the baseline. The baseline is generally split into two components, namely perpendicular $B_\perp$ (or normal B_n) and parallel $B_\parallel$. Referring to the equation of the phase in Relation (3.17), the phase difference between two different measurements can be described by a simple differentiation

$$\Delta\varphi = \varphi_2 - \varphi_1 = -(\frac{4\pi}{\lambda})(R_2 - R_1) + \Delta\varphi_{\mathrm{scatterer}} + \Delta\varphi_{\mathrm{delay}} \,. \qquad (3.29)$$

The first component refers to the path length difference $R_2 - R_1$, which also corresponds to the parallel baseline. The second element refers to variations due to the position of the targets. In this case, the slightly different look angle between

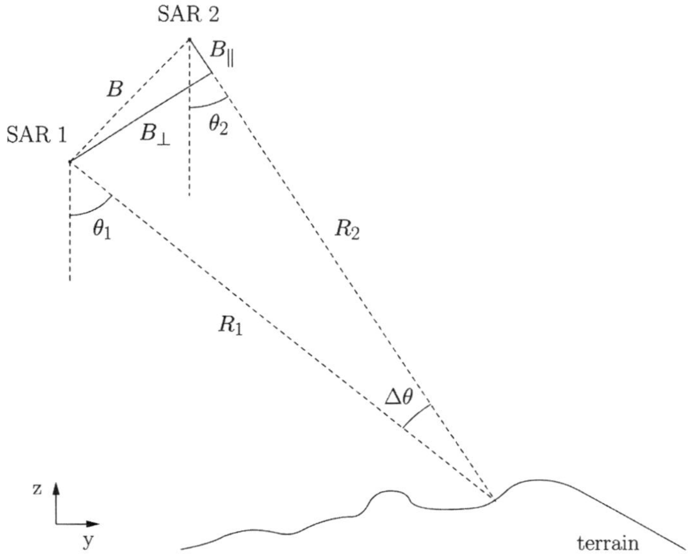

Fig. 3.2 SAR interferometric configuration

the interferometric images changes the speckle patterns, thus introducing noise. The third term is related to the potentially varying conditions (atmosphere, ionosphere) lying between the sensors and the targets. Considering two pixels on an image, the phase difference can be further developed as follows (Askne 2003; Santoro 2003):

$$\phi = \Delta\varphi = \frac{4\pi B_{\mathrm{n}}}{\lambda R \tan(\theta)}\Delta R + \frac{4\pi B_{\mathrm{n}}}{\lambda R \sin(\theta)}\Delta z + \frac{4\pi}{\lambda}\Delta\eta + \frac{4\pi}{\lambda}\Delta\rho + \Delta\varphi_{noise} + n \cdot 2\pi\,. \tag{3.30}$$

The first component is the difference in slant-range distances between the first and second acquisitions. This difference is influenced by the curvature of the Earth, which can be easily compensated for by flat Earth removal. For the different locations on the images, the second term includes a measure of the altitude variation Δz. Thus, the phase difference is proportional to the terrain elevation. In this respect, it should be noted that the elevation refers to the scattering phase center. For example, in the case of forests, the latter may be found inside the canopy instead of at the top of the canopy and may, therefore, not coincide with the forest height. The third element indicates the potential displacements of the terrain or coherent movements of the surface. The fourth component indicates phase differences due to varying atmospheric and ionospheric conditions (phase delays), and the two last terms ($\Delta\varphi_{\mathrm{noise}}$ and $n \cdot 2\pi$) indicate phase noise contribution and a factor n of 2π electromagnetic waves cycles, which is essential for unwrapping the interferometric phase and deriving physical parameters. Further details of the interferometric phase can be found in Hansson (2001).

3.2.2.1 Coherence

As mentioned previously, the SAR signal can be described as a complex number. Assuming a SAR image $g = A\exp(-j\varphi)$, the phase difference is obtained by considering the product of the first image g_1 with the conjugate of the second image g_2. This operation results in a complex image called interferogram, which represents a phase component given by the phase difference (Eq. 3.30) and a magnitude component given by the product of the SAR amplitudes. In practice, an interferogram is computed from the average of all samples within a defined window. The calculation involves normalizing the expected value of the interferogram by the expectation of each SAR product separately. It results in a complex quantity, which includes the interferometric phase and a measure of noise which is provided by the degree of correlation between the two acquisitions. The latter quantity is referred to as interferometric coherence and is described by the following relation:

$$\gamma = |\gamma|\, e^{j\phi} = \frac{\mathrm{E}\{g_1 g_2^*\}}{\sqrt{\mathrm{E}\{|g_1^2|\}\mathrm{E}\{|g_2|^2\}}}\,. \tag{3.31}$$

Here γ is the complex coherence, $|\gamma|$ is the interferometric coherence (also called coherence or coherence magnitude), ϕ is the interferometric phase and E{} is the expected value. The interferometric coherence ranges between 0 and 1 and describes total decorrelation when approaching 0. To compute the expected value, a maximum likelihood (ML) estimator is usually considered. In this case, the coherence magnitude of Eq. (3.31) is expressed as follows:

$$\left|\hat{\gamma}_{\text{ML}}\right| = \frac{\left|\sum_{i=1}^{N_s} g_{1,i} g_{2,i}^* e^{-j\varphi_i}\right|}{\sqrt{\sum_{i=1}^{N_s} \left|g_{1,i}\right|^2 \sum_{i=1}^{N_s} \left|g_{2,i}\right|^2}}. \tag{3.32}$$

The added exponential term $\exp(-j\varphi_i)$ in Eq. (3.32) corresponds to the topographic phase. As the name suggests, it takes corrections for topography into account. This term may be derived from a DEM or may be directly estimated from the interferogram. It was shown that the ML estimation was biased, in particular for low coherence and/or small estimation windows. The bias can be calculated as follows:

$$\sigma_\gamma = \frac{1 - |\gamma|^2}{\sqrt{2L}}. \tag{3.33}$$

Relation (3.33) is referred to as Cramér-Rao bound approximation. L denotes the number of independent looks and γ the coherence. To minimize the bias of low coherence values, the number of looks must be sufficiently large. Similar to Eq. (3.33), the standard deviation of the phase can be calculated as follows:

$$\sigma_\phi = \frac{\sqrt{1 - |\gamma|^2}}{|\gamma|}. \tag{3.34}$$

According to Eq. (3.34), the phase accuracy improves when the coherence or number of looks increases.

3.2.2.2 Decorrelation Factors

The degree of coherence can be decomposed into six important contributions, each expressing a decorrelation source

$$\gamma = \gamma_{\text{proc}} \cdot \gamma_{\text{SNR}} \cdot \gamma_{\text{az}} \cdot \gamma_{\text{rg}} \cdot \gamma_{\text{vol}} \cdot \gamma_{\text{temp}}. \tag{3.35}$$

The first decorrelation factor is related to the interferometric processor, which includes errors due to co-registration, resampling, interpolation, data quantisation, etc. Modern SAR processors allow accurate computations, and, therefore, this term can be neglected. The second decorrelation factor refers to noise in the SAR backscatter signal. Several noise contributions (such as thermal noise) affect SAR systems.

This decorrelation can be accounted for with the following equation:

$$\gamma_{\mathrm{SNR}} = \frac{1}{\sqrt{1 + SNR_1^{-1}}\sqrt{1 + SNR_2^{-1}}} = \frac{1}{1 + SNR^{-1}}, \tag{3.36}$$

where SNR_1 and SNR_2 refer to the *SNR* computed for the first and second interferometric acquisitions, respectively. In the case where the interferometric acquisitions are performed with the same or an equivalent sensor (i.e. TSX and TDX), a common *SNR* value can be assumed and a simplified calculation of the γ_{SNR} can be envisaged.

The aforementioned first two decorrelation quantities depend solely on the sensor and processing characteristics. As for the third and fourth components, namely γ_{az} and γ_{rg}, the decorrelations can be expressed in terms of sensor characteristics and target properties. For the azimuth Doppler and range spectra, the contributions γ_{az} and γ_{rg} refer to frequency shifts induced by a non-overlapping spectrum between the two SAR images. This effect is produced by the different positions of the sensor during the two interferometric SAR acquisitions, which cause differences in the antenna squint angle and incidence angle. The non-overlapping spectra can be well-compensated by applying common band filtering. However, this technique, which is also known as wavenumber shift filtering, only applies to surface scattering, and its performance may decrease with the inclination of the terrain, in particular with azimuth slopes.

The range frequency shift is linked to the spatial configuration of the interferometric acquisitions, and the relation between this parameter and the normal baseline can be expressed by Gatelli et al. (1994)

$$\Delta f_{\mathrm{r}} = \frac{cB_{\mathrm{n}}}{\lambda r \tan(\theta - \alpha)}. \tag{3.37}$$

Equation (3.37) presents the range frequency shift Δf_{r} as a function of the normal baseline B_{n}. It can be seen that as the baseline increases, the frequency shift also increases. However, there is a limit for which the perpendicular baseline produces complete decorrelation in the signal (Lee and Liu 2001)

$$\gamma_{\mathrm{spatial}} = 1 - \left|\frac{\Delta f}{W}\right|. \tag{3.38}$$

According to Eq. (3.38), this limit appears where the frequency shift Δf_{r} equals the radar bandwidth W and is known (in terms of baseline) as critical baseline. Consequently, the critical baseline can be inferred from Eq. (3.37) as follows:

$$|(B_{\mathrm{critical}})| = 1 - \left|\frac{WR\lambda \tan(\theta - \alpha)}{c}\right|, \tag{3.39}$$

where λ is the wavelength, W is the bandwidth, R is the slant-range distance, θ is the incidence angle and α is the topographic slope defined positive for slopes facing away from the sensor (Hansson 2001). The critical baseline given in Eq. (3.39) may also be explained with reference to speckle. Indeed, by increasing the normal baseline, the speckle patterns become more distinct between the two acquisitions. Thus, the normal baseline for which the variation in speckle patterns produces a complete decorrelation corresponds to the critical baseline. In Eq. (3.37), the variable α refers to the local slope. The sloping terrain affects the frequency shift and induces spatial decorrelation (see Fig. 3.3). The slopes facing the radar induce the greatest decorrelation. When the local slopes are close to the incidence angle, the blind-angle region, which exhibits complete decorrelation, appears (Cloude 2009). From this point, the increase in the local slope facing the radar decreases the spectral shift, but a layover configuration arises and decorrelates the signal. On the backslopes, the frequency shift is reduced with terrain inclination, thus enhancing the spatial correlation. However, the loss of the signal due to the non-common spectra reoccurs in the shadow regions. The frequency shift (and therefore the decorrelation) becomes more prominent on a sloping terrain with the increase in normal baseline. Also, as discussed with Eq. (3.15), the bandwidth determines the range resolution. In this respect, by reducing the overlapping spectra of the two acquisitions, the frequency shift reduces the spatial resolution.

The coherence term γ_{rg} indicates the baseline decorrelations for surface scattering. The consideration of vertically extended distributed targets such as forests introduces another baseline decorrelation contribution known as volume decorrelation γ_{vol}. In this case, as in Eq. (3.28) for volume scattering, the volume decorrelation can be expressed as a function of height

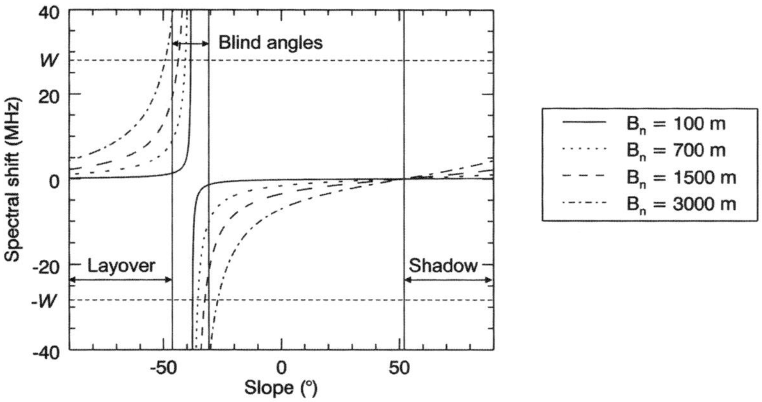

Fig. 3.3 Relation between frequency shifts and terrain slopes for ALOS PALSAR system with $\theta = 39°$. *Positive slopes* refer to slopes facing away from the radar. *Negative slopes* correspond to slopes facing radar

$$\gamma_{vol} = \frac{\displaystyle\int_0^{h_v} \sigma_{veg}(z)\exp\left(-j\frac{4\pi B_n}{R\lambda\sin(\theta)}z\right)\,\mathrm{d}z}{\displaystyle\int_0^{h_v} \sigma_{veg}(z)\,\mathrm{d}z}, \tag{3.40}$$

where

$$k_\mathrm{z} = \frac{4\pi\Delta\theta}{\lambda\sin(\theta)} \approx \frac{4\pi\Delta B_\mathrm{n}}{\lambda R\sin(\theta)} = \frac{2\pi}{h_\mathrm{a}}. \tag{3.41}$$

Volume decorrelation γ_vol is the normalised complex coherent sum of all backscatter echoes σ_veg along the canopy height h_V. It is determined from the normal baseline component B_n and radar wavelength λ, both of which partly define the interferometric wavenumber k_z given after range spectral filtering. The parameter k_z represents the sensitivity of the interferometric system and may be related to the height of ambuigity h_a which is defined as the elevation difference corresponding to a 2π difference of the interferometric phase (see Eq. 3.41). The scattering model describing the vertical distribution of the effective scatterers $\sigma_\mathrm{veg}(z)$ depends on the extent of penetration of the waves into vegetation and may take different forms. The most common relationship is provided by the RVoG model. This is described within the context of the PolInSAR theory discussed later in this chapter. The speckle patterns discussed with the critical baseline (Eq. 3.39) represented surface scattering. Similar to this case, the vertically distributed scatterers can be visualised as producing a vertical speckle (Papathanassiou 2011). This speckle also changes with different perpendicular baselines, which signifies that the volume decorrelation is also related to the normal baseline. In this respect, an increase in the perpendicular baseline causes an increase in the volume decorrelation. Similar to spectral-shift decorrelation, volume decorrelation represents a case of spatial decorrelation.

The last decorrelation factor is temporal decorrelation. This contribution involves physical changes between the first and second acquisitions, which may occur in repeat-pass acquisitions. For example, temporal changes are influenced by the dielectric properties of the target, weather conditions occurring during acquisitions or stability of the scatterers. The temporal effects may be found to be more pronounced in high-frequency systems and for long temporal baselines. However, these statements may not always be true as temporal decorrelation is the result of many physical processes.

3.2.3 SAR Polarimetry

As mentioned earlier in this section, polarisation is an important wave parameter investigated in SAR remote sensing. The polarisation of an EM wave denotes the orientation of the electric field intensity. This orientation is determined by the direction of the waves' propagation plane. In this respect, referring to Eq. (3.13), the EM wave can be decomposed into three components in a Cartesian coordinate system, each referring to a propagation plane

$$\vec{E}(z,t) = E_{0x}\cos(\omega t - kz + \delta_x)\vec{e}_x + E_{0y}\cos(\omega t - kz + \delta_y)\vec{e}_y\,, \tag{3.42}$$

where E_{0x} (resp. E_{0y}) and δ_x (resp. δ_y) are the amplitude and the phase components of the EM wave. The latter are described by the Jones vector defined later in this section. In Eq. (3.42), the wave propagates along the z-axis with electric fields components $\vec{e}_x$ and $\vec{e}_y$ oriented on planes x and y, respectively. In this case, $\vec{e}_x$ and $\vec{e}_y$ are orthogonal. The phase of the wave is described by δ_x and δ_y and the state of the polarised wave can be calculated using the relative phase shift $\delta = \delta_x - \delta_y$.

The trajectory of an EM wave is known to be helical. To more conveniently characterise this particular trajectory and the polarisation state, different representations of the waves were proposed. One of them deals with the wave polarisation within an ellipse. The considered ellipse is characterised by two angles, namely ellipticity and orientation. The ellipticity angle χ is the ratio between the minor (axis b) and major (axis a) semi-axes of the ellipse. The orientation angle ψ is the inclination angle of the major axis. The ellipticity χ varies between $-\pi/4$ and $\pi/4$, while ψ varies between $-\pi/2$ and $-\pi/2$.

The polarisation ellipse takes different forms according to χ and ψ. The most common shapes are linear and circular polarisations. The former represents the ellipse as a line ($\chi = 0$), while the latter describes the ellipse as a circle ($\chi = -\pi/4$ or $\chi = \pi/4$). In addition to the polarisation ellipse or the use of the wave parameters E_{0x}, E_{0y} and δ, the polarisation state can be represented with the Stokes vector

$$g = \begin{bmatrix} I \\ Q \\ U \\ V \end{bmatrix} = \begin{bmatrix} \left|E_y\right|^2 + |E_x|^2 \\ \left|E_y\right|^2 - |E_x|^2 \\ 2\Re(E_y E_x^*) \\ -2\Im(E_y E_x^*) \end{bmatrix}. \tag{3.43}$$

Here I corresponds to the total intensity, Q is the difference between the vertical and horizontal intensities and U and V are the phases, with $*$ denoting the complex conjugate operation, and R and C are the real and imaginary parts of the electric fields, respectively. All parameters in Eq. (3.43) are real and depict the same physical dimensions. Also, three of these parameters are independent, which results for completely polarised waves in to the following expression:

$$I^2 = Q^2 + U^2 + V^2\,. \tag{3.44}$$

Considering the set of parameters defined within Eq. (3.43) and Fig. 3.4, the Stokes vector can be rewritten as follows:

Fig. 3.4 Polarisation ellipse

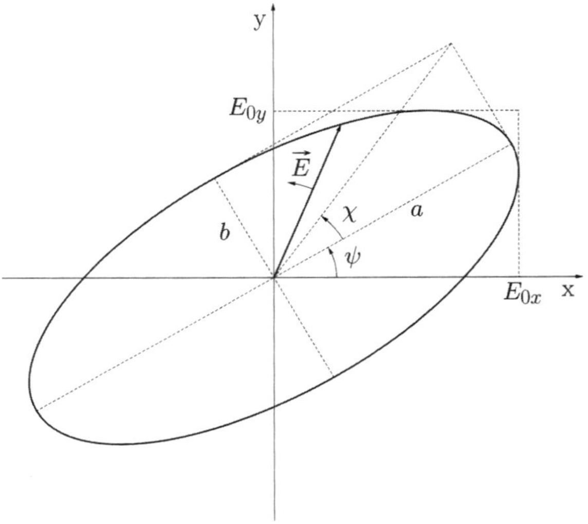

$$g = \begin{bmatrix} I \\ Q \\ U \\ V \end{bmatrix} = \begin{bmatrix} a_x^2 + a_y^2 \\ a_x^2 - a_y^2 \\ 2a_x a_y \cos(\Delta\phi) \\ 2a_x a_y \sin(\Delta\phi) \end{bmatrix} = I \begin{bmatrix} 1 \\ \cos 2\psi \cos 2\chi \\ \sin 2\psi \cos 2\chi \\ \sin 2\chi \end{bmatrix} . \tag{3.45}$$

Equation (3.45) shows that the polarisation state can be described on a sphere of radius I with 2χ and 2ψ representing the latitude and longitude, respectively. This representation is commonly known as the Poincaré sphere.

In the past, only non-coherent radar systems were available. With the development of coherent radar systems, a new mathematical expression was established for considering amplitude and phase. This characterisation is performed by the so-called Jones vector, which can be inferred from Eq. (3.42)

$$\vec{E}(0) = \begin{bmatrix} E_{0x} e^{i\delta_x} \\ E_{0y} e^{i\delta_y} \end{bmatrix} . \tag{3.46}$$

As discussed earlier in this chapter, the transmitted EM waves are scattered back from the targets and are recorder by the receiver. In polarimetry, a SAR transmitter emits linear EM waves with horizontal or vertical polarisations, and the receiver records the waves with either horizontal or vertical polarisation. These different polarisations produce various interactions with the targets, which can be coherently measured and interpreted. These interactions are synthetised under a 2 × 2 complex scattering matrix, which describes how the scatterers modify the incident electric field vector

$$\vec{E}^{\mathrm{R}} = [S] \cdot \vec{E}^{\mathrm{t}} = \frac{e^{ikR}}{R} \cdot \begin{bmatrix} S_{\mathrm{HH}} & S_{\mathrm{HV}} \\ S_{\mathrm{VV}} & S_{\mathrm{VH}} \end{bmatrix} \cdot \begin{bmatrix} E_{\mathrm{h}}^{\mathrm{t}} \\ E_{\mathrm{v}}^{\mathrm{t}} \end{bmatrix} . \tag{3.47}$$

Here matrix $[S]$ is the complex scattering matrix (also known as Sinclair matrix) and the electric fields $\vec{E}^{\mathrm{t}}$ and $\vec{E}^{\mathrm{r}}$ represent the Jones vectors of the transmitted and received fields, respectively. The scattering matrix denotes the relationship between the polarisation states of these two electric fields. R involves the slant-range distance. It is factored in the coefficient R^{-1} and in the exponential term $\exp(ikR)$, which are referring to the attenuation and delay of the wave induced by the distance separating the target and the antenna. The 'HH', 'VV', 'HV' and 'VH' polarisations refer to horizontal transmission and reception, vertical transmission and reception, horizontal transmission and vertical reception and vertical transmission and horizontal reception, respectively.

In most conditions of SAR remote sensing operating in the backscatter mode, reciprocity can be assumed. In such cases, $S_{\mathrm{VH}} = S_{\mathrm{HV}}$, and the scattering matrix is reduced to only three scattering components. Also, SAR systems deal mainly with far-field zones, as the distance between the sensor and the target is proportionally longer than the size of the studied areas. In this case, the scattering described by the matrix in Eq. (3.47) is assumed to be valid only for far-field conditions.

In radar, waves are transmitted and received in the propagation plane, which refers to the direction of the antenna. To provide standards and define the coordinate systems for the propagation directions of waves, two IEEE conventions were created, namely forward scatter alignment (FSA) and back scatter alignment (BSA). In the case of the FSA norm, the coordinate system 'follows' the wave propagation direction. Taking into account the Cartesian coordinate system introduced earlier in this section, the z-axis direction is positive for both transmitted and scattered waves. In the case of the BSA standard, the coordinate system 'follows' only the incident waves. The z-axis direction of the coordinate system is positive for the incident and negative for the backscattered waves.

It is possible to rewrite the radar cross-section by considering the scattering matrix for polarised waves

$$\sigma_{pq} = 4\pi \left|S_{pq}\right|^2 . \tag{3.48}$$

The backscatter cross-section σ_{pq} for any arbitrary combination of transmitted p and received q polarisations is obtained from the square of the magnitude of the scattering matrix.

The designation of completely polarised waves was used from the definition of the Stokes vector in Relation (3.44). In the ideal case, waves are completely polarised, which means that they are independent of time and spatial position. However, this is rarely the case, and thus, the difference between the term 'completely polarised waves' and from partially or quasi-polarised waves, which refer to time- and space-varying EM waves, is clarified. The partially polarised waves introduce second-order statistical terms known as covariance and coherency matrices.

Assuming reciprocity ($S_{\mathrm{VH}} = S_{\mathrm{HV}}$) (Borgeaud et al. 1978), the covariance matrix can generally be written as a 3×3 complex Hermitian (self-adjoint) matrix. Its formulation is given by the expected value of the product of the scattering vector $\vec{k}_{\mathrm{L}}$ with its complex conjugate $\vec{k}_{\mathrm{L}}^*$

$$[C] = \left\langle \vec{k}_{\mathrm{L}} \cdot \vec{k}_{\mathrm{L}}^{*T} \right\rangle,$$
$$= \begin{bmatrix} \left\langle |S_{\mathrm{HH}}|^2 \right\rangle & \sqrt{2}\langle S_{\mathrm{HH}} S_{\mathrm{HV}}^* \rangle & \langle S_{\mathrm{HH}} S_{\mathrm{VV}}^* \rangle \\ \sqrt{2}\langle S_{\mathrm{HV}} S_{\mathrm{HH}}^* \rangle & \left\langle |S_{\mathrm{HV}}|^2 \right\rangle & \sqrt{2}\langle S_{\mathrm{HV}} S_{\mathrm{VV}}^* \rangle \\ \langle S_{\mathrm{VV}} S_{\mathrm{HH}}^* \rangle & \sqrt{2}\langle S_{\mathrm{VV}} S_{\mathrm{HV}}^* \rangle & \left\langle |S_{\mathrm{VV}}|^2 \right\rangle \end{bmatrix}, \tag{3.49}$$

where

$$\vec{k}_{\mathrm{L}} = \left[S_{\mathrm{HH}}\ S_{\mathrm{HV}}\ S_{\mathrm{VH}}\ S_{\mathrm{VV}} \right]^T. \tag{3.50}$$

The scattering vector $\vec{k}_{\mathrm{L}}$ indicates the lexicographic representation of the polarimetric data. The $\langle\rangle$ of the covariance components denotes ensemble averaging, which involves loss of spatial resolution.

As an alternative to Relation (3.49), the coherency matrix based on the Pauli vectorisation $\vec{k}_{\mathrm{P}}$ can be calculated. With respect to the covariance matrix, assuming that the reciprocity theorem holds, the coherency matrix can be represented by a 3×3 Hermitian matrix, which is determined by the expected value of the product of the scattering vector $\vec{k}_{\mathrm{P}}$ with its complex conjugate $\vec{k}_{\mathrm{P}}^*$

$$[T] = \left\langle \vec{k}_{\mathrm{P}} \cdot \vec{k}_{\mathrm{P}}^{*T} \right\rangle,$$
$$= \frac{1}{2} \begin{bmatrix} \left\langle |S_{\mathrm{HH}} + S_{\mathrm{VV}}|^2 \right\rangle & \langle (S_{\mathrm{HH}} + S_{\mathrm{VV}})(S_{\mathrm{HH}} - S_{\mathrm{VV}})^* \rangle & 2\langle (S_{\mathrm{HH}} + S_{\mathrm{VV}}) S_{\mathrm{HV}}^* \rangle \\ \langle (S_{\mathrm{HH}} - S_{\mathrm{VV}})(S_{\mathrm{HH}} + S_{\mathrm{VV}})^* \rangle & \left\langle |S_{\mathrm{HH}} - S_{\mathrm{VV}}|^2 \right\rangle & 2\langle (S_{\mathrm{HH}} - S_{\mathrm{VV}}) S_{\mathrm{HV}}^* \rangle \\ 2\langle S_{\mathrm{HV}} (S_{\mathrm{HH}} + S_{\mathrm{VV}})^* \rangle & 2\langle S_{\mathrm{HV}} (S_{\mathrm{HH}} - S_{\mathrm{VV}})^* \rangle & 4\left\langle |S_{\mathrm{HV}}|^2 \right\rangle \end{bmatrix}, \tag{3.51}$$

where

$$\vec{k}_{\mathrm{P}} = \frac{1}{\sqrt{2}} \left[S_{\mathrm{HH}} + S_{\mathrm{VV}}\ S_{\mathrm{HH}} - S_{\mathrm{VV}}\ 2S_{\mathrm{HV}} \right]^T. \tag{3.52}$$

The covariance $[C]$ and coherency $[T]$ matrices present the same information. However, $[T]$ is arranged such that the terms of the matrix represent elementary scattering mechanisms which can be physically interpreted. Therefore, this matrix is usually preferred over the covariance matrix. The choice of studying the $[C]$ and $[T]$ matrices at the expense of the scattering matrix $[S]$ depends on the nature of the targets. In the case of distributed targets such as forests, $[C]$ and $[T]$ matrices are examined, while for coherent (pure) targets such as man-made dihedrals, the scattering matrix is considered.

3.2.3.1 Yamaguchi Decomposition

As introduced in Chap. 2, the Yamaguchi decomposition algorithm is an incoherent decomposition technique. ‘Incoherent’ stands for non-deterministic or distributed targets and 'decomposition' refers to the separation of the SAR signal into a sum of elementary scattering from which physical interpretations can be drawn.

The Yamaguchi decomposition is an extension of the Durden–Freeman decomposition algorithm, which is dedicated to forested areas. Its general idea is based on the assumption that the coherency matrix can be described as a combination of individual matrices, each representing distinct physical scattering mechanisms. While Durden–Freeman decomposition deals with a three-component model, the algorithm proposed by Yamaguchi adds a new component which takes the cases where reflection symmetry does not hold into account. Equation (3.53) presents Yamaguchi decomposition:

$$\begin{aligned}\langle|[T]|\rangle &= f_s \langle|[T]|\rangle_{\text{surface}} + f_d \langle|[T]|\rangle_{\text{dbl}} + f_v \langle|[T]|\rangle_{\text{vol}} + f_c \langle|[T]|\rangle_{\text{helix}}, \\ &= f_s \begin{bmatrix} 1 & \beta^* & 0 \\ \beta & |\beta|^2 & 0 \\ 0 & 0 & 0 \end{bmatrix} + f_d \begin{bmatrix} |\alpha|^2 & \alpha & 0 \\ \alpha^* & 1 & 0 \\ 0 & 0 & 0 \end{bmatrix} + \frac{f_v}{4} \begin{bmatrix} 2 & 0 & 0 \\ 0 & 1 & 0 \\ 0 & 0 & 1 \end{bmatrix} + \frac{f_c}{2} \begin{bmatrix} 0 & 0 & 0 \\ 0 & 1 & \pm j \\ 0 & \pm j & 1 \end{bmatrix}\end{aligned} \tag{3.53}$$

Here the coherency submatrices represent the surface, double bounce, volume and helix scattering mechanisms, and the corresponding expansion coefficients f_s, f_d, f_v and f_c are to be determined; also, the parameters α and β are to be defined using the Fresnel reflection coefficients. It should be noted that Yamaguchi decomposition may be similarly described in terms of covariance matrices. This section focuses on the decomposition of the coherency matrix.

The parameters of the decomposed coherency matrices represent the physical properties of the forest. As introduced above, the principal contribution made by Yamaguchi involved assuming the non-symmetry of certain targets. For this, the use of the VV to HH cross-sectional ratio was suggested to determine which canopy model was to be applied. The conditions were defined as follows:

$$\langle|[T]|\rangle_{\text{vol}}^{\text{dipole}} = \frac{1}{30} \begin{bmatrix} 15 & 5 & 0 \\ 5 & 7 & 0 \\ 0 & 0 & 8 \end{bmatrix}, \quad \text{for } 10 \log \left(\frac{\langle |S_{\text{VV}}|^2 \rangle}{\langle |S_{\text{HH}}|^2 \rangle} \right) < -2\,\text{dB}, \tag{3.54}$$

$$\langle|[T]|\rangle_{\text{vol}}^{\text{dipole}} = \frac{1}{4} \begin{bmatrix} 2 & 0 & 0 \\ 0 & 1 & 0 \\ 0 & 0 & 1 \end{bmatrix}, \quad \text{for } -2\,\text{dB} < 10 \log \left(\frac{\langle |S_{\text{VV}}|^2 \rangle}{\langle |S_{\text{HH}}|^2 \rangle} \right) < 2\,\text{dB}, \tag{3.55}$$

$$\langle|[T]|\rangle_{\text{vol}}^{\text{dipole}} = \frac{1}{30} \begin{bmatrix} 15 & -5 & 0 \\ -5 & 7 & 0 \\ 0 & 0 & 8 \end{bmatrix}, \quad \text{for } 10 \log \left(\frac{\langle |S_{\text{VV}}|^2 \rangle}{\langle |S_{\text{HH}}|^2 \rangle} \right) > 2\,\text{dB}. \tag{3.56}$$

In the case where the ratio falls within ± 2 dB, the canopy is assumed to be a cloud of uniformly distributed dipoles. Assuming the reflection azimuthal symmetry of the forest canopy, $\langle|[T]|\rangle_{\text{vol}}^{\text{dipole}}$ becomes diagonal (Eq. 3.55). When the ratio falls below -2 dB, a horizontally structured canopy is conceived (Eq. 3.54), and when the ratio is greater than 2 dB, a vertically oriented canopy is considered (Eq. 3.56). In these conditions, reflection symmetry can be assumed; the components $\langle|[T_{12}]|\rangle_{\text{vol}}^{\text{dipole}}$ and $\langle|[T_{21}]|\rangle_{\text{vol}}^{\text{dipole}}$ are also defined in the $\langle|[T]|\rangle_{\text{vol}}^{\text{dipole}}$ matrices (Richards 2009). With reference to Eq. (3.53), if the case of a uniformly distributed cloud of dipoles applies, the relation represents a four-component decomposition. If the dipoles are horizontally or vertically oriented, the equation becomes a three-component decomposition. For uniformly distributed dipoles, five equations with six unknowns (α, $\beta, f_{\text{s}}, f_{\text{d}}, f_{\text{v}}$ and f_{c}) can be derived:

$$P_{\text{s}} = f_{\text{s}}\left(1 + |\beta|^2\right), \tag{3.57}$$

$$P_{\text{d}} = f_{\text{d}}\left(1 + |\alpha|^2\right), \tag{3.58}$$

$$P_{\text{v}} = f_{\text{v}} = 4T_{33} - 2P_{\text{c}} = 8\left\langle|S_{\text{HV}}|^2\right\rangle - 2P_{\text{c}}, \tag{3.59}$$

$$P_{\text{c}} = f_{\text{c}} = 2\left|\Im\left\langle S_{\text{HV}}^{*}\left(S_{\text{HH}} - S_{\text{VV}}\right)\right\rangle\right|, \tag{3.60}$$

$$P_{\text{t}} = P_{\text{s}} + P_{\text{d}} + P_{\text{v}} + P_{\text{c}} = \left\langle|S_{\text{HH}}|^2 + 2\,|S_{\text{HV}}|^2 + |S_{\text{VV}}|^2\right\rangle. \tag{3.61}$$

Here P_{s}, P_{d}, P_{v} and P_{c} describe the scattering powers for surface, double bounce, volume and helix components, respectively, and P_{t} is the total power, which represents the span of the polarimetric data. According to Relation (3.60), P_{c} can be directly calculated from the measurable quantities S_{HV}, S_{HH} and S_{VV}. P_{v} can then be derived from S_{HV} and P_{c}. At this stage, three equations remain with four unknowns (α, $\beta, f_{\text{s}}, f_{\text{d}}$). After making assumptions regarding the dominant scattering between surface or double bounce scatterers and a few mathematical manipulations, the two power components, namely P_{d} and P_{s}, can be inferred.

As mentioned in Chap. 2, the use of FDD decomposition showed negative powers in vegetated areas, which is physically inconsistent. This problem is related to the definition of volume scattering, which may lead to an overestimation of the volume component and, therefore, to an underestimation of the remaining components, namely surface and double bounce scattering, under certain conditions. To deal with this issue, Yamaguchi's solution involved the use of an amplitude ratio, as discussed above, to choose between three different canopy models. This solution reduced the nonnegative power effect, but did not remove it completely.

3.2.3.2 Azimuth-Slope Correction

The polarimetric orientation of EM waves is affected under topographic conditions by the slopes facing the azimuthal direction of the radar. This effect can be compensated for by rotating the polarimetric matrices (scattering matrix, coherency matrix, etc.) along the line-of-sight by the negative value of the induced azimuth slopes. This correction requires the calculation of the orientation angle (*OA*), which is geometrically related to the sloping terrain and to the radar look angle

$$\tan\theta = \frac{\omega}{-\tan\gamma\cos\phi + \sin\phi}\,. \tag{3.62}$$

The parameter θ represents the *OA*, $\tan\omega$ is the azimuth slope, $\tan\gamma$ is the slope in the ground range direction and ϕ is the radar look angle. In practice, several methods can be used to derive the *OA*. One of the most common approaches is the circular polarisation method. By defining the left–left and right–right circular polarisation responses, the *OA* can be directly extracted from the scattering-matrix components

$$\nu = Arg(\langle S_{\mathrm{RR}} S^{*}_{\mathrm{LL}}\rangle) = \frac{1}{4}\left[\tan^{-1}\left(\frac{-4\Re(\langle (S_{\mathrm{HH}} - S_{\mathrm{VV}})S^{*}_{\mathrm{HV}}\rangle)}{-\langle |S_{\mathrm{HH}} - S_{\mathrm{VV}}|^2\rangle + 4\langle |S_{\mathrm{HV}}|^2\rangle}\right) + \pi\right], \tag{3.63}$$

and

$$\eta = \begin{cases} \nu, & \text{if } \nu \le \pi/4\,, \\ \nu - \pi/2, & \text{if } \nu > \pi/4\,. \end{cases} \tag{3.64}$$

Here ν describes the OA derived from the polarimetric radar data and $\Re$ characterises the real part of the complex quantities. The condition expressing the replacement of η by $\nu - \pi/2$ for $\nu > \pi/4$ takes the case of negative-*OA* into account. Similar to phase unwrapping for interferometry, a factor π is considered in Eq. (3.63) to match the *OA* to the azimuth slopes, thus allowing the extraction of physical information. Once the *OA* is derived, it can be applied to the polarimetric matrices. The rotation of the scattering matrix is performed as follows:

$$[S(\eta)]_{\mathrm{corr}} = [R_S(\eta)]\,[S]\,[R_S(\eta)]^T\,, \tag{3.65}$$

where

$$[R_S(\eta)] = \begin{bmatrix} \cos\eta & \sin\eta \\ -\sin\eta & \cos\eta \end{bmatrix}.$$

Here $[S(\eta)]_{\mathrm{corr}}$ is the scattering matrix compensated for the azimuth slopes. It can be used to calculate the covariance or coherency matrices, as shown in Eqs. (3.49) and (3.51), respectively. The coherency matrix may also be directly corrected using the following relations:

$$[T(\eta)]_{\mathrm{corr}} = [R_T(\eta)]\,[T]\,[R_T(\eta)]^T\,, \tag{3.66}$$

where

$$[R_T(\eta)] = \begin{bmatrix} 1 & 0 & 0 \\ 0 & \cos 2\eta & \sin 2\eta \\ 0 & -\sin 2\eta & \cos 2\eta \end{bmatrix}.$$

All terms of the compensated coherency matrix $[T(\eta)]_{\text{corr}}$ are modified, except $\langle |S_{\text{HH}} + S_{\text{VV}}| \rangle$. The T_{11} component is roll invariant. Comparison of the $[T]$ matrix before and after corrections showed that the volume scattering power consistently decreases, while the double bounce component increased. These observations are further confirmed with the application of the Yamaguchi decomposition to the compensated coherency matrix. A similar rotational transformation can be applied to the covariance matrix.

3.2.4 Polarimetric SAR Interferometry

Sections 3.2.2 and 3.2.3 described the theory of InSAR and PolSAR, respectively. By combining both techniques, it is possible to rewrite the coherence defined in Eq. (3.31) as a function of polarisation. The formulation becomes (Richards 2009)

$$\gamma_{\text{u}} = \frac{\left\langle \vec{w}_1^{*T} \left[\Omega_{12}\right] \vec{w}_2 \right\rangle}{\sqrt{\left\langle \vec{w}_1^{*T} \left[T_{11}\right] \vec{w}_1 \right\rangle} \sqrt{\left\langle \vec{w}_2^{*T} \left[T_{22}\right] \vec{w}_2 \right\rangle}}, \tag{3.67}$$

where

$$\begin{aligned}
[\Omega_{12}] &= \left\langle \vec{k}_1 \cdot \vec{k}_2^{*T} \right\rangle, \\
&= \frac{1}{2} \begin{bmatrix} \langle (S_{1\text{HH}} + S_{1\text{VV}})\left(S_{2\text{HH}}^* + S_{2\text{VV}}^*\right) \rangle & \langle (S_{1\text{HH}} + S_{1\text{VV}})\left(S_{2\text{HH}}^* - S_{2\text{VV}}^*\right) \rangle & 2\langle (S_{1\text{HH}} + S_{1\text{VV}}) S_{2\text{HV}}^* \rangle \\ \langle (S_{1\text{HH}} - S_{1\text{VV}})\left(S_{2\text{HH}}^* + S_{2\text{VV}}^*\right) \rangle & \langle (S_{1\text{HH}} - S_{1\text{VV}})\left(S_{2\text{HH}}^* - S_{2\text{VV}}^*\right) \rangle & 2\langle (S_{1\text{HH}} - S_{1\text{VV}}) S_{2\text{HV}}^* \rangle \\ 2\langle S_{1\text{HV}}\left(S_{2\text{HH}}^* + S_{2\text{VV}}^*\right) \rangle & 2\langle S_{1\text{HV}}\left(S_{2\text{HH}}^* - S_{2\text{VV}}^*\right) \rangle & 4\langle S_{1\text{HV}} S_{2\text{HV}}{}^* \rangle \end{bmatrix}.
\end{aligned} \tag{3.68}$$

Here γ_{u} is the complex polarimetric interferometric coherence, T_{11} and T_{22} are the coherency matrices of the two interferometric acquisitions, $\vec{w}_1$ and $\vec{w}_2$ are unitary filter vectors representing different polarisation combinations from the target vectors, respectively, Ω is the new joint image complex coherency matrix and $\vec{k}_1$ and $\vec{k}_2$ are Pauli-scattering vectors for the first and second image pairs, respectively. Complex polarimetric interferometric coherence is a generalisation of polarimetric data on the basis of the complex InSAR coherence. It contains phase and amplitude information. The phase denotes the phase difference between the first and second interferometric acquisitions in addition to variations related to polarisation differences. The amplitude is used to measure the correlation between the two interferometric images. Similar to the InSAR system described earlier in this section, the maximum amplitude of interferometric phase correlation is unity, and the decorrelation factors described in Eq. (3.35) apply to the complex polarimetric interferometric coherence.

The complex polarimetric interferometric coherence γ_{u} can be plotted on an Argand diagram with the coherence magnitude described by the radius and the phase by the angle starting counter-clockwise from the positive real axis (see Fig. 3.5).

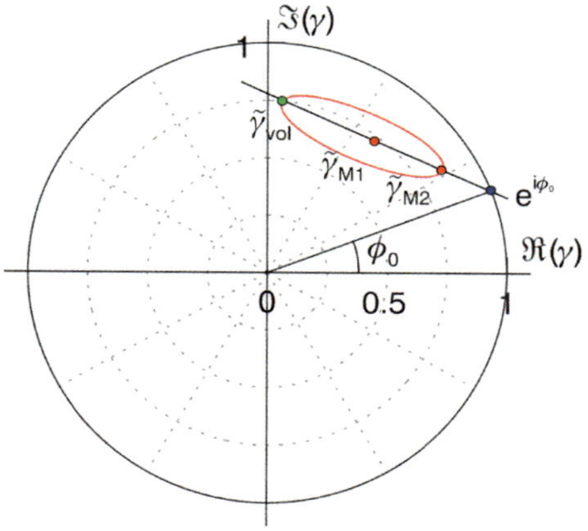

Fig. 3.5 Loci representation of the complex polarimetric interferometric coherence (Cloude 2009)

The range of the complex polarimetric interferometric coherence function γ_u can be displayed as the loci obtained for all polarisations. It refers to loci representation and denotes a specific area known as the coherence region. The size and shape of the region are determined by the nature of the scattering processes. In the case of forests, the phase angles of the canopy and the ground differ significantly. This characteristic provides information on the location of the scattering phase centers. If the microwaves partially reach the ground, the signal is mixed by ground and crown returns, leading to significant phase variations (i.e. increase of phase angle). In such conditions, the coherence region stretches over the Argand diagram and generally takes an elliptical form. If the canopy completely attenuates the signal or is completely transparent to the microwaves, the variations in the phase angle reduce, and the coherence region shrinks to a single point. Generally, suitable complex coherence regions are found near the circumference of the plot, where the coherence is high. In such optimal conditions, on the basis of the loci coherence, two extreme points can be identified in the coherence elliptical region, one assimilated to the ground and the other considered at the top of the trees. With further mathematical considerations, this configuration enables a straightforward derivation of the forest height.

Equation (3.40) represents the volume decorrelation for an interferometric SAR system. In the case of PolInSAR systems, the RVoG model describes the joint polarimetric and interferometric information. This model connects complex polarimetric interferometric coherence with the forest height, canopy extinction, ground reflection and topographic phase. Its principal concept involves balancing the contribution of two different layers, one referring to the ground and the other to the forest canopy. The RVoG model ignores double bounces and other higher-order scattering mechanisms. Equation (3.69) introduces the RVoG model

$$\tilde{\gamma}_m(\vec{w}) = exp(i\phi_0)\frac{\tilde{\gamma}_{vol} + M(\vec{w})}{1 + M(\vec{w})}, \qquad (3.69)$$

$\gamma_m(\vec{w})$ represents complex coherence as a function of polarisation, with $\tilde{\gamma}_{vol}$ being the complex coherence of the random homogenous volume, $M(\vec{w})$ the ground-to-volume amplitude ratio and ϕ_0 the ground interferometric phase. M is the only parameter which is polarisation-dependent and describes the relative contribution of the ground over the canopy and vice-versa. It is further formulated as follows:

$$M(\vec{w}) = \frac{m_G(\vec{w})}{m_V(\vec{w})} \exp\left(-\frac{2\sigma_v h}{\cos\theta}\right). \tag{3.70}$$

Here the parameter m_G is the scattering contribution from the ground, m_V is the direct volume scattering from the canopy, h is the volume height, σ_v is the mean extinction coefficient and θ is the incidence angle. The ratio $m_G(\vec{w})/m_V(\vec{w})$ is the ground-to-volume amplitude ratio $m(\vec{w})$. The variable σ_v denotes the density and the dielectric constant of the scatterers present in the volume, and it is not affected by the polarisations of the waves.

The volume decorrelation $\tilde{\gamma}_{vol}$ given in Eq. (3.69) can be further developed into the following relation:

$$\tilde{\gamma}_{vol} = \frac{\int_0^{h_V} \exp\left(\frac{2\sigma_V z}{\cos(\theta)}\right) \exp(i\kappa_z z)\, dz}{\int_0^{h_V} \exp\left(\frac{2\sigma_V z}{\cos(\theta)}\right) dz}. \tag{3.71}$$

Here k_z, which is the vertical wavenumber, was defined in Eq. (3.41). The complex coherence $\tilde{\gamma}_{vol}$ is a complex quantity. As suggested by its relation, volume decorrelation increases with increasing volume height as well as with decreasing mean extinction coefficient. The RVoG model given in Eq. (3.69) can be represented in the complex plane of the Argand diagram as a straight line. In practice, by fitting the RVoG model over optimum coherences, this line would be drawn through the major axis of the loci coherence ellipse. This characteristic is used as an inversion scheme to retrieve the forest height, ground topography or canopy extinction parameters defined within the RVoG model.

3.3 Sensors Fusion

3.3.1 Definitions of Fusion

The concept of data fusion appeared in the early years of remote sensing. However, the exact definition had only been discussed at the end of the 1990s. A definition for fusion became necessary because the interpretations and descriptions of fusion differed considerably between researchers. Indeed, each researcher interpreted this term differently according to his/her own research interests. For instance, Hall (1992) suggested that 'Multisensor data fusion seeks to combine data from multiple sensors to perform inferences that may not be possible from a single sensor alone'. Pohl and van Genderen (1994, 1998) proposed image fusion as a 'combination of two or more different images to form a new image by using a certain algorithm'. Mangolini (1994) defined fusion as a 'set of methods, tools and means using data coming from various sources of different nature, in order to increase the quality (in a broad sense) of the requested information'. While these definitions were accurate, they either lacked information or placed excessive emphasis on a specific issue. Therefore, Wald (1998) suggested a general definition which appeared to be universally accepted by the scientific community. Wald (1998) expressed 'data fusion' as 'a formal framework in which are expressed means and tools for the alliance of data originating from different sources, in order to obtain information of greater quality'. In addition to the term 'fusion', studies have addressed SAR information fusion commonly in relation to other close terms for 'fusion' such as 'merge', 'combination', 'synergy', 'integration'. The two former terms were defined by Wald (1998) as 'any process that implies a mathematical operation performed on at least two sets of information'. He emphasised that merging is broader than fusion, and combination can be considered in an even broader sense than the terms 'merge' and 'fusion'. Wald (1998) also mentioned that merging and combination denote certain 'operations', while fusion designates a 'framework'. Synergy was not clearly defined with regard to remote sensing. With reference to the Oxford Dictionary, Cracknell (1998) inferred his own definition as 'the utilisation of two or more data sources together in order to extract more environmental information from their combination than one could extract from the separate data sources individually'. 'Integration', which is also closely related to fusion, was defined by the Canada Centre for Remote Sensing (CCRS 2007) as 'the combining or merging of data from multiple sources in an effort to extract better and/or more information'.

3.3.2 Fusion Levels

Fusion methods involve a wide range of existing fusion techniques, which can be considered at different levels of fusion (Gamba and Chanussot 2008; Pohl and van Genderen 1998) (Fig. 3.6).

At one extreme, there is signal-level fusion, which concerns mainly data pre-processing. Then, pixel-level fusion appears. In such cases, a pixel-based approach

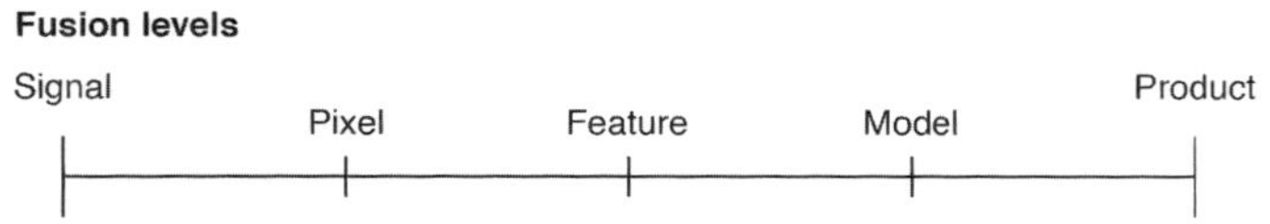

Fig. 3.6 General fusion processing levels

combining two images with raw information can be realised. The arithmetic combinations, Brovey transformation or intensity–hue–saturation (IHS) transformation methods are typical examples of pixel-level fusion. The third level refers to segmented data. In this case, decision trees or quad-trees are often cited. The fourth fusion level is a combination of different models, including statistical and physical models. At the other extreme of the signal level, the final level is the fusion of the products derived from the data. The fusion processes generally involve one fusion level; however, a combination of techniques belonging to different levels may also be realised.

References

Abadia A, Gil E, Morales F, Monates L, Montserrat G, Abadia J (1996) Marcescence and senescence in a submediterranean oak (Quercus subpyrenaica E.H. del Villar): photosynthetic characteristics and nutrient composition. Plant Cell Environ 19(6):685–694

Arner SL, Woudenberg S, Waters S, Vissage J, MacLean C, Thompson M, Hansen M (2001) National algorithms for determining stocking class, stand size class, and forest type for forest inventory and analysis plots. Technical report, US Forest Service, USA

Askne J (2003) Remote sensing using microwaves. Technical report, University of Technology, Gothenburg

Attema EPW, Ulaby FT (1978) Vegetation modeled as a water cloud. Radio Sci 13(2):357–364

Badeck F-W, Bondeau A, Bottcher K, Doktor D, Lucht W, Schaber J, Sitch S (2004) Responses of spring phenology to climate change. New Phytol 162(2):295–309

Bannan MW, Bindra M (1970) The influence of wind on ring width and cell length in conifer stems. Can J Bot 48(2):255–259

Bonan GB, Shugart HH (1989) Environmental factors and ecological processes in boreal forests. Annu Rev Ecol Syst 20(1):1–28

Bonnor GM (1985) Inventory of forest biomass in Canada. Technical report, Petawawa National Forestry Institute, Chalk River, Ontario

Borgeaud M, Shin RT, Kong JA (1987) Theoretical models for polarimetric radar clutter. J Electromagn Waves Appl 1(1):73–89

Brack C (2000) Forest measurement and modelling. http://fennerschool-associated.anu.edu.au/mensuration/home.htm

Bradley RT (1963) Thinning as an instrument of forest management. Forestry 36(2):181–194

CCRS (2007) Tutorial: Fundamentals of remote sensing. http://www.nrcan.gc.ca/earth-sciences/geomatics/satellite-imagery-air-photos/satellite-imagery-products/educational-resources/9347

Cloude S (2009) Polarisation: applications in remote sensing. Oxford University Press, Oxford

Cochard H, Granier A (1999) Fonctionnement hydraulique des arbres forestiers. Revue Forestière Française. 2:121–133

Cochard H, Lemoine D, Améglio T, Granier A (2000) Winter embolism, mechanisms of xylem hydraulic conductivity recovery and springtime growth patterns in walnut and peach trees. Oxford J 21(1):27–33

Cochard H, Lemoine D, Améglio T, Granier A (2001) Mechanisms of xylem recovery from winter embolism in Fagus sylvatica. Tree Physiol 21(1):27–33

Cornu S (1998) La circulation de l'eau dans l'arbre ascension de la sève et transpiration foliaire. Technical report, Université Claude Bernard Lyon I, Lyon, France

Cracknell AP (1998) Review article synergy in remote sensing—what's in a pixel? Int J Remote Sens 19(11):2025–2047

Cruiziat P, Cochard H, Améglio T (2002) Hydraulic architecture of trees: main concepts and results. Ann For Sci 59(7):723–752
Curtis RO, Marshall DD (2000) Why quadratic mean diameter. W J Appl For 15(3):137–139
del Rio M, Montero G, Bravo F (2001) Analysis of diameter-density relationships and self-thinning in non-thinned even-aged Scots pine stands. For Ecol Manag 142(1–3):79–87
Ellis CR, Pomeroy JW, Essery RLH, Link TE (2011) Effects of needleleaf forest cover on radiation and snowmelt dynamics in the Canadian Rocky Mountains. Can J For Res 41(3):608–620
Escudero A, del Arco JM (1987) Ecological significance of the phenology of leaf abscission. Oikos 49(1):11–14
Fiedler H, Krieger G, Jochim F, Kirschner M, Moreira A (2002) Analaysis of satellite configurations for spaceborne SAR interferometry. In: International symposium formation flying missions and technologies, CNES, Toulouse, France
Gamba P, Chanussot J (2008) Foreword to the special issue on data fusion. IEEE Trans Geosci Remote Sens 46(5):1283–1288
Gatelli F, Guamieri AM, Parizzi F, Pasquali P, Prati C, Rocca F (1994) The wavenumber shift in SAR inferometry. IEEE Trans Geosci Remote Sens 32(4):855–865
Global forest resources assessment (2000) main report. Technical report, FAO, Rome, Italy 2000
Global forest resources assessment (2005) main report. Technical report, FAO, Rome, Italy 2005
Gray HR (1956) The form and taper of forest-tree stems. Imperial forestry institute paper, Imperial Forestry Institute, vol 32, p 79
Hall DL (1992) Mathematical techniques in multisensor data fusion. Artech House, Inc., Norwood
Hansson R (2001) Radar interferometry: data interpretation and error analysis. Baker and Taylor, New York
Hewson EW, Wade JE, Baker RW (1978) Vegetation as an indicator of high wind velocity. Technical report, prepared for the United States Department of Energy division of Solar Energy Federal Wind Energy Programme, Corvallis, Oregon, USA
Husch B, Beers TW, Kershaw JA Jr (2003) Forest mensuration, 4th edn. Wiley, Hoboken
Kellogg RM, Arber FJ (1981) Stem eccentricity in coastal western hemlock. Can J For Res 11(3):714–718
Kim S (2009) Antenna mask design for SAR performance optimization. IEEE Geosci Remote Sens Lett 6(3):443–447
Köhl M, Magnussen SS, Marchetti M (2006) Sampling methods, remote sensing and GIS: multiresource forest inventory. Springer, Heidelberg
Lapointe L (2001) How phenology influences physiology in deciduous forest spring ephemerals. Physiol Plant 113(2):151–157
Lebourgeois F (2006) Réponse au climat du Sapin (Abies alba Mill.) et de l'Épicéa (Picea abies (L.) Karst.) dans le réseau RENECOFOR. Revue Forestière Française, 53(6):419–432
Lee H, Liu JG (2001) Analysis of topographic decorrelation in SAR interferometry using ratio coherence imagery. IEEE Trans Geosci Remote Sens 39(2):223–232
Levine JS (1996) Remote sensing and modeling of biomass burning, and biomass burning in the boreal forest. In: Biomass burning and global change, vol 1. The MIT Press, USA
Liang L, Schwartz MD (2009) Landscape phenology: an integrative approach to seasonal vegetation dynamics. Landscape Ecol 24(4):465–472
Littell JS, Peterson DL, Tjoelker M (2008) Douglas-fir growth in mountain ecosystems: water limits tree growth from stand to region. Ecol Monogr 78(3):349–369
Mangolini M (1994) Apport de la fusion d'images satellitaires multicapteurs au niveau pixel en télédétection et photo-interprétation. Ph.D. thesis, Université Nice - Sophia Antipolis
Marino A (2012) The theoretical equivalent number of looks. Personal communication
Massonnet D, Souyris J-C (2008) Imaging with synthetic aperture radar. EPFL Press, Lausanne
Mayr S, Schwienbacher F, Bauer H (2003) Winter at the alpine timberline. Why does embolism occur in Norway spruce but not in stone pine? Am Soc Plant Biol, 131(2):780–792
McElhinny C, Gibbons P, Brack C, Bauhus J (2005) Forest and woodland stand structural complexity: its definition and measurement. For Ecol Manag 218(1–3):1–24

Mette T (2007) Forest biomass estimation from polarimetric SAR interferometry. Technische Universität München, Doktorarbeit

Miller RB (1999) Structure of wood. In: Wood handbook: wood as an engineering material. USDA Forest Service, Forest Products Laboratory, Madison (Chapter 2)

Northcote TG, Hartman GF (2004) Fishes and forestry: worldwide watershed interactions and management. Blackwell, Oxford

Oliver CJ, Quegan S (2004) Understanding synthetic aperture radar images. SciTech Publishing Inc, Boston

Oxford University Press (2013) Definition of structure. http://www.oxforddictionaries.com/definition/english/structure

Papathanassiou K (2011) Vertical speckle noise. Personal communication

Penman J, Gytarsky M, Hiraishi T, Krug T, Kruger D, Pipatti R, Buendia L, Miwa K, Ngara T, Tanabe K, Wagner F (2003) Good practice guidance for land use. Land-use change and forestry. Technical report, IPCC, Japan

Pohl C, van Genderen JL (1994) Multisensor fusion: optimization and operationalization for mapping applications. In: Ivan K, Vibeke L (eds) Proceedings of SPIE, pp 17–25

Pohl C, van Genderen JL (1998) Review article multisensor image fusion in remote sensing: concepts, methods and applications. Int J Remote Sens 19(5):823–854

Pommerening A (2002) Approaches to quantifying forest structures. Forestry 75(3):305–324

Richards JA (2009) Remote sensing with imaging radar (signals and communication technology). Springer, Heidelberg

Santoro M (2003) Estimation of biophysical parameters in boreal forests from ERS and JERS SAR interferometry. Doctor of philosophy, Friedlich-Schiller University Jena

Santoro M, Beer C, Cartus O, Schmullius C, Shvidenko A, McCallum I, Wegmüller U, Wiesmann A (2011) Retrieval of growing stock volume in boreal forest using hyper-temporal series of Envisat ASAR ScanSAR backscatter measurements. Remote Sens Environ 115(2):490–507

Snoeij P, van der Valk N, Boom E, Hoekman D (2001) Effect of the ionosphere on P-band spaceborne SAR images. In: IGARSS 2001. Scanning the present and resolving the future. Proceedings of IEEE 2001 international geoscience and remote sensing symposium (Cat. No.01CH37217), vol 1, pp 132–134(IEEE 2001)

Spies TA (1998) Forest structure: a key to the ecosystem. Northwest Sci 72(2):34–39

Telewski FW (1995) Wind and trees. In: Wind-induced physiological and developmental responses in trees. Cambridge University Press, Cambridge, pp 237–263 (Chapter 14)

Tolunay D (2011) Total carbon stocks and carbon accumulation in living tree biomass in forest ecosystems of Turkey. Turkish J Agric For 35:265–279

van Laar A, Akça A (2007) Forest mensuration (managing forest ecosystems), 2nd edn. Springer, Gottenberg

Wade JE, Hewson EW (1979) Trees as a local climatic wind indicator. J Appl Meteorol 18(9):1182–1187

Wald L (1998) Data fusion: a conceptual approach for an efficient exploitation of remote sensing images. In: Processing of EARSeL conference on fusion of earth data, Sophia Antipolis, France

West P (2009) Tree and forest measurement, 2nd edn. Springer, Heidelberg

Williamson L (1975) Out-of-roundness in douglas-fir stems. For Sci 21(4):365–370

Woodhouse IH (2005) Introduction of microwave remote sensing. CRC Press, Glasgow (Taylor & Francis edition)

Young HE (1980) Biomass utilization and management implications. In: Weyerhaeuser science symposium 3, forest-to-mill challenges of the future, Tacoma, Washington, pp 65–80

Zianis D, Muukkonen P, Mäkipää R, Mencuccini M (2005) Biomass and stem volume equations for tree species in Europe. Silva Fennica Monogr 4:63

Chapter 4
Test Site, Data and Methods

To fulfill the objectives of this study, the Thuringian Forest, located at the centre of Germany, was selected for the experimental processes. In addition, various datasets were collected and pre-processed, and methods were developed for the elaboration of this study. This chapter intends to provide a detailed description of these different steps. The first section introduces the selected test site. In particular, information on the Thuringian Forest and some of the relevant characteristics of the dominant tree species composition is provided. The site and species descriptions provide a good basis for interpreting the results presented in Chap. 5. The second section introduces the available data used in this study. The ground reference data, SAR acquisitions and ancillary data are described. The data delivered by different institutions required several pre-processing steps to become comparable. In this respect, the third section focuses on the various pre-processing techniques applied in this study. Pre-processing data comprised the forest inventory, regions of interest (ROIs), weather and SAR data. Finally, the last section focuses on the main components of the methodologies developed for this study. In particular, the procedures applied to the exploration of the data are briefly discussed. The processing approaches related to data modelling with non-parametric and parametric models are highlighted, and the proposed data fusion approach is described.

4.1 Test Site

4.1.1 Thuringian Forest

The Thuringian Forest covers an area of approximately 520,000 ha in the state of Thuringia and is divided into four categories of owners, namely 'Staatswald' (41 %), 'Privatwald' (41 %), 'Körperschaftswald' (16 %) and 'Treuhandwald' (2 %) (Reinholz 2011). Usually, the Thuringian Forest also refers to the 50-km-wide belt

N. Ackermann, *Growing Stock Volume Estimation in Temperate Forested Areas Using a Fusion Approach with SAR Satellites Imagery*,
Springer Theses, DOI 10.1007/978-3-319-13138-2_4

Fig. 4.1 Location of Thuringia and Thuringian Forest in Germany

of forested areas which extend from the northwest to southeast direction (50° 18′N–51° 00′N and 10° 00′E–11° 20′E) in the South of Thuringia. Figure 4.1 provides an overview of the location of the Thuringian Forest in Germany.

The overall area of the forest belt is estimated to be approximately 5,500 km^2, 75 % of which is forested. The remaining portion consists of arable and pasture land, but agriculture has relatively low economic importance. From a geological point of view, the Thuringian Forest is the point of convergence of two subduction plaques, namely the Saxo-Thuringian upper plate and the Rheno-Hercynian lower plate. Consequently, the South Region is essentially characterised by hilly terrain, ranging from 300 to 800 m above sea level (ASL) for the most part, while the North Region represents mainly the Thuringian Plateau with altitudes in the range 150–400 m. The steepest slope gradients do not generally exceed 35°, and the highest point, namely the Great Beerberg, is 982 m ASL (an overview of the Thuringian topography is provided in Appendix A.1). The Thuringian Forest is managed and exploited to meet the requirements of the timber industry. Thus, the forest is generally not allowed to naturally regenerate itself and consists primarily of pure homogeneous stands with Norway Spruce (*Picea abies*) (44 %), European Beech (*Fagus sylvatica*) (21 %) or Scots Pines (*Pinus sylvestris*) (15 %) tree species composition (Reinholz 2011). Because its intended purpose is production of timber and pulpwood, diverse logging activities are conducted in the forest, which entails the forest in a large range of *GSV*, in some cases reaching more than 700 $m^3\,ha^{-1}$. With respect to the climate of the Thuringian Forest, the region experiences cool weather and frequent rainfall (see the climatic diagrams in Appendix A.1). During winter, snowfall is abundant, while in summer, there are local weather variations depending on the topography of the area, exposure to the sun and degree of exposure to wind. The topography determines the climate of the Thuringian Forest. When the prevailing southwesterly winds reach the topographic areas, they are transported to high altitudes; the winds then accumulate in the cloudy regions to the South of Thuringia and cause rainy showers, while in the North (and particularly on the Thuringian Plateau), the weather conditions remain warm and dry. On 18th and 19th January 2007, a violent storm (Kyrill) passed over the Thuringian Forest, partially or completely damaging a large number of forest stands. The volume of fallen timber was estimated to be 2.5 million m^3, and the damaged area was estimated to be approximately 11,000 ha, which represents more than 2 % of the forest (Reinholz 2011). The latter were delineated and digitalised into a geoinformatic system.

4.1.2 Tree Species Description

As mentioned above, the Thuringian Forest consists mainly of three tree species, namely the Norway Spruce, European Beech and Scots Pine. Figure A.7 in Appendix A.1 depicts a picture of each of these species.

Norway Spruces are referred to as evergreen trees. They are categorised as strong soft wood and are economically valuable. Owing to their high market value, they are well-distributed over the temperate zone, especially in the European Alps, Scandinavia and North America, where plantation of this species is common. In the Thuringian Forest, spruces are grown in general as pure stands. Their foliage composition consists of needles that are 1–2 cm long, and their pyramidal crowns usually present a narrow conical form which limits resistance to the wind. However, strong wind conditions can easily damage the stands of the spruces owing to shallow roots of the trees. At a tree level, the branches are orthogonally connected to the trunk which leads for spruces to a dominant horizontal branching structure. While growing, new generation of branches take place on the upper part of the tree and dead branches usually remain along the trunk under the canopy. In addition to being shade tolerant, spruces may grow in acidic soil conditions. However, they require good soil properties in terms of nutrient and water supply, and both preconditions are generally well-satisfied in the Thuringian Forest. The production of a litter with high concentration of manganese is typical in spruce forests, especially for mature stands. This high manganese concentration generally prevents the regrowth of spruces as well as other species. Thus, the forest floor and understory of the spruce forest stands generally exhibits little ground vegetation. As a result of the topographic terrain, the Thuringian region is regularly covered by snow during winter. The accumulation of snow on spruce branches increases the load on them which must be released before the tree branches break. To release the intercepted snow, spruces have adapted strategies using the shapes of their branches; instead of a straight and rigid 'plate' form, the spruce branches are deflected in 'brush' or 'comb' form with pendulous distributions of needles. Consequently, a high diversity of spruce-branching structures can be observed in the Thuringian Forest. Spruce has a greater resistance to low temperatures than beech. Its metabolic activities such as photosynthesis reduce during winter, especially when the air temperature falls near or below 0 °C. In the Thuringian Forest, these conditions are particularly met in January.

The European Beech, which is categorised as a hardwood, is a deciduous broad-leaved tree. As the name suggests, although it can be found in temperate regions such as Asia, North America or New Zealand, it is a species native to Europe and is, consequently, commonly observed in European countries. A beech requires high humidity and can grow in both acidic and basic soil conditions as long as the soil provides good drainage. This last specificity explains the reason for beeches in the Thuringian Forest being mostly observed on sloped areas instead of the bottom of a valley, which would be frequently waterlogged. The European Beech shows a radial branching structure with branches directed mainly towards the sky at 45°–90°. They present dense foliage at maturation, with leaves having sizes of 5–15 cm and breadths of about 4–10 cm and a wide network of roots. Owing to its efficient use of light, this beech is normally shade tolerant, but its dense foliage and root networks usually prevent the growth of an understory in beech forest stands. As mentioned above, beeches are deciduous trees; they are characterised by the senescence phenomenon in autumn and marcescence during winter until foliation in the new spring season. The senescence phenomenon in beeches also marks the beginning of their dormancy period which causes a reduction in all photosynthetic activities.

Similar to spruces, Scots Pines are known as evergreen conifers. Owing to their easy adaptation to different soils and conversion into products with high economical value, they are the most widely distributed pines in the world. Their habitat extends from Europe to Asia (including boreal forests) and North of America. A comparison of Scots Pines and Norway Spruce shows that young pines have the same conical shape as young spruces, but mature pines present flat- or round-topped branches and an open and spreading crown, unlike spruces which retain the conical form even after maturation. Scots Pines are also characterised by high intolerance to shade in comparison to spruces, limiting their growth in understories. Therefore, pine forest stands in the Thuringian Forest consist mainly of a unique pine canopy layer.

4.2 Data

4.2.1 Thuringian Forest Inventory

4.2.1.1 Forest Inventory Description

The reference data used in this study is the forest inventory provided by the 'Thüringer Landesanstalt für Wald, Jagd und Fischerei' (TLWJF), which is the institution responsible for forest-management in Thuringia. The provided inventory played a key role in this study, as it served as a reference for the observations conducted using remote sensing systems. The inventory of the Thuringian Forest comprised 57,840 polygons, representing forest stands, in the vector format, which included different forest attributes. The data were based on the 'Deutsches Hauptdreiecksnetz Zone 4' (DHDN 4) (i.e. Postdam) datum and was referred to as the Gauss–Krüger projection. Table A.1 in Appendix A.3 depicts a sample of the main parameters.

To organise and supervise the different forest-management approaches for the entire Thuringian Forest, TLWJF established different forest divisions. As shown in Table A.1, the largest division is the 'Forstamt' (*FOA*), followed by 'Eigentumsart' (*ETF*), 'Revier' (*REV*), 'Abteilung' (*ABT*), 'Unterabteilung' (*UABT*) and 'Teilfläche' (*TFL*). Each vector polygon of the forest inventory represents the smallest established forest division, namely the TLF. The latter can be assimilated into a forest stand but is not always as it appeared to hold different species or uneven-aged forests in some cases. In other words, the TFL can be considered as a forest stand, but certain precautions must be taken. This detail was considered in the preparation of the ground data (Sect. 4.3). In addition to forest divisions, Table A.1 in Appendix A.3 depicts the main available forest parameters evaluated for this study. Among the forest attributes, the stand area (*AREA*), tree heights (*HÖHE*), *DBH* (*BHD*), Age (*ALT*), dominant species percentage (*BAAN*), relative stocking (*VG*), tree species composition (*HBA_S*), number of understory layers (*BEART*) and growing stock volume (*GSV*) (*HAVOR_GES*) are included in the available forest inventory. The acquisition specificities of these parameters and derivation of the *GSV* are described below. It should be noted that the *GSV* investigated with SAR data was referenced in this study to the provided *HAVOR_GES* attribute.

The earliest measured stands implemented in the forest inventory date back to 1st January 1989. The forest stands were measured annually during summers and were implemented into a geographic information system (GIS) during winters. It should be noted that the forest inventory comprised a large number of stands, and thus, updating all stand measurements in a single year was difficult. Therefore, the measurements of the stands were operated by regions. For instance, the stands at the centre of the Thuringian Forest were evaluated in 2009, while the stands to the South of the forest were evaluated in 2002. Using this procedure, a complete update of the forest stands with new measurements was achieved every 10 years. To keep the entire forest inventory updated each year, the measurements based on logging data and growth models were updated annually. The latter process was executed using the Datenspeicher Wald 2 (DSW2) software.

As mentioned in Sect. 4.1, the Thuringian Forest is owned by several entities, such as the government, societies, private individuals and other particular owners. It may be mentioned that TLWJF undertook the responsibility of forest management only in the forest stands owned by the German authority. Therefore, the forests controlled by the government were regularly updated, and the corresponding data were much more reliable than the forests belonging to private individuals or other owners. Owing to data policies, the forest inventory delivered by the TLWJF consisted of only 'Staatswald', i.e. forest stands owned by the German government.

In addition to the forest inventory, information about Kyrill-damaged areas (see Sect. 4.1) was collected and provided in the form of a vector map.

4.2.1.2 Forest Inventory Acquisition Specificities

The TLWJF institution defined measurement methodologies for the foresters for operation in the forest stands. The procedure followed on the terrain was as follows. The forester first located a forest stand and selected the number and size of circular plots to be measured in that stand. The plots represent the samples of the stand. Their number and size depend on the homogeneity of the stand and visibility in the stand. Visibility plays an important role, as for each tree, the forester needs to observe several parameters from the central location of the plot. For example, owing to its dense branches, a young regenerating stand provides reduced visibility, thus leading to the necessity of increasing the number of plots and also reducing their size. Generally, two or three plots with radii of 25–30 m are subjectively chosen for intermediate sizes (1–2 ha) and relative homogenous stands. The number and size of plots are required to be such that a surface of approximately 0.25 and 1 ha is covered for young and mature stands, respectively. After reaching a given plot, the forester located and marked the plot centre and delimitated the trees out of the circular plot. Distance measurements were performed with a laser Vertex hypsometer instrument.

The *DBH* and height are mainly required to evaluate the *GSV* in a stand (see Chap. 3). In the Thuringian Forest inventory, the *DBH* parameter was derived from the basal area. The method used by the foresters in the Thuringian Forest is based on the Bitterlich approach, in reference to Walter Bitterlich who developed it. It involves the use of a gauge having a known width; this gauge is held at a fixed distance from the eye. The forester turned around from the centre of the plot and counted each tree which appeared wider (at breast height) than the width of the gauge. If a tree appeared to be wider, it was considered as 'in' and counted as '1'. If a tree appeared to have exactly the same width as that of the gauge, it was then counted as '1/2'. Trees appearing to be smaller than the gauge width were ignored. The total count was multiplied by the 'factor' of the gauge in order to obtain the basal area per hectare (see Eq. 3.5) and determine the quadratic mean diameter of the plots. This method was used because it is fast to set up and meets the accuracy required for forestry activities. To retrieve the stand height, the foresters selected 15–20 trees in a plot and measured their height. The selection usually included at best the smallest and largest diameter trees. The height mensurations were performed with a clinometer, which enabled rapid and accurate height measurements. In a further stage, the trees' heights were plotted against their respective *DBH* values for regression analysis and in order to compute the height curve, from which the mid-height of the plot can be subsequently obtained. Once the *DBH* and mid-height were known for the different plots, the average plot values for each parameter were computed, and the *GSV* was evaluated using the allometric relation provided in Eq. (3.3).

In addition to the stand surface, basal area, *DBH* and height, which are the parameters used for deriving the stand *GSV*, the Thuringian Forest inventory possess considerable additional information relevant to gaining an understanding of the stands in the forest. The age, relative stocking, tree species composition, tree species fraction, acquisition date and vertical-layer structuration of the stands were all terrain-observed attributes, which had been registered by foresters. This information was directly inscribed in a working protocol, which was served at the office for the implementation of the data on a GIS system.

As mentioned above, the foresters' field campaigns for data acquisition were conducted during the vegetative period. Subsequently, foresters started the digitalisation task for the delineation of the forest stand. Vectorisation involved correcting or adding the limits of the stands. Several additional bits of information such as aerial digital images and terrestrial GPS measurements were used. The aerial photos were analysed before the field surveys were performed so that the potential new stand boundaries were easily validated and/or adjusted.

4.2.1.3 Forest Inventory Accuracy

The forest inventory *GSV* value accuracy was estimated for approximately 20 % of the measured *GSV* (Chamara 2009), and the georeferencing accuracy of the stands was up to 10 m (Chamara 2009). The inaccuracies in the *GSV* values were essentially sources of basal area measurements. Indeed, the Bitterlich method mainly depends on the subjective appreciation of the forester and can therefore lead to over- or underestimation of the basal area. Other errors contributing to the 20 % error in the *GSV* were related to post-treatment procedures. For example, while the forest height was measured at mid-height, the DSW2 software in this study was based on the top height for actualizing the stands parameters. Further information about the errors of the forest inventory is provided in Grassmann (2012).

4.2.2 Satellite Data

The spaceborne data for this study were collected by L-band systems with ALOS PALSAR and X-band systems with TSX, TDX and CSK. Table 4.1 briefly summarises the available radar data used in this study.

As seen in Table 4.1, 222 radar acquisitions were available for the investigations. It should be noted that the datasets covered different regions of the test site. Among available acquisitions, time-series data were requested. A visual representation of the data frames over the test site and a temporal representation of the data are provided in Appendix A.2.

The time-series acquisitions pertained to vegetation–phenology investigations and were also used for the application of interferometric techniques. Table 4.2 summarises the available data, which could be used for the study of InSAR in the Thuringian Forest.

Table 4.1 Summary of SAR spaceborne acquisitions

Sensor	Frequency	Beam mode	Polarisation	Incidence angle	Nbr. of scenes
PALSAR	L-band	FBS	HH	39°	51
PALSAR	L-band	FBD	HH/HV	39°	60
PALSAR	L-band	PLR	HH/HV/VH/VV	24°	13
TSX	X-band	HS	HH, VV, HH/VV	20°–50°	41
TSX	X-band	SL	HH, VV, HH/VV	20°–50°	9
TSX	X-band	SM	HH/HV, HH/VV, VV/VH	20°–50°	18
CSK	X-band	SM	HH	20°–50°	26
TDX	X-band	SM	HH	38°	4
				Total	222

Table 4.2 Summary of interferometric SAR spaceborne acquisitions

Sensor	Beam mode	Polarisation	Incidence angle	Temporal baseline	Normal baseline (m)	Critical baseline (km)	Nbr. of pairs (-)
PALSAR	FBS	HH	39°	46 days	11–1706	15.7	12
PALSAR	FBD	HH/HV	39°	46 days	67–3874	7.8	24
PALSAR	PLR	HH/HV/VH/VV	24°	46 days	359–362	3.7	2
TSX	HS	HH, VV	20°–50°	11 days	28–209	3.1–8.8	10
CSK	SM	HH	20°–50°	1 day	2–296	1.8–8.6	13
TDX	SM	HH	38°	Single-pass	259	7.9	4
						Total	65

Table 4.2 indicates that six different acquisition configurations were available for examining the interferometric phase. These configurations included 46 days repeat-pass L-band frequency with PALSAR sensor as well as single-pass, 1 day and 11 days repeat-pass X-band frequency configurations with TDX, CSK, and TSX instruments, respectively. The perpendicular baseline ranges for PALSAR were between 11 and 3,874 m, while for TSX, CSK and TDX the normal baseline ranges were between 2 and 296 m. When the perpendicular baselines were compared with the critical baselines, it was observed that the L-band baselines consisted of more than 10 % of the critical baseline, reaching up to 50 % with the normal baseline measured at 3,874 m in the FBD configuration. In contrast to the L-band, the X-band baselines remained below 10 % of the critical baseline. It should also be noted that owing to the varying incidence angles, the critical baselines for TSX and CSK were in the range 3–9 and 2–9 km, respectively. The different datasets summarised in Tables 4.1 and 4.2 are further described below, and a detailed representation of the geometrical and temporal baselines for the available interferometric acquisitions is provided in Appendix A.2.

4.2.2.1 ALOS PALSAR Scenes

Following an acquisition strategy elaborated before the launch of the satellite, the L-band ALOS PALSAR sensor has been systematically acquiring images spanning the Thuringian Forest. The data were acquired between 2006 and 2009 in three different operation modes, namely FBS, FBD and PLR. The three modes were operated during the ascending pass. During this study, 51 FBS, 60 FBD and 13 PLR scenes had been provided by the Japan Aerospace Exploration Agency (JAXA). FBS data were acquired during winter in HH polarisation at an incidence angle of about 39°, while FBD data were measured during summer in HH and HV polarisations with an incidence angle of approximately 39°. It should be noted that FBS data acquired at an incidence angle of about 47° were also collected by ALOS PALSAR. However, this data was declared by JAXA to be unusable because of ambiguities in the range direction (ESA 2007). This piece of data was not considered in this study and was not accounted for in Tables 4.1 and 4.2. Given the consistent acquisition plan elaborated by JAXA, FBS and FBD footprints consisted of eight equivalent frames distributed over the test site in a 2 lines × 4 columns configuration (see Appendix A.2). Each frame, which is 65 km wide and 55 km long, covered a part of the test site and overlapped in the longitude direction of about 30 km (see Appendix A.2). The PLR scenes were acquired during spring at an incidence angle of about 24° with full polarisation (i.e. HH, VV, HV and VH). The footprints with length and size of 25 and 65 km, respectively, were distinguished in 5 different frames (see Appendix A.2). ALOS PALSAR images were ordered pre-processed to level 1.1 in the single look complex (SLC) format. The bandwidths of the data were 14, 28 and 14 MHz for FBD, FBS and PLR modes, respectively.

4.2.2.2 TerraSAR-X Scenes

TSX acquisitions were performed over the Thuringian Forest spanning two different years, namely 2009 and 2010. The TSX dataset consisted of 41 HS, 9 SL and 18 SM scenes acquired in different passes and look angle configurations. Because the TSX is capable of measuring a spatial resolution of up to 1 m, emphasis was placed on the HS mode with dual polarisation. This configuration allowed investigations of different polarisations at extremely high spatial resolutions. It should be noted that the HS mode allowed acquisitions only for HH and VV polarisations, while the SM mode permitted cross-polarisation acquisitions. The TSX satellite was developped for both scientific and commercial utilisation. The performed acquisition orders were in some cases cancelled owing to order conflicts with users having a higher priority. Consequently, the acquisition plan was updated during the survey period with a consideration of the cancelled orders and the availability of limited relative orbits. The positions of the TSX frames were chosen on the basis of the objectives of this study. The coverage of the dominant species, recent forest stands and wide spatial extension over the test site (see TSX frames in Appendix A.2) were given primary importance. Furthermore, to analyze the phenological characteristics of the forest, time-series acquisitions were performed. TSX datasets were ordered in the SLC format at processed level 1A with precise orbit computation and bandwidth of 150 MHz.

4.2.2.3 Cosmo-SkyMed Scenes

CSK acquisitions were undertaken over the selected test site. With the benefit of the CSK constellation system (see Sect. 1.2), 26 datasets representing 13 interferometric pairs with a 1-day repeat-pass were ordered. The data were acquired in HH polarisation with the Himage Ping Pong SM mode. The acquisitions were performed between 2010 and 2011. The scenes were delivered for various incidence angle and pass direction configurations. The frame centre of the acquisitions was defined near Schmiedefeld am Rennsteig in the Thuringian Forest (see Appendix A.2). With a footprint of 30 km × 30 km, this location was shown to have a large amount of spruces, beeches and pines in the same frame. CSK interferometric data were delivered in SLC format at level 1A—SCSB and with a bandwidth of 72 MHz.

4.2.2.4 TanDEM-X Scenes

With the recent launch of TDX, interferometric acquisitions using both TDX and TSX in a single pass were possible. Thus, when TDX acquisitions are mentioned below, if not explicitly mentioned, they refer to TSX and TDX interferometric pairs. Four interferometric acquisitions were undertaken with TDX over the Thuringian test site. The acquisitions consisted of four scenes measured successively on 30th August 2011 with an incidence angle of approximately 38° in the ascending pass and bistatic mode. The first frame was located in the South of the Thuringian Forest, while the last frame was acquired from the North of the site. TDX acquisitions were ordered in the SM mode, HH polarisation and with a bandwidth of 150 MHz. The interferometric data were delivered in level 1A in the SLC format.

4.2.3 Ancillary Data

To support processing and analysis of the satellite data, some auxiliary datasets have been ordered. The Landesamt für Vermessung und Geoinformation (TLVermGEO) delivered LiDAR ground/non ground Earth points data which were acquired during an Airborne Laser scanning campaign operated in 2003 and 2004. Several series of the Airborne Laser Terrain Mappers (ALTM) instrument were used for the surveys. With a flight altitude of 1,000 m, the density of the points ranged between 0.69 and 1.09 pts m^{-2}, leading to a grid sample dimension (GSD) approaching 1 m. The precision of the footprints was 50 cm for flat areas. The acquisition of laser points permitted to derive a 5-m digital elevation model (DEM) and digital surface model (DSM) over the entire test site. In addition to LiDAR data, the TLVermGEO delivered digital orthophotos, which were acquired between 2007 and 2008. The data consisted of RGB images covering regions with an area of 2 $km^2 \times 2$ km^2 with a spatial resolution of 20 cm. The orthophotos were used at a further stage for collecting ROIs over different landcovers. Five landcover classes were established, namely *crop*, *forest*, *open area*, *urban* and *water* (see Sect. 4.3). As mentioned in this chapter, the topography of the Thuringian Forest played a significant role in recording the weather conditions in the test site. Thus, several hydrological studies were conducted in the region, and many weather stations were distributed all over the site. In the framework of this study, approximately 200 weather stations were used to obtain information about the weather conditions in different parts of the test site. The weather stations consisted of individual or climate (i.e. including several measured parameters) stations, which measured precipitation, air temperature, wind speed, sunshine duration, air relative humidity, snow depth and equivalent snow water. The acquired weather data had the same time interval range as that of the satellite data (2006–2012) and was obtained at a daily and hourly frequency. The delivered weather data was operated by the 'Deutscher Wetterdienst' (DWD) and was in the form of .xml files. Further information regarding the weather data is provided in Sect. 4.3. To assess the quality of the forest inventory and obtain a better insight into the forest, forest campaigns were undertaken. Totally, 21 selected forest stands were measured and 20 other selected stands were visited for ground-level observations. Each measured and visited stand was documented using both photographs and GPS measurements. The measurements included specific forest height and *DBH* for computing the *GSV* values. The results of the measured stands are provided in Sect. 4.3.

4.3 Data Pre-processing

4.3.1 Forest Stands

The pre-processing of the forest stands involved three topics: selection of reliable stands, preliminary investigations of the available biophysical parameters and assessment of the forest inventory quality. These different elements are discussed in the present section.

4.3.1.1 Selection of Reliable Forest Stands

The forest inventory was created by foresters for forest-management and industrial activities. To fulfil the aims of this study, it was necessary to perform pre-processing on the inventory. The steps and criteria applied for processing the forest inventory data are summarised below:

- Projection conversion and datum transformation;
- Kyrill-damaged areas;
- Stands erosion (25 m);
- Acquisition date $\geq$ 1.01.2000;
- $GSV \geq 750\,\text{m}^3\,\text{ha}^{-1}$;
- Single canopy layer ($BEART = 1$);
- Forest division ($BESTAND = 1$).

First, the inventory was converted from Potsdam datum and Gauss-Krüger projection to UTM 32 (North) projection, which is based on the WGS84 ellipsoid. Then, the stands intersecting with the Kyrill-damaged areas were removed using the digitalised damage map delivered by the foresters. A comparison of the stands with the available digital orthophotos showed that the damaged areas appeared to affect not only the underlying stand but also the surrounding ones (Fig. 4.2), justifying the choice of an intersect function for this process. Owing to the spatial inaccuracy of the forest inventory, the border of the stands may not correspond to the real forest stand limits, leading to potential artefacts when comparing the stands with the satellite data. After overlapping the forest stands with the satellite images and additional data sources (i.e. digital orthophotos), it was established that 25-m erosion of the stand border should be sufficient to avoid this effect. It should be noted that 25 m correspond to 1 pixel for the least-processed spatial resolution of the available satellite datasets (PALSAR FBD and PLR data). It should also be noted that selection of 50-m erosion would have substantially limited the number of available stands. The forester commenced recording measurements in 1989. Although updates of the old measurements using growing models and logging data could be performed (see Sect. 4.2), it was preferable to limit the investigations to the forest stands measured after the year 2000. Consequently, forest stands acquired before the millennium were removed from the forest inventory. This criterion limited the number of potential errors associated with

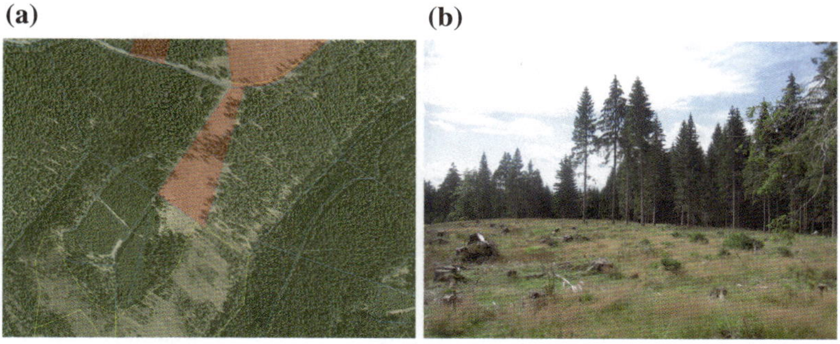

Fig. 4.2 Visualisation of Kyrill-damaged areas with **a** a digital orthophoto (*red areas*) and **b** a corresponding ground level picture. The forest stands' borders are overlaid on the orthophoto in *blue* and *yellow* for spruce and beech tree species compositions, respectively. The orthophoto was acquired on the 28th April 2008 after the passage of Kyrill. The damaged area is located nearby Frauenwald (50° 36.233′N, 10° 52.900′E)

the stands acquired much earlier than the satellite data. The forest inventory contained non-systematic errors introduced particularly during the digitalisation of the forest measurements. As a consequence, stands with a *GSV* lying outside the range 0–750 $m^3\,ha^{-1}$ were considered to be unrealistic and were, therefore, removed from the inventory. Some forest stands included *GSV* values of 0 $m^3\,ha^{-1}$. These stands referred to forested areas which effectively contained forests but were not measured because the *DBH* of the trees was below 7 cm. Forest stands with $GSV = 0$ were removed from the selection, as no information about these forests were available. With respect to the *BEART* attribute, most of the forest stands in the Thuringian Forest were found to present no tree undergrowth, but a few stands clearly depicted two or more canopy layers. To avoid the mixing of the radar signal from the primary canopy with that of the understory, stands with multiple layers were removed from the selection of stands. As mentioned above, the smallest established forest division refers to the *TFL*, which may not always be assimiled into a forest stand. To remove potential uneven-aged forests, forest stands with $BESTAND = 1$ were selected.

In addition to the selected stand parameters defined above, it was necessary to consider two final criteria, one referring to the homogeneity of the stands and another to the processed spatial resolution of the radar data. The criteria needed to be considered with a tradeoff between the number of available stands and the quality of the stands. Taking into account that each species presented a different number of stands and each scene covered a different part of the forest inventory and the radar datasets were processed at different spatial resolutions, several forest inventories with different limited stand sizes and homogeneity parameters were created. Table 4.3 presents the different processed inventories and their related criteria.

Table 4.3 Definition of the stand size (*AREA*), dominant species percentage (*BAAN*) and relative stocking (*VG*) parameters for the three species and different remote sensing datasets investigated in this study

Sensor—beam mode	Processed spatial resolution (m)	Tree species composition	Stand size (*AREA*) (ha)	Dominant species percentage (*BAAN*)	Relative stocking (*VG*) (-)
TSX, CSK, TDX	10	Spruce	0.5	0.8	0.8–1.3
		Beech, pine	0.25	0.4	0.4–1.3
PALSAR FBS	12.5	Spruce	1	0.8	0.6–1.3
		Beech, pine	0.5	0.4	0.4–1.3
PALSAR FBD, PLR	25	Spruce	2	0.8	0.6–1.3
		Beech, pine	1	0.4	0.4–1.3

As shown in Table 4.3, the parameters used for representing the quality of the stands were minimum stand size, dominant species percentage (*BAAN*) and relative stocking (*VG*). These two parameters were considered to determine the homogeneity of the stands.

To determine the size of the stands, the forest inventory and satellite data were compared by extracting the pixels given within the stands (see Sect. 4.4). To improve the radar statistics derived from the stands, it is generally recommended to work with large forest stands, especially when the spatial resolution of the radar data is low. The size of the forest inventory stands ranges between 0.1 and 35 ha, with majority of the stands having an area below 5 ha. Thus, selecting forest stands larger than 5 ha would seriously limit the number of available stands. As shown in Table 4.3, the SAR datasets were pre-processed at 10, 12.5 and 25 m for the X-band sensors TSX, TDX and CSK, the L-band PALSAR FBS data and the L-band PALSAR FBD and PLR data, respectively. On the basis of the size of the stand and the radar pre-processed spatial resolutions, in the case of spruce, the minimum stand sizes of 0.5, 1 and 2 ha were set for spatial resolutions of 10, 12.5 and 25 m, respectively, and for beech and pine 0.25, 0.5 and 1 ha for spatial resolutions of 10, 12.5 and 25 m pre-processed, respectively. These sizes corresponded to at least 30 pixels for spruce stands and 15 pixels for beech and pine stands and thus led to a good tradeoff between the reliability of the statistics and the available number of stands. It should be noted that the number of pixels was separately considered for spruce, beech and pine because the number of stands varied between the species. In particular, a greater number of spruce stands were available compared to beech and pine. It should also be noted that the stand size did not refer to the original size of the stands but to the buffered size, as the buffered stands were used for the comparison between the forest parameters and satellite data.

Considering the *BAAN*, this attribute required selection of values close to 1. Indeed, a *BAAN* of 1 would indicate a single tree species composition within a stand, while a *BAAN* of 0.5 would indicate the occupation of half of the stand by secondary species.

As shown in Table 4.3, the *BAAN* was fixed to 0.8 for spruce species, while it was 0.4 for beech and pine species. As for the minimum stand size parameter, the values assigned to spruce were different from those assigned to beech and pine in order to obtain an adequate number of stands while maintaining the reliability of the stands.

With reference to the relative stocking parameter, this parameter may be used to control the relative distribution of the trees in a stand. For a full stocked forest ($VG = 1$), the stands are per definition (see Sect. 3.1) at their optimal *GSV* capacity, indicating that the stands' conditions are optimal for tree growth. Thus, in this case, it can be assumed that the spaces between the trees would tend to be equal and the forest would be homogenously distributed. However, for an understocked forest ($VG < 1$), the optimal capacity of the stands in terms of *GSV* is not achieved, suggesting that the stands' conditions can be potentially improved. Thus, it is assumed that the number of trees present in the stand can be increased in this case, or the distribution of the trees is not optimal. Therefore, to improve the reliability of the stands, it is recommended to choose a relative stocking close to the fully stocked conditions, i.e. *VG* close to 1. As described in Table 4.3, the relative stocking parameter was limited to the range 0.6–1.3 for spruce and 0.4–1.3 for beech and pines. The relative stocking was more relaxed for European Beech and Scots Pines as compared to spruce so that a greater number of these stands could be included in the analysis. It should be noted that the inventory corresponding to the spatial resolution of 10 m showed a large number of stands compared to the other inventories. In the latter case, the relative stocking of spruce was chosen to be between 0.8 and 1.3.

After the application of the different criteria, the number of forest stands initially available was considerably reduced. At this stage, to preserve the characteristics of the Thuringian Forest, it is important to ensure that the statistics, especially the relative *GSV* distribution of the initial inventory, were not altered. Figure 4.3 depicts the percentage of available stands as a function of *GSV* for the composition of three tree species investigated in this study.

Figure 4.3 suggests the large range of *GSV*, especially for spruce and beech with values up to 750 $m^3\,ha^{-1}$, compared to pine, which has values in the range 0–400 $m^3\,ha^{-1}$. Comparing the *GSV* ranges from the Thuringian Forest with the ones from other forests both in and outside Europe, they are uncommonly large (Pulliainen et al. 1996). This large range may be explained by the optimal growing conditions found in the Thuringian Forest as well as by the management practices followed in Germany. With the comparison of the selected and the overall available stands for each species, the selected stands appear to have a distribution equivalent to that of the stands of the original forest inventory (solid dotted curves). The shape of the curves appears to follow a normal distribution with a slight break around 150 $m^3\,ha^{-1}$, which showed a relative decrease in the number of stands. The mean values were approximatively 350, 350 and 300 $m^3\,ha^{-1}$ for spruce, beech and pine, respectively, which correspond to values of mature stands. For all species, the percentage of young forest being regrown (0–100 $m^3\,ha^{-1}$) appeared to be lower than the proportion of mature stands (100–400 $m^3\,ha^{-1}$). However, the number of stands with low *GSV* remained sufficient for the investigations.

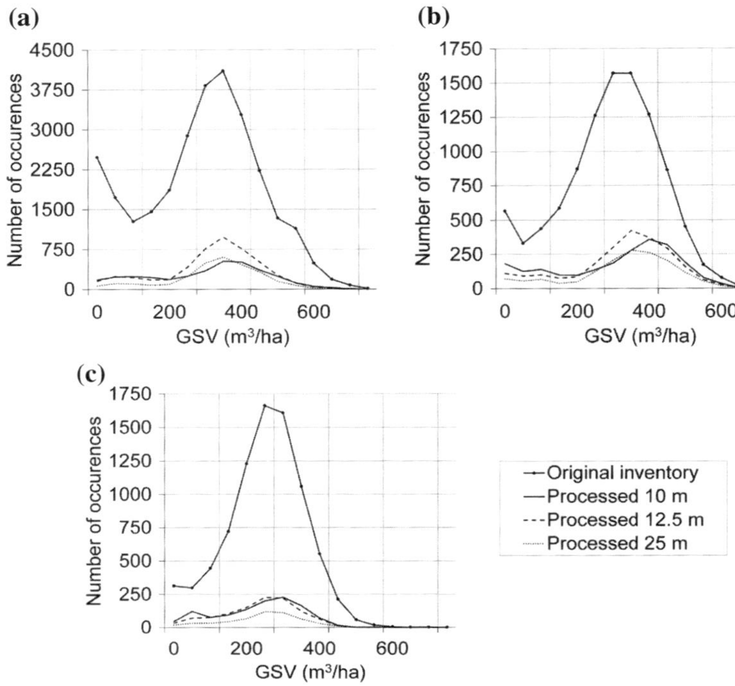

Fig. 4.3 Comparison of the *GSV* distribution between the delivered forest inventory and the different processed inventories for spruce, beech and pine species. *Solid dotted curve* depicts the original inventory, *dotted curve* represents the stands selected for the remote sensing data processed at a spatial resolution of 25 m, *dashed curve* represents the stands selected for a 12.5-m resolution and *solid curve* the stands selected for a 10-m spatial resolution

4.3.1.2 Biophysical Properties of the Thuringian Forest

As described in Chap. 3, the derivation of *GSV* from field measurements follows an allometric relation which includes forest height and basal area. This relation primarily depends on the characteristic of the tree species. Thus, it may be significant to establish a comparison among these biophysical parameters and among each species on the basis of these parameters. Figure 4.4 depicts the relationship between the different forest biophysical parameters of the Thuringian Forest.

In Fig. 4.4, it was observed that the parameters are generally well-correlated. Plot (a) shows that the forest height and forest age are exponentially increasing. Comparing the different species present in this plot, it is also observed that beech becomes older and higher than spruce and pine. However, as shown by the yellow cloud relative to the blue and red clouds of points, in the first 50 years, compared to the other two conifers, beech requires more time to grow. In considering plot (b) from Fig. 4.4, the increase in *GSV* becomes limited at around 75 years for the three

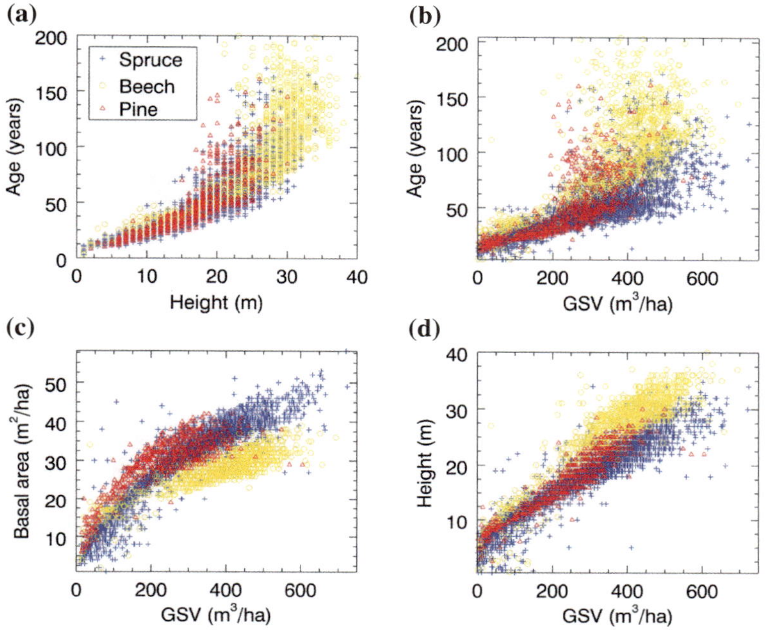

Fig. 4.4 Comparisons of the Thuringian Forest biophysical parameters for different compositions of tree species with **a** forest height versus forest age, **b** forest *GSV* versus forest age, **c** forest *GSV* versus forest basal area and **d** forest *GSV* versus forest height. The spruce, beech and pine tree species compositions are depicted in *blue*, *yellow* and *red*, respectively

species under investigation. In other words, the species attain their mature height and *DBH* after around 75 years. From plot (c), it was observed that the basal area depicts a logarithmic relation with *GSV*. The species on plot (c) must be compared with those on plot (d). Indeed, the basal area and height show clearly opposing trends when compared with *GSV* for the different species. The basal area of pine and beech rapidly increases in the first stages of development ($GSV < 100\,\text{m}^3\,\text{ha}^{-1}$). The basal area of spruce increases at a moderate rate compared to the two other species. In terms of height, the composition of the three tree species depicts equivalent growth rates in the initial stages of developpment. After reaching $100\,\text{m}^3\,\text{ha}^{-1}$, spruce gains a larger basal area compared to beech; however, the increase in the basal area of spruce is offset by a decrease in height compared to beech. Thus, while beech gains in volume, its density (i.e. basal area) decreases and its height increases. Conversely, while spruce gains in volume, its height decreases and its density increases. As a consequence, mature beeches and spruces present different structures. Section 5.2 shows that this information is key to interpreting the radar signal.

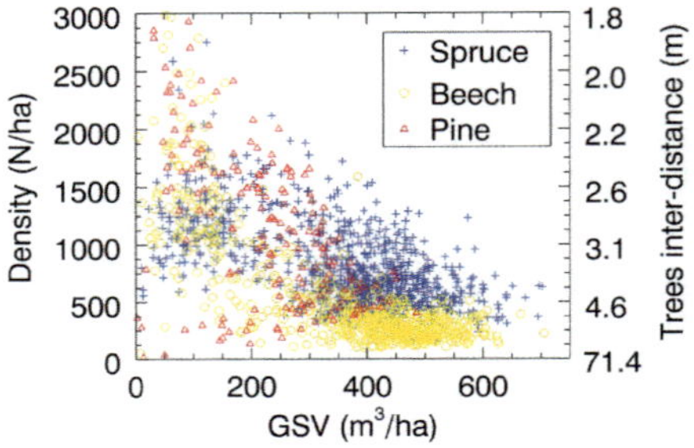

Fig. 4.5 Comparison of forest *GSV* with stem density. Spruce, beech and pine tree species compositions are depicted in *blue*, *yellow* and *red*, respectively

Although the basal area provides information about the density of the trees in a stand, it may be easier to interpret the density parameter of the stand. Stand density (*N*), expressed in trees/ha, was not available in the forest inventory, but it could be calculated using the basal area and *DBH*. The density was then plotted against *GSV*, as shown in Fig. 4.5.

Figure 4.5 shows that the density generally decreases with *GSV*. The highest density is mainly shown for beech and pine with more than 2,000 trees ha^{-1}, while the lowest density is essentially depicted by beech with less than 250 trees ha^{-1}. Assuming a uniform distribution of the trees on the stand, 2,000 and 500 trees ha^{-1} would be equivalent to the respective distances of 2.2 and 6.3 m between each tree. In the range 400–600 $m^3\ ha^{-1}$ (mature stands), beech shows a smaller density than spruce.

The physiology of a forest may be significantly influenced by the topography of the terrain (see Chap. 3 and Sect. 4.1). In this context, the parameters of the forest stands of the Thuringian Forest were analysed with the rough terrain of the site. For this purpose, polar plots displaying the *GSV* on the radius, the aspect of the terrain on the circumference and slope classes represented by different colours were computed. Figure A.18 in Appendix A.6 depicts the polar plots created for (a) spruce, (b) beech and (c) pine species. On comparing the polar plots, it is generally observed that pines are located mainly on flat areas, while spruce and particularly beech are mostly present on hilly terrain. As introduced in Sect. 4.1, the beech species require well-drained soils, which is why they are predominately found on topographic areas. Also, the pine plot indicates that pines growing over sloped areas are located mainly at orientations ranging from 60° to 330°. This indicates the lower tolerance of pines to the shaded areas described in Sect. 4.1. Polar plots were also obtained using forest height, basal area and other forest parameters, but no specific phenomena could be observed.

As introduced in Chap. 3, both topography and wind modify the cross-sectional shape of the tree trunks. To investigate wind loading and slope effects on the eccentricity of trunks, the *DBH* parameter of trees should be retrieved with two orthogonal measures, one parallel and another perpendicular to the prevailing wind and slope orientation. Unfortunately, such information was not retrieved and was also not available in the forest inventory.

4.3.1.3 Quality Assessment of *GSV* in Forest Inventory

After selecting the stands on the basis of the criteria defined earlier in this section, comparisons with orthophotos were established. The comparisons were performed for three parameters, forest height, basal area and forest homogeneity. The observations showed first that the inventory border of forest stands appropriately delimited the stands. The comparison results showed that the observed homogeneity almost completely agreed with the *BAAN* parameter provided in the forest inventory. However, forest height and basal area were different from the forest inventory; this difference amounted to 40 % in some cases. Although orthophotos provided reliable observations of the forest, it should be noted that these observations remain subjective. Moreover, the acquisition time between the orthophotos and the forest stands differed in some cases, ranging from 1 to 5 years. Therefore, although the comparisons provided a good first indication about the quality of the forest inventory, caution should be taken while considering these results.

To obtain a better evaluation of the forest inventory quality, forest campaigns with external measurements were organised. Two types of campaigns were performed. One campaign concentrated on gaining an understanding of the Thuringian Forest, especially young regenerative forests, and the other campaign focused on measurements of the forest stands.

The first campaign involved a 2-day visit of 20 forest stands. Appendix A.3 describes a list of the visited stands. The study involved gaining an insight into the forest by identifying and documenting relevant characteristics of young regenerative forest stands. The observations essentially provided two sets of information. First, the young forest stands in particular presented extremely dense forests, and in such conditions, light penetration was significantly reduced. Second, the forest floor of the young forests consisted mainly of bare soils with dispersion of dead woods, which was due to the forest density and reduced penetration of light as well as species-specific tolerances to shadow and soil conditions. Finally, on comparing the species, it was noted that the branches of spruces were directed orthogonally towards the trunk, while the beeches' branches were directed mainly towards the sky at 45°–90°. It was also noted that dead branches were clearly present along spruces' trunks, as opposed to the case of pines and beeches.

The project which aimed to evaluate forest stands involved three different campaigns, which were performed in 2009, 2010 and 2011. These campaigns were organised in collaboration with the Fachhochschule Erfurt (FHE) and constituted a Bachelor Thesis in the field of forestry and ecosystems management (Grassmann 2012). The measurements involved determining the *GSV* for the selected forest stands. For this purpose, the procedure described in Sect. 4.2 was followed considering a few modifications. The differences concerned essentially the *DBH* and basal area determinations. In the case of the forest campaign, *DBH* was measured for each tree inside the delimited plot by taking samples using a calliper, and the basal area was then calculated from the *DBH* measurements. In contrast, the foresters measured the

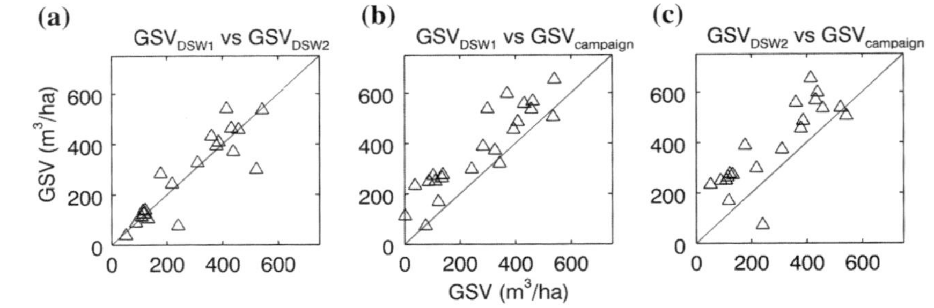

Fig. 4.6 Comparisons of forest stands' *GSV* values among the provided forest inventory excluding actualisation of the stands (GSV_{DSW1}), that includes updates on the stands (GSV_{DSW2}) and measurements undertaken in forest campaigns (GSV_{campaign}). The actualisation of the stands with DSW2 included corrections using growing models and logging data

basal area using the Bitterlich method introduced in Sect. 4.2 and calculated the *DBH* parameter. The calliper method was chosen for the campaigns, as it was more accurate than the Bitterlich method. Once the required forest parameters were measured, post-treatments were performed, particularly by computing *GSV* using the BWIN pro v.6.3 software. The resulting *GSV* measurements are summarised and compared to two different forest inventories, namely one with (DSW2) and the other without (DSW1) actualisation of the forest stands. Figure 4.6 depicts the different *GSV* values as scatterplots with (a) GSV_{DSW1} versus GSV_{DSW2}, (b) GSV_{DSW1} versus GSV_{campaign} and (c) GSV_{DSW2} versus GSV_{campaign}.

On comparing the *GSV* values for the two provided forest inventories in Fig. 4.6a, it is observed that with a few exceptions, the *GSV* values from the updated forest inventory are slightly higher than those from the non-actualised inventory. However, the difference remains insignificant. Plots (b) and (c) in Fig. 4.6 indicate that the *GSV* values measured during the campaigns are approximately 100 $\mathrm{m}^3\,\mathrm{ha}^{-1}$ higher than those provided by the inventories. This difference is significant. The fact that most of the *GSV* values in the campaigns are higher than those in the inventories may indicate a systematic error. This error may be related to the slightly different methods used for obtaining the *GSV* of the forest stands. It should be noted that the measurements of the stands for the campaigns and inventories were undertaken at various epochs and for various species of trees (see Appendix A.3). Therefore, some points on the scatterplots in Fig. 4.6 are expected to be dispersed.

According to the conclusions in Grassmann (2012), the relevant differences shown in Fig. 4.6b, c between the campaigns and *GSV* inventories may have been related to the inclusion of several inaccuracies in the measurements. In particular, the forest campaign performed in 2010 excluded the slope inclination for the correction of the height measurements on a hilly terrain and evaluation of the surface of the plots. These parameters were assessed during post-treatments with a local DEM and by considering the size and homogeneity of the stands, respectively. Moreover, as mentioned in Sect. 4.2, forest reference heights were based on the top stand heights obtained from the DSW2 software, while foresters measured the mid-height. Further details are provided in Grassmann (2012).

4.3.2 Regions of Interest (ROIs)

As introduced in Sect. 4.2, on the basis of digital orthophoto observations, ROIs of several landcover classes were collected in a database. The landcover information was required in this study to investigate the identification of forested areas and compare the radar signals obtained from forests and land classes, such as open areas. Overall, five different landcovers were selected, crop, forest, open area (grass), urban and water. The five selected landcovers were distributed throughout the test site. Crops were mostly located on the Thuringian Plateau, while forests were found mainly in the South, where the topographic areas covered Thuringia. Open areas were characterised by grasslands such as meadows. Urban regions consisted of small villages in the South and cities in the North of the site such as Erfurt, Illmenau, Jena and Weimar. Finally, the class water essentially described the supply of drinking water and electricity by artificial lakes to the Thuringian region. It should be noted that the forest class was not based on orthophoto observations but on the pre-processed forest inventory, as the latter data provided information about forests. An overview of the selected ROIs is given in Appendix A.4.

4.3.3 Weather Data

The weather data acquired in the Thuringian Forest was delivered for each weather parameter in the form of XML files with measurements given for every station. The objective of weather data pre-processing was to be able to compare the satellite data with the weather conditions occurring at the time of the SAR acquisitions. For this purpose, the J2k module developed on the Jena Adaptable Modelling System (JAMS) software at the University of Jena as well as implemented IDL scripts was used. The procedure included several processing steps leading to the regionalisation of weather data over the entire Thuringian state. The regionalisation process consisted of a simple inverse distance weight (IDW) kriging, which considered a DEM (SRTM) to correct the regionalisation of the weather parameters with the topography. To provide boundary conditions to the regionalisation process, an area larger than Thuringia was considered. Also, regionalisation was processed at a pixel size of 90 m. Considering the distances between the weather stations and the spatial resolution of SRTM, this spatial resolution was optimal in terms of the processing time and accuracy of the regionalisation.

After regionalizing the weather data, statistics such as mean and standard deviation could be extracted from the forest stands for each satellite acquisition. The statistics were then stored in a database, which could be used for the investigations. The weather conditions may influence the radar signal at different temporal scales. For example, wind is a variable which may change every few hours, while precipitation is a variable which may affect soil moisture conditions for several days. Thus, to analyse and consider weather effects at different temporal scales, for each satellite acquisition, weather data obtained from 4 hourly and 4 daily measurements were to be processed. Also, the weather measurements of the daily weather stations were performed every day at 7.30 a.m. Because the satellite data were acquired at different times each day, the last day of the 4-day weather measurements was selected to be as close as possible to the satellite acquisition time. In these terms, ALOS PALSAR was programmed for all ascending acquisitions to survey over the Thuringian Forest at 9.30 p.m. on a specific day (d). The period considered for the weather measurement then included the 2 days before satellite acquisition ($d - 2$) until 1 day after satellite acquisition ($d + 1$). Moreover, the ascending acquisitions of TSX were performed at around 6 a.m. Thus, in this case, the chosen period was 3 days before satellite acquisition ($d - 3$) up to the day of acquisition (d). This approach helps in obtaining the weather data which are as close as possible to the actual weather conditions for each satellite acquisition.

A summarised version of the processed weather database is provided in Appendix A.5.

4.3.4 SAR Data Pre-processing

Radar information such as backscatter intensity, SAR polarisation and interferometric coherence were investigated in this study. As given below, the pre-processing of these different data is described. An evaluation of the noise level on the processed datasets is provided.

4.3.4.1 Backscattering Intensity

As introduced in Sect. 4.2, the different spaceborne datasets were delivered in the SLC format. To extract the backscatter intensity from the SLC data and make the radar data comparable with each other and with the forest inventory, several steps were required. These different steps conducted with the GAMMA software are presented below and are depicted as a processing flowchart in Appendix A.7.

Calibration of the data was first processed using the calibration constant provided by the different space agencies. In the case of ALOS PALSAR, the updated calibration constants were considered for computing the calibration of the different PALSAR acquisition modes (Shimada et al. 2009). In addition to calibration, the radiometry of the data was converted from beta nought (β^0) to sigma nought (σ^0) projection using the sine of the incidence angle. Multi-looked intensities were then generated by applying multi-looking factors. The latter were selected on the basis of the final spatial resolution to be achieved and the imaging SAR configuration (i.e. incidence angle, range and azimuth resolution). The objectives of multi-looking are as follows: sufficiently reduce speckles while preserving an adequate spatial resolution and get as close as possible to the defined final range and azimuth spatial resolution for the geocoding process (see below). Appendix A.7 provides a summary of the multi-looking factors used for the different datasets. It should be noted that the range of multi-looking factors were calculated with a $\sin(\theta)^{-1}$ component, which enabled the consideration of the geometric projection from the slant to the ground range with the SAR incidence angle θ. The final spatial resolution was defined on the basis of the intrinsic SAR systems' spatial resolutions. PALSAR FBD and PLR modes were processed at 25 m, PALSAR FBS mode at 12.5 m and TSX, TDX and CSK acquisitions at 10 m. Geocoding was finally executed using a lookup table and a simulated SAR image generated from the available LiDAR 5 m DEM and SAR geometry. The lookup table was refined by generating a cross-correlation analysis between the SAR and the simulated SAR image (Wegmuller 1998). After reaching sub-pixel accuracy, the refined lookup table was used for geometric transformation and resampling of the SAR image to the DEM geometry. The resampling was based on bilinear interpolation. The geocoding process, which uses a simulated SAR image, is a fully automatic approach. To control the accuracy of the process, the geocoding range and azimuth offsets were examined, and the standard deviations of the range and azimuth were verified.

As mentioned in Sect. 4.1, the forested areas in the Thuringian Forest were located mainly on hilly regions. Because SAR data encounter radiometric and geometric distortions with high topographic variations, some radiometric and geometric corrections, including a DEM, were required. A normalisation process was considered by applying corrections for the local incidence angle and the ground scattering area. Equation (4.1) exposes the different parameters involved in the topographic normalisation process (Ulander 1996; Castel et al. 2001)

$$\gamma^0 = \sigma^0 \frac{A_{\mathrm{flat}}}{A_{\mathrm{slope}}} \left(\frac{\cos\theta}{\cos\theta_{\mathrm{l}}} \right)^n . \tag{4.1}$$

Equation (4.1) describes the relative dependence of the backscattering coefficient γ^0 on several topography-related parameters. The cosine of the local incidence angle θ_{l} corrects the radiometry for local slopes and converts the data from sigma nought (σ^0) to gamma nought (γ^0), and the cosine of the incidence angle of reference θ scales the data to the mid swath incidence angle. The terms A_{flat} and A_{slope} represented a normalisation factor which corrected for slopes towards or away from the radar (Ulander 1996). The exponential parameter n reflected the potential varying scattering contribution of the investigated targets over hilly areas. In the case of bare surfaces, the n coefficient would take the value 1. However, when considering a volume medium such as a forest canopy, the terrain slope would modify the path length of the radar microwaves through the canopy, which requires the consideration of a specific n coefficient value which would depend on the environmental conditions as well as on the properties of the canopy.

Castel et al. (2001) suggested an approach for deriving the exponent n. The approach involved relating the n coefficient as a function of the SAR acquisition geometry and the optical depth of the canopy on the basis of RT equations and WCM. The relation is described as follows:

$$n = 1 + a\log\left(\frac{1 - \exp(-b\tau)}{1 - \exp(-c\tau)}\right), \tag{4.2}$$

where

$$\tau = K_e H_c . \tag{4.3}$$

The parameters a, b and c are empirical variables derived for different SAR configurations. They depend on the SAR acquisition parameters (i.e. wavelength) and local geometry. The optical path length τ is derived from the product of the canopy coefficient of extinction K_e and the canopy crown height H_c. These two biophysical parameters may in turn be related to other variables. In particular, H_c may depend on the local slope and the tree height, while K_e can be related to the tree structure and the environmental weather conditions occurring during acquisition.

Although the computation of n may be performed directly from Eq. (4.2), information about the canopy, such as height and extinction coefficient, were required. These data were not available in the Thuringian Forest inventory; thus, another approach for the derivation of the n parameter was considered. The approach involved testing different n coefficients by evaluating the coefficient of variation ($CV = \sigma/\mu$) of the amplitude until its minimum was obtained. The n coefficient corresponding to the minimum CV could then be successfully applied to the correction of the radar data.

As the final step of the topographic corrections, the areas presenting geometrical distortions (layover, foreshortening and shadows) were masked because these areas did not possess valuable information.

4.3.4.2 Polarimetry

The two specific computations performed with polarimetric data were decomposition of scattering mechanisms and azimuth-slope correction. The former was performed using the Yamaguchi decomposition algorithm presented in Chap. 3. Assuming a volume filled with dipoles, this decomposition approach was designed for forest targets and should therefore be suitable for the investigations conducted in this study. Azimuth-slope correction was performed with the method proposed by Lee et al. (1999) (see Chap. 3). The *OA* was estimated from the polarimetric data using the circular polarisation technique. This approach was chosen because it showed a straightforward relationship with the investigated data, and its effectiveness was demonstrated in several studies (Lee et al. 2004). The decomposition of the scattering mechanisms and azimuth-slope corrections was performed using the PolSARpro software (Pottier et al. 2009).

4.3.4.3 Interferometric Coherence

The radar data presented in Sect. 4.2 included interferometric acquisitions. In this regard, the interferometric coherence was examined in this study. The process of the interferometric phase magnitude was performed with the GAMMA software using two main sequences:

1. SLC data co-registration
2. InSAR processing

The first step consisted of perfectly registering the two SLC images together. For this purpose, a processing scheme using a lookup table was established. The lookup table linked the geometries of the two SLC images and included offset corrections computed from a DEM. Thus, the offsets due to the topography could be incorporated into the SLC data registration. Similar to the geocoding of the SAR intensities, calibrated and multi-looked images (MLIs) were first computed. The lookup table could be then generated between the MLIs and refined using cross-correlation analysis between the two SAR images in a further stage. The refined lookup table could be finally applied to resample the slave SLC image to the master one. Although the lookup table was refined and included corrections for topography, some inaccuracies due to the DEM, orbit state vectors and geocoding still remained. Therefore, a further refinement step consisting of estimating the offsets between the master SLC image and the resampled SLC image was also conducted. This co-registration processing methodology led to a variation which ranged from 0.05 to 0.1 subpixels, which represented a shift ranging between 50 cm and 1 m for the data processed at a spatial resolution of 10 m. This very high accuracy was necessary for reducing the spatial decorrelation from the end coherence magnitude product (Hansson 2001; Massonnet and Souyris 2008). The standard deviation constitutes an average value encompassing the total offsets. The offsets may be large at some locations and low in other locations of the images. Thus, although the standard deviations of the offsets are good indicators of the general accuracy, caution must be paid while interpreting these values. The second stage of the phase coherence magnitude computation included three steps. The first one involved calculating the geometrical baselines between the two satellite acquisitions. The estimations of the parallel and perpendicular baselines were performed using the orbit states vector. The orbits of ALOS PALSAR, TSX, CSK and TDX were sufficiently accurate for use in the calculation of the geometrical baseline. Once the baseline was computed, an interferogram based on the co-registered SLC data was computed. Additionally, a simulated (unflattened) interferogram was derived from the available LiDAR DEM. The simulated interferogram could then be subtracted from the interferogram generated from the co-registered SLC images, thus resulting in a differential interferogram that considers the topography-induced phase variations. The interferometric acquisitions were performed with two measurements taken at slightly different angles, thus leading to spatial decorrelation (see Chap. 3). To deal with this phenomenon, a common band filtering approach, applied for the range and azimuth directions, was used prior to the generation of the interferogram (Gatelli et al. 1994). It should be noted that the latter corrections were performed adaptively using topographic information. At this stage of the process, the differential interferogram presented residual systematic fringes resulting from errors in the orbital state vectors. The spatial baseline was thus recalculated on the basis of the differential interferogram fringes and was used to compute a new simulated interferogram. The latter could finally be subtracted from the differential interferogram to create a differential interferogram free of topography-induced phase variations and systematic errors. In the last stage of the InSAR processing, the magnitude of the interferometric phase was estimated. The calculation was performed using an adaptive window which was adjusted to the local degree of coherence (Wegmuller 1998). More specifically, the window size of the estimator was enlarged over low coherence values and reduced over high coherence values. In the case of PALSAR data, the window was set to 3×3 and 5×5 (range and azimuth) for the minimum and maximum sizes, respectively. For TSX, TDX and CSK data, the window sizes were fixed to 3×3 and 7×7. In addition to the window size, texture information derived from the SAR intensity was used to improve the estimation of the coherence. The texture was computed with the *CV*, allowing the distinction between small inhomogeneous (high texture) and large homogenous (low texture) areas. The texture could be used together with the estimator window size, constraining small windows over small texturised areas. By referring to the window sizes, the choice of a maximum size of 5×5 pixels for PALSAR and 7×7 for the X-band datasets was made in order to not only achieve a good estimate of the coherence but also consider a sufficient number of samples for limiting the coherence bias (see Sect. 3.2).

4.3.4.4 Noise Evaluation

SAR systems are affected by speckle and several other noise sources such as thermal noise. To avoid misinterpretation of the results, it is necessary to quantify noise. A broad evaluation of the noise was performed in this study for the backscatter intensity and interferometric coherence. For this, the SNR was considered. The equations for the SNR calculations were introduced in Sect. 3.2. Moreover, as described in Sect. 3.2, the ENL was used to evaluate the signals' accuracy.

The *SNR* and ENL values were computed for the pre-processed SAR datasets showing different acquisition configurations. The *SNR* was based on the *NESZ* provided by the different space agencies in sigma nought projection (refer to JAXA 2012; Eineder et al. 2009; Fritz et al. 2012; ASI 2007). Thus, to compare the signal of the pre-processed datasets with the *NESZ*, the pre-processing procedure described in Eq. (3.21) was applied to the provided *NESZ* values. The *SNR* computation led to *SNR* images with *NESZ* values converted in gamma nought projection and included topographic corrections based on the local incidence angle and pixel area.

The ENL values were computed using two different approaches. The first approach, which was named for the purpose of this study, ENL_S, included statistics inferred from the data (see Eq. 3.24). It should be recalled that this calculation assumed uniform distributed areas and a normal distribution (Oliver and Quegan 2004). SAR data usually do not assume a Gaussian distribution, and some of the targets considered, such as crops, may not fit the uniformity assumption, as they represent different growing stages. However, ENL_S still provided a good way of comparing the signal accuracy between classes and radar datasets. A varying number of pixels may influence the statistical ENL estimation. Hence, to provide comparable ENL_S values between classes and SAR datasets, a general fixed number of 50 pixels per stand or ROIs were considered in this study for evaluating the number of independent looks. Also, the number of available ROIs varied from class to class. For instance, water or urban ROIs had a smaller number of ROIs than crops, forests or open areas. When possible, a number of 100 ROIs were used for calculating the ENL. When the number of ROIs available was smaller, the ENL_S estimation accuracy should be reduced. The second ENL calculation approach, referred to as ENL_t, was a purely theoretical approach. In this case, as presented in Eq. (3.23), the principle involved dividing the initial radar pixel spacing by the spatial resolution provided by the space agencies and then multiplying it with the multi-looking factors and the number of pixels used during pre-processing of the data. The ENL_t values may be considered as a reference for the evaluation with ENL_S of the accuracy of the radar signals.

In theory, the noise should be equivalent for different targets covering an image. However, the various scattering mechanisms often lead to great variations between the targets. To analyse the noise level, 2D scatterplots with the *SNR* (dB) of the backscatter and interferometric coherence (-) on the abscissa and ordinate, respectively, were obtained. The investigations were focused on the selected ROIs showed in Appendix A.4. Figure 4.7 presents an illustration of the computed scatterplot for PALSAR FBD HH and TDX SM HH data acquired on 10th June/26th July 2010 and on 30th August 2010, respectively. In addition to *SNR* values, the theoretical curve (red function) based on Eq. (3.23) is displayed on the graphs to support the interpretations.

The scatterplots in Fig. 4.7 indicate that the noise level varies significantly between the different landcovers. The noise also appears to differ between sensors. On further examination, it is observed that for PALSAR data, the noise is the greatest for water ($SNR = -5$ to 0 dB) and then for crops and open areas ($SNR =$ 0–10 dB) followed by forest and urban areas ($SNR =$ 10–20 dB). In contrast to PALSAR, TDX depicted a high level of noise for water ($SNR =$ 0–5 dB) and then for crops, open areas and forest ($SNR =$ 10–15 dB) followed by urban areas ($SNR =$ 15–20 dB). With reference to these observations, the noise level for PALSAR and TDX is negligible for forests. This result may be explained by the high power caused by volume scattering in the forested areas. In terms of decorrelation, according to the modelled *SNR* (red curve), the noise decorrelation for forests is also of minor importance in PALSAR and TDX configurations. However, with respect to the difference between the measured and the modelled coherence, for PALSAR and TDX, PALSAR, which has a larger difference,

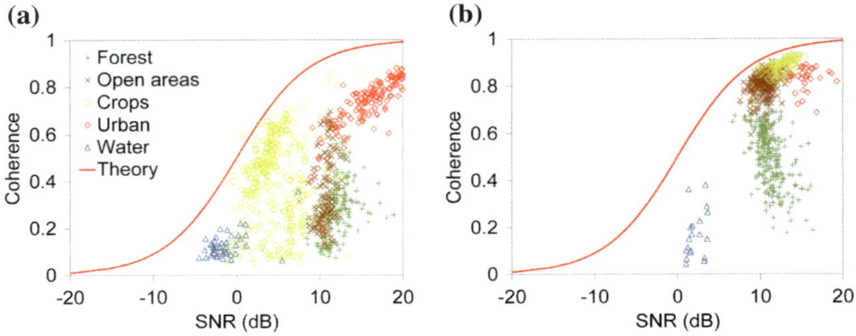

Fig. 4.7 *SNR* noise versus interferometric coherence with **a** ALOS PALSAR FBD HH 39° Asc. $B_n = 363$ m (frame FBD3) and **b** TDX SM HH 38° Asc. $B_n = 258$ m (frame TDX3) data. The PALSAR and TDX interferometric acquisitions were acquired on 10th June/26th July 2010 and on 30th August 2010, respectively. The *red curve* describes the theoretical noise decorrelation (γ_{SNR}). The *SNR* values were calculated with the first acquisitions of the interferometric pairs. The landcovers represent crops (*yellow*), forest (*green*), open areas (*brown*), urban (*red*) and water (*blue*)

appears to be significantly affected by other decorrelation sources such as temporal or volume decorrelations. These elements will be addressed in more detail in Sects. 5.1 and 5.3.

As mentioned above, general noise evaluation was performed on the overall available sensor and acquisition configurations. Table 4.4 summarises the average *SNR* and ENL_s values for the forest class as well as the ENL_t values calculated for the available sensor configurations. The average computations were performed over 100 spruce stands with 50 pixels selected for each stands.

The ENL values in Table 4.4 indicate that the ENL_s and ENL_t values are significantly different. The ENL_t value ranges between 250 and 4,000, while ENL_s extends between 3 and 30. As compared to the theoretical ideal ENL values, the ENL_s inferred from the statistics are pessimistic. However, comparing ENL_t and ENL_s for the various incidence angles, it is clearly observed that both follow the same trend, namely a high ENL for large incidence angles and a low ENL for steep angles. Then, with respect to the *SNR* values, it is generally seen that with an *SNR* ranging mainly between 5 and 15 dB, forests present an acceptable noise level. The decorrelation due to noise also appeared to be sufficient, especially for PALSAR HH data, which depicted a γ_{SNR} value of above 0.9. Among these observations, one exception could be mentioned, namely CSK, which appeared to present a lot of noise with a *SNR* for large or steep incidence angle configurations remaining below 1 dB. In the latter case, the *NESZ* value of −19 dB provided by ASI (2007) appeared to be slightly high and may have not been suitable for CSK sensor configurations. Then, on comparing the polarisations, it is observed that HH polarisation data was less noisy than HV data. In L-band and X-band systems, the *SNR* values were found to be approximately 6 dB lower for HV than those for HH. Like-polarisations generally presented higher power than cross-polarisations, which explains why higher noise was found with the HV

Table 4.4 Noise evaluation for spruce forest stands with ENL_s, ENL_t, SNR and γ_{SNR} measures for different acquisition configurations

Sensor	Beam mode	Polarisation	Incidence angle	ENL_s (-)	ENL_t (-)	SNR (dB)	γ_{SNR} (-)
PALSAR	FBS	HH	39°	1279	16	11.9	0.93
PALSAR	FBD	HH	39°	1320	30	11.9	0.94
PALSAR	FBD	HV	39°	1320	24	6	0.79
PALSAR	PLR	HH	24°	722	22	11.1	0.91
TSX	HS (single-pol)	HH	23°	862	12	14.9	0.97*
TSX	HS (single-pol)	HH	48°	3505	22	10.9	0.93*
TSX	HS (single-pol)	HH	37°	2052	20	5.2	0.76
TSX	HS (dual-pol)	HH	37°	2052	24	6.1	0.77
TSX	HS (dual-pol)	VV	37°	2052	21	6.1	0.77
TSX	SL (dual-pol)	HH	41°	2949	24	11.4	0.93*
TSX	SM (dual-pol)	HH	37°	3073	19	14	0.91*
TSX	SM (dual-pol)	HV	37°	3073	22	7.2	0.84*
CSK	HS	HH	20°	353	3	0.17	0.52
CSK	HS	HH	37°	1266	17	0.8	0.53
CSK	HS	HH	50°	2012	33	0.13	0.52
TDX	SM (dual-pol)	HH	38°	257	19	11.4	0.93

The ENL_s and SNR values were calculated with the first acquisitions of the interferometric pairs. For the γ_{SNR} values marked by '$*$', no interferometric data were available, and therefore, they were processed with the SNR of one image instead of two

channel. Finally, on observing various incidence angles, it was expected that there would be a greater noise with a steep incidence angle, especially due to the greater returns occurring in such configuration. However, in this case, the *SNR* evaluation did not show any such trend.

4.4 Methodology

4.4.1 Data Exploration

As presented in Sect. 4.2, a large number of SAR and reference data are available. To automate the investigations, an IDL data exploration routine was developed. This comprised first the initialisation and input-data preparation. The inputs consisted of the reference polygons shapefile (forest inventory, ROIs), a raster copy of the shapefile and the satellite data to be investigated. The raster copy was converted beforehand from the shapefile and contains the polygons ID values. The raster copy was used here to establish the link between the shapefile and satellite data. This raster, also called the raster ID, needed to be processed at the same spatial resolution as the input satellite data. Each set of input data was then read, and the process of extracting satellite data pixels included in each reference polygon could be started. Statistics such as the polygon mean or number of pixel values were implemented to summarise the information. Four additional optional processes were also developed: (1) requesting specific criteria such as the tree species, tree height or a defined slope inclination, (2) modelling the observations, (3) plotting the data and (4) saving the plots and statistics. For the comparison of the satellite data with each other and with the reference data (forest inventory, ROIs), particular attention was paid to the following elements. First, it was verified whether the scenes were overlapping and their spatial resolutions corresponded with each other. Then, subsetting of the common area of the datasets was performed so that each scene had the same area. Additionally, geometrical distortions for radar systems are specific to the radar acquisition configurations. Thus, the nodata information over each considered scene was commonly masked. Finally, with respect to the parameter under investigation, it was verified that the remaining parameters presented equivalent configurations, namely same incidence angle, polarisation, spatial resolution, frequency, etc.

4.4.2 Parametric Modelling

Parametric regression modelling is one of the techniques applied in this study to estimate *GSV*. The modelling approach was distinguished into four different topics, namely selection, training, inversion and testing of the models. Before describing each of these processes, a general overview of the different processing steps can be presented. Figure 4.8 depicts a flowchart describing the performed modelling phases.

As illustrated in this figure, parametric modelling included two sets of data, the forest inventory serving as a reference and the satellite data from which the *GSV* could be estimated. The entire modelling process was repeated using bootstrapping techniques so that the estimations became statistically more reliable. Overall, only 10 bootstraps were considered, as it appeared to be a good tradeoff between computation

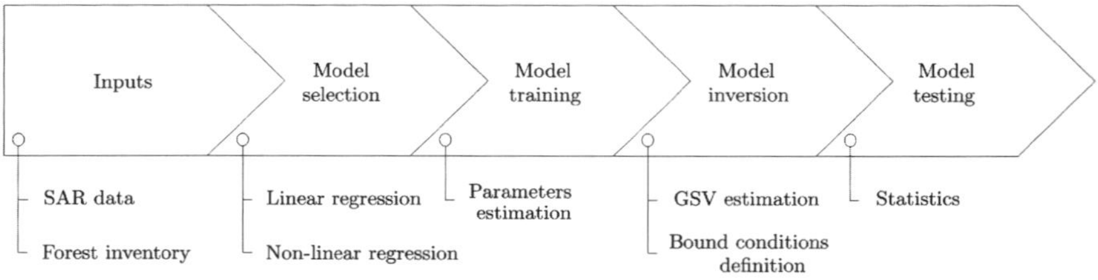

Fig. 4.8 Flowchart describing the parametric modelling process

time and statistics reliability. It should be noted that the training and testing stands random selection described later in this section was repeated for each bootstrap so that configurations of different stands may be used for the estimation of modelling accuracy. The different steps of the modelling process, shown in Fig. 4.8, are detailed below.

4.4.2.1 Model Selection

In Sect. 5.3, by studying the relation between the available radar data and *GSV*, it could be shown that in the Thuringian Forest, interferometric coherence decreased with increasing *GSV*. More specifically, it could be shown that in some cases, this negative relationship was linear, and in other cases, it exponentially decreased with *GSV*. In this context, simple linear and nonlinear regression models were defined for the estimation of *GSV*. For the nonlinear models, the exponential function already used by Eriksson et al. (2003) was chosen for the purpose of this study. The linear and nonlinear models considered for the parametric modelling of this study are provided in Eqs. (4.4) and (4.5), respectively:

$$\gamma(GSV) = A \cdot GSV + B\,, \tag{4.4}$$

$$\gamma(GSV) = A \cdot e^{B \cdot GSV} + C\,. \tag{4.5}$$

Both models described coherence γ as a function of *GSV*. The linear model (Eq. 4.4) comprised two unknown parameters (i.e. *A*, *B*), while the nonlinear model (Eq. 4.5) held three different unknowns (i.e. *A*, *B*, *C*). In the case of the linear relation, *A* denoted the slope of the function and *B* represented the y-intercept. *A* and *B* could be interpreted as the decreasing rate and the dynamic range of the line, respectively. In the case of the nonlinear model, *A* and *B* depicted the dynamic range and the decreasing rate of coherence, respectively. The remaining parameter, *C*, indicated the coherence offset.

Equations (4.4) and (4.5) are empirical relations. In this study, the choice of empirical functions instead of semi- or physical functions was motivated by different reasons. First, the Thuringian Forest, as shown in the scattering and decorrelation mechanisms analysis (Sects. 5.2 and 5.3), presents specific structural characteristics such as large *GSV* ranges and very dense canopies. In this case, the application of a semi-empirical model such as the IWCM would imply modifications of the model, which is beyond the scope of this study. Second, the sensitivity of coherence and intensity to *GSV* was shown to be rather limited, particularly with the backscatter intensity. Thus, as a good starting point for future studies, the application of empirical models may provide a first approximation of the retrieval accuracy.

4.4.2.2 Model Training

The training of the two models presented above was performed by least-squares minimisation using the forest stands as a reference from the available forest inventory. The stands were randomly selected and set into different groups so that both, training and testing of the models could be executed. The following three groups were considered:

- 40 % training;
- 30 % testing—model selection;
- 30 % testing—final accuracy.

The first 40 % of the stands were dedicated to the training process. This percentage was chosen to be higher relative to the other ones in order to guarantee (with sufficient forest stands) a reliable estimation of the model parameters. The second and third fractions of stands referred to the testing of the model. These two fractions each accounted for 30 % of the overall stands. Two different sets of testing stands were defined to take into consideration the evaluation of the most suitable model between the abovementioned models. Thus, the 30 % testing set was allocated to the selection of the model, while the final 30 % of the stands were dedicated to the final accuracy calculation of the selected model. For each of the three sets of data, the stands were randomly selected using uniform distribution. Also, for the selection of the model, the same training and testing stands were considered so that the accuracy could be estimated on the basis of the configurations of the same datasets.

The good fit of a model depends on the model itself, the number of available training stands and their repartition over the *GSV* range. In the case under study, the number of stands and their repartition varied significantly from one scene to another; the statistics were not always consistent. In this context, an iteration process was performed, which enabled reaching at least 300 stands for each acquisition. The iterations were based on the parameters which describe the homogeneity of the stands (*BAAN*, *VG*) (see Sect. 4.3). When the number of available stands was not sufficient, the *VG* was first relaxed, after which the same was followed for *BAAN*. This procedure implied that for the scenes covered by few stands, the quality of the selected stands would decrease because with the relaxing of *VG* and *BAAN*, the stands would tend to be more heterogeneous. The choice of $N = 300$ stands was shown to be a good tradeoff between the available number of stands for the different datasets ($N_{\text{train}} = 120$, $N_{\text{test}}^{\text{model}} = 90$, $N_{\text{test}}^{\text{accuracy}} = 90$) and the quality of the stands. The issue regarding obtaining sufficient forest stands for the modelling process was partly resolved by fixing the number of stands. For the Thuringian Forest, the proportion of stands in the high *GSV* range was larger than that in the low *GSV* range (see Sect. 4.3). Thus, the problem related to small number of available stands in the low *GSV* range remained for some of the acquisitions. The solution to this problem would be to control the stand quality for the low and high *GSV* ranges using *BAAN* and *VG* criteria differently. This idea was not implemented, as the forest structure between mature and young forests would have been modified and supervision of the forest parameters would have been limited.

Although the most reliable stands were selected for the training and testing of the model (see Sect. 4.3), outliers were still observed during the training process. By comparing the outlier stands with orthophotos and Google Earth images in addition to weather data, they were mainly identified as logged or potentially Kyrill-damaged forest stands which were not updated in the actual forest inventory or changing local weather conditions. To remove these outliers, a bootstrap process was performed. More specifically, it involved systematic removal of forest stands lying above and below 2σ from the model and repeating this process until all the stands ranging outside the 2σ interval could be discarded from the training process (Santoro et al. 2011).

4.4.2.3 Model Inversion

After estimation of the model parameters with the training process, model inversion was performed. The inversed functions were expressed for the linear (Eq. 4.6) and nonlinear (Eq. 4.7) functions in the following form:

$$G\hat{S}V = \frac{\gamma - B}{A}\,, \tag{4.6}$$

$$G\hat{S}V = \frac{1}{B}\ln\left(\frac{\gamma - C}{A}\right)\,. \tag{4.7}$$

The estimated *GSV* ($G\hat{S}V$) was calculated from Eqs. (4.6) and (4.7) as a function of the interferometric coherence and the estimated parameters. Accordingly, a pixel-based *GSV* image may be derived for each available coherence image. However, the model was suitable for the range of coherence values for which it was fitted. Outside this range, the retrieved *GSV* should theoretically not make any sense. Thus, on the basis of the approach used in Santoro et al. (2011), some conditions were considered for the coherence pixel values lying outside the minimum and maximum value of the regression curve. The defined conditions can be summarised as follows:

$$G\hat{S}V = \begin{cases} 0, & \text{if } \gamma > \max\left(\gamma_{\text{model}}\right) + 2\sigma\,, \\ 0, & \text{if } \max\left(\gamma_{\text{model}}\right) < \gamma < \max\left(\gamma_{\text{model}}\right) + 2\sigma\,, \\ \max(GSV), & \text{if } \min\left(\gamma_{\text{model}}\right) - 2\sigma < \gamma < \min\left(\gamma_{\text{model}}\right)\,, \\ \max(GSV), & \text{if } \gamma < \min\left(\gamma_{\text{model}}\right) - 2\sigma\,. \end{cases} \tag{4.8}$$

The idea of the above conditions is fixing the unrealistic $G\hat{S}V$ values. For the case where the coherence lies above the maximum of the modelled coherence plus two standard deviations (condition 1), $G\hat{S}V$ was set as zero, which is the minimum possible value. Theoretically, nodata value should be defined for this range of values because it should correspond to the outliers range. However, zero was preferred because as shown in Sect. 5.3, the very high coherence observed in forested areas is physically attributed to the high stability of the very young regenerative trees. Similar to the above condition, the $\hat{GSV}$ for the coherence values ranging between the maximum modelled coherence and the maximum modelled coherence plus two standard deviations were fixed to zero (condition 2). In the case where the coherence lies between the minimum modelled coherence and the minimum modelled coherence minus two standard deviations, the $\hat{GSV}$ was set to the maximum stand *GSV* value (condition 3), as it is the theoretically modelled maximum value. Finally, when coherence values were shown to be lower than the minimum of the modelled coherence minus two standard deviations, as in the case for condition 1, the $\hat{GSV}$ was set to the maximum modelled *GSV* (condition 4) instead of nodata, as the coherences in such ranges referred mainly to large forests canopies showing considerable instability and/or volume decorrelation.

4.4.2.4 Model Testing

The determinant part of the modelling procedure concerned the testing of the models. As presented above, two different testing sets were defined for this process: one for the selection and the other for the final accuracy assessment of the model. To perform both processes, several statistics such as the *RMSE*, mean absolute error (*MAE*), bias and coefficient of determination (R^2) were computed (see Appendix C.1). The *RMSE* relied on the difference for stand i between the estimated $\hat{GSV}_i$ and the ground measured GSV_i. This difference, which constituted the error, was squared and averaged for the N available forest testing stands. Finally, the root of this given amount was considered in order to obtain the *RMSE*. It should be noted that in some studies (Santoro et al. 2002), the sampling error was removed in the *RMSE* computation. In the case of the Thuringian Forest, the sampling error intrinsic to the field measurements was shown to be approximately 20 % of the measured *GSV* (see Sect. 4.2). If this error is removed from the *RMSE* calculation, *RMSE* would not be representative of the overall error, namely the radar and ground-measurement errors. Therefore, in this study, the sampling error was considered in the final *RMSE* estimation.

The statistics computation using the first testing dataset (30 %) was performed for the two models and each interferometric dataset. For each coherence scene, the best model was selected on the basis of the derived *RMSE*. After selecting the model, the same statistics were then recomputed using the selected model and the independent testing set of stands (30 %) which provided the final accuracy of the *GSV* estimation.

4.4.3 Non-parametric Modelling

For comparison with the abovementioned regression modelling method, a second technique was tested in this study for estimating forest *GSV*. This technique referred to the non-parametric k-nearest neighbour (k-NN) algorithm. The k-NN approach is described below in two different steps, training and testing.

4.4.3.1 Model Training

k-NN is a data-driven approach used for retrieval of *GSV* and is based solely on information provided by the forest inventory. The basic principle of the k-NN method is assuming that forest stands with similar forest characteristics should hold similar spectral properties. Accordingly, each pixel of a satellite image is assigned a specific *GSV* value according to its spectral signature.

The simplest case of the k-NN operator is $k = 1$, which signifies that only the first nearest forest stand is used for determining the estimated *GSV*. To improve the estimation, it is possible to consider multiple k numbers. In such a case, a weighting of the k nearest forest stands selected would be required. The important relationships in the k-NN approach are shown below:

$$G\hat{S}V_p = \sum_{i=1}^{k} w_{p_i,p} \cdot GSV_i \,, \tag{4.9}$$

$$w_{p_i,p} = \frac{\dfrac{1}{d^s_{p_i,p}}}{\sum_{j=1}^{k} \dfrac{1}{d^s_{p_j,p}}} \,. \tag{4.10}$$

According to Eq. (4.9), $G\hat{S}V_p$ was calculated by multiplying the inventory GSV_i for the forest stand i with ponderation factor $w_{p_i,p}$. The ponderation factor calculation is described in Eq. (4.10). It includes the spectral distance d between the pixel p to estimate and the pixel p_i for forest stand i as well as exponent value s. The exponents $s = 0$, 1 or 2 were considered for the mean, inverse and inverse-square distance, respectively. Of the common existing spectral distances, this study evaluated the Euclidean, Mahalanobis and Manhattan distances. The following equations provide the relations corresponding to these distances

$$D_{\text{Euclidian}} = \sqrt{(P - P_{\text{r}})^T \, (P - P_{\text{r}})} \,, \tag{4.11}$$

$$D_{\text{Mahalanobis}} = \sqrt{(P - P_{\text{r}})^T \left[M^{-1}\right] (P - P_{\text{r}})} \,, \tag{4.12}$$

$$D_{\text{Manhattan}} = |P_1 - P_{\text{r}1}| + |P_1 - P_{\text{r}2}| + \cdots + |P_1 - P_{\text{r}n}| \,. \tag{4.13}$$

In the aforementioned relations, the multivariate vector P is the spectral value of the pixel to be estimated (target), while P_r is the spectral value of the pixels averaged for the forest stand being referenced. In the case of processing n multichannels, P and P_{r} would be holding the dimension $1 \times n$. The Mahalanobis distance differs from the Euclidean distance by considering a covariance matrix M which may consider the potential correlation between different channels when several datasets are used. In the specific case where M equals the identity matrix, the Mahalanobis distance would reduce to the Euclidean distance. The Manhattan distance consists of the sum of the absolute difference between the target and reference spectral values.

4.4.3.2 Model Testing

Similar to the regression modelling technique presented in Sect. 4.4.2, the evaluation of the *GSV* estimation using the k-NN algorithm was based on statistics such as *RMSE* and bias. However, instead of selecting a specific percentage of the available forest stands, the overall stands were used in this case for the training as well as testing of the model. In this study, the applied evaluation method, known as the leave-one out cross-validation (LOOCV) technique, involved the removal of one forest stand from the reference dataset for use in the testing (i.e. evaluation) process, while the rest of the stands were considered for the training (i.e. estimation) process. This sequence was repeated for each forest stand such as when each of them was used once as a testing dataset. Although this method was computationally intensive, it provided for the k-NN method a good estimate of the *GSV* accuracy without requiring any specific subdivisions of the datasets.

4.4.4 Fusion Approach

This section describes the fusion approach developed in this study for the creation of a forest *GSV* map. As it may be conceived, the processing of a *GSV* map not only involves estimating the *GSV* but may also involve the identification of forested areas and separation of tree species because such information is required for delivering forest and species-specific *GSV* maps. The modelling of the interferometric coherence presented in Sects. 4.4.2 and 4.4.3 allowed the derivation of *GSV* quantities for the available remote sensing data. However, the issues related to identifying forested areas and separating tree species were not treated in this study. In this context, two topics were examined for the fusion methodology: one related to the derivation of forest/non-forest as well as tree species composition masks and the second dedicated to the fusion of the derived biophysical products. A brief introduction to the selected fusion approach and the motivation behind the chosen approach is provided below. Then, the strategy for separating forest/non-forest and different tree species composition is presented. Finally, a detailed description of the fusion methodology leading to the end product *GSV* map is given. As shown in Sect. 3.3, the terminologies ‘fusion’, ‘merge’, ‘combination’, ‘synergy’ and ‘integration’ are closely related. Thus, these terms simply refer to the combination of data.

4.4.4.1 Approach and Motivation

According to the theory presented in Sect. 2.2, remote sensing data can be combined with many different approaches. In the framework of this study, it was decided to focus the combination of the SAR information on derived high level products such as forest/non-forest masks, tree species masks or *GSV* images. The main idea was to process a *GSV* map which could be derived from different sources of data constituting an efficient combination of the available datasets. Data combination refers not only to the fusion of the *GSV* information but also to the inclusion of forest and tree species information.

The suggested fusion approach introduced above involved the comination of several secondary products in a single *GSV* map. Different reasons motivated the conception of such an approach. First, the techniques used to retrieve forest parameters recently reached maturity. Thus, the fusion of various biophysical products may now be envisaged for deriving updated national and global *GSV* maps. Second, the development of new SAR sensors has significantly increased over the past few decades, and a large amount of data is actually available. This large amount of data constitutes a prominent source of information, which may contribute to the derivation of forest parameters. In this regard, the availability of different datasets may increase the accuracy of the *GSV* maps as well as their spatial extension and updated frequency. Third, by proposing the combination of established products, the proposed approach can provide considerable flexibility. Indeed, the combined products may not only be derived from one single technique but from several techniques such as a *GSV* map estimated from the regression and *k*-NN modelling methods described earlier in this section. Finally, most of the fusion approaches reported in literature until now depended on two or three different datasets, limiting the forestry applications to the availability of these datasets (see Sect. 2.1). With the proposed approach, the ingestion in the computation of various acquisitions measured by several sensors and acquisition modes is proposed. In terms of usability, this approach would enable new perspectives for the user, such as the possibility of refining the *GSV* map specifically to some defined objectives. In this sense, the proposed approach would be user-oriented.

4.4.4.2 Mask Establishment and Accuracy Assessment

As mentioned in the introduction to this section, a forest *GSV* map intends to show not only the *GSV* of the forested areas but also the *GSV* specific to the different species composing the forested areas. Thus, to create a *GSV* map, it is necessary to know the location of the forested areas and the different species. In this regard, before working on the combination of the remote sensing data, the derivation of thresholds for the forest and species classes was performed. The developed methodology involved extracting the numerical values from different landcovers and species on a scene and deriving the threshold values as well as statistics for the separation of the classes on the basis of bimodal frequency-histogram analyses.

In the present study, the classes crop, forest, open area, urban and water for the different landcovers and spruce, beech and pine for the composition of different species were considered. The information on landcovers was based on the defined ROIs (see Sect. 4.2), while the species information was established from the forest inventory. In the case of landcovers, the objective was to extract the forest class from the other classes. Thus, in this case, the histogram analyses always referred to the forest class with one of the other classes. In the case of species composition, the goal was to distinguish between the species under study. The frequency histogram analyses were performed by considering the normalised difference between two functions, with each function representing a different class. Equation (4.14) describes the analytical function used to extract the classes' thresholds

$$H_{\text{norm}}(x) = \frac{f(x)}{\max(f(x))} - \frac{g(x)}{\max(g(x))} = 0\,. \tag{4.14}$$

In Eq. (4.14), $H_{\text{norm}}(x)$ represents the normalised difference between the histogram of the first class $f(x)$ and the second class $g(x)$, where the function $H_{\text{norm}}(x)$ is set to zero, as the two functions $f(x)$ and $g(x)$ intersect. The numerical x-value at this point of intersection corresponds to the threshold value. To automatically determine this position, a process based on the principle of the bisection method was performed. The extraction of a class usually requires the calculation of two thresholds. The first threshold would correspond to the abovementioned point of intersection, while the second threshold is found either at the extremity of the histogram representation or at a second intersection point, depending on the potential separation of the classes. It should be noted that an optimal definition of the number of bins may improve the derivation of the threshold. In the present case, $n = 100$ bins was chosen. This is a good tradeoff between the histogram smoothness and the number of pixels used in each histogram. In addition, to derive the position of the thresholds more accurately, a smoothing function was used. It should also be noted that the thresholds were computed for each available dataset. To obtain a consistent threshold value for each SAR configuration, the average of thresholds was evaluated. In terms of radar configurations, parameters such as incidence angle, polarisation, frequency and perpendicular baseline or acquisition period were distinguished. The configurations showing a deficient number of pixels were not considered in the calculation of the average values.

The histogram analysis chosen for identifying forests and classifying forest species is close to the Bayesian ML classification approach. In the present case, the histogram was directly used, while in the Baysian method, probability distribution functions (PDFs) are considered. Although the method based on the raw histogram is hardly transferable to a different test site, this simple method was preferred for the following reasons. First, the statistical distributions were initially unknown, particularly after applying ratios (see fusion process below) from the different SAR datasets. Second, forest and species mapping were not the main objective of this thesis. Finally, classifications obtained from the histogram analysis are more accurate than those obtained by assuming PDFs.

In addition to deriving the thresholds, information about the separability of the classes is required to evaluate the potential of extracting only one signal. For this purpose, normalised cross-correlation analyses were performed, thus enabling the quantification of the correlation between two classes. Equations (4.15) and (4.16) present the calculation of the cross-correlation and the normalised cross-correlation, respectively,

$$R(h) = \int_{-\infty}^{+\infty} f(x)g(x+h)\,\mathrm{d}x\,, \tag{4.15}$$

$$R_{\mathrm{norm}}(h) = \frac{R(h)}{\max(R(h))}\,. \tag{4.16}$$

In Eq. (4.15), $R(h)$ represents the cross-correlation coefficient calculated as a function of the lag h. The cross-correlation indicates the correlation of the functions $f(x)$ and $g(x)$ at different lags h. When $h = 0$, the correlation coefficient indicates the separability of the two functions $f(x)$ and $g(x)$ at their initial position. The higher the correlation, the less separable the two functions, i.e. the two classes. Cross-correlation at lag $h = 0$ represents the Pearson's correlation coefficient r. Although this coefficient is the information of interest in the present case, its calculation by cross-correlation is more appropriate as cross-correlation is widely used in signal processing. Similar to thresholds, consistent values of the correlation coefficient were obtained by averaging the coefficients for each equivalent radar configuration. Configurations showing an insufficient number of pixels were also not considered.

4.4.4.3 Fusion Process

Having introduced the fusion approach and the methodology for identifying forested areas and separating the tree species, the fusion processing scheme leading to the final *GSV* map may now be described. The fusion approach was split into four different steps: inputs, initialisation, core processing and end products generation. Each of these steps was distinguished by subprocesses. Figure 4.9 provides an overview of the latter processes.

As depicted in Fig. 4.9, the first steps involved the determination of input data and initialisation of the process. Then, core procesing included temporal merging and ratio fusion processes, threshodling computations and synergy processes. Finally, the end products were obtained by masking and mosaicking of the tiled products. The entire processing steps were performed at the pixel level and implemented in an IDL routine. The IDL routine allowed the automatic processing of the proposed fusion approach. The end products were *GSV*, forest/non-forest and species maps, each of which was represented on a corresponding accuracy flag map. Further details of the fusion processing scheme introduced in Fig. 4.9 are presented.

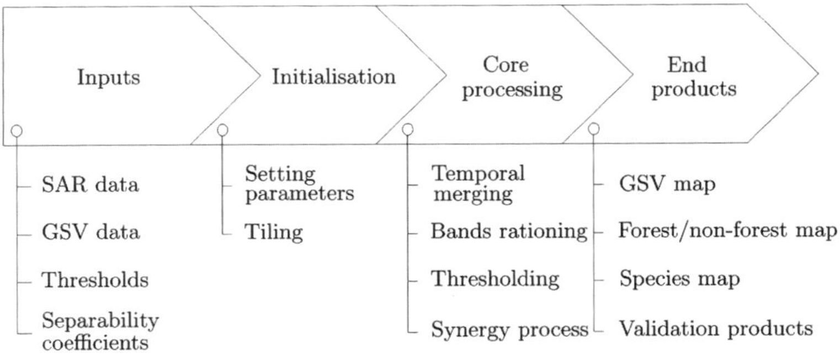

Fig. 4.9 Flowchart describing the fusion process

Inputs

To compute the *GSV*-map product, the developed algorithm required a minimum of two different datasets. First, it required a *GSV* image containing the estimated *GSV* of a forest type or a specific forest species for an entire scene. In the present study, this information would be directly provided by the regression and *k*-NN modelling techniques described in Sects. 4.4.2 and 4.4.3, respectively. Then, it required a backscatter intensity or interferometric coherence image to establish forest/non-forest and species masks with the thresholds derived from the histogram analysis presented above. The intensity and coherence information may be pre-processed on the basis of the techniques described in Sect. 4.3.

Initialisation

With the developed algorithm, a few parameters which can be set by the user were proposed. One of them is the desired spatial resolution of the final products. It should be noted that all processing steps are performed for all datasets at the same defined resolution. If the remote sensing data have a lower spatial resolution than the user-defined data, the data would be downsampled; on the other hand, if the spatial resolution of the remote sensing data was higher, the data would be oversampled. Another parameter proposed to the user is the size of the processed tiles. This size should be defined on the basis of the computer memory and CPU capacity. The last parameter is the specification of the minimum accuracy to be processed. This parameter allows the user to select the accuracy to be achieved for a certain product. For the forest/non-forest and species masks, the accuracy is given as a score (see Eq. 4.18), while for the *GSV* map, it is provided as an *RMSE* value. The dataset which does not enable the attainment of the specified accuracy would not be considered when processing the final product and would be flagged in the validation products as no information. An example is presented in Sect. 5.5.

The tiling process or block operation considered the overall data provided in the input by the user. It was performed for two main reasons. First, the ingestion of hundred or more datasets requires high-memory-capacity computing systems. To solve the memory-capacity problem, the datasets are computed in tiles, which are saved in a temporary folder. Another reason for applying tile processing is the simplification of the masking process. The size of the tiles is known and is the same for all data, thus facilitating the overlaying of the data and masking processes.

Core Processing

As suggested by its name, core processing is a relevant step in the proposed fusion approach. To properly understand the processes involved, an illustration of the main processes is provided in Fig. 4.10. It should be noted that core processing involved two distinct methods of data combination, fusion and synergy. These two terms are distinguished below. This step was performed iteratively over each tile, and the results could be saved in temporary folders.

As suggested in Fig. 4.10a, a fusion process was envisaged for multitemporal data. The fusion approach involved merging of the multitemporal datasets when such data configuration was available. Multitemporal averaging was performed on the backscatter intensity and interferometric coherence, which were used for the establishment of forest/non-forest and species masks. The temporal aggregation of the data enabled reduction of the intrinsic noise of the backscatter intensity and improvement of the interferometric coherence estimates. The fusion processes were performed at the tile level. When the available data presented band-rationing capabilities, a simple ratio between two selected channels was considered (Fig. 4.10b). Band rationing was performed by referring to the results of the histogram analysis described in Sect. 5.5. Similar to the multitemporal merging process, band rationing enabled noise reduction, and this procedure was performed at the tile level. After performing the above fusion stages, the processing of the masks was conducted using the thresholds and correlation values established in the histogram analysis. Equation (4.17) describes the thresholding process:

$$g_{ij} = \begin{cases} 1, & \text{for} f_{ij} \geq T_{\text{h}} , \\ 0, & \text{for} f_{ij} < T_{\text{h}} . \end{cases} \tag{4.17}$$

Here g_{ij} is the masked image, f_{ij} the SAR intensity or coherence image and ij the corresponding radiometric pixel location. Pixels with values greater than or equal to the defined threshold T_{r} are assigned the value 1. The remaining pixels are assigned the value 0. The threshold and correlation values were stored in a database for each investigated radar dataset. Computing tile by tile, the algorithm searched for the available radar backscatter intensity and interferometric coherence data and computed the masks and correlation images using the stored threshold and coefficient

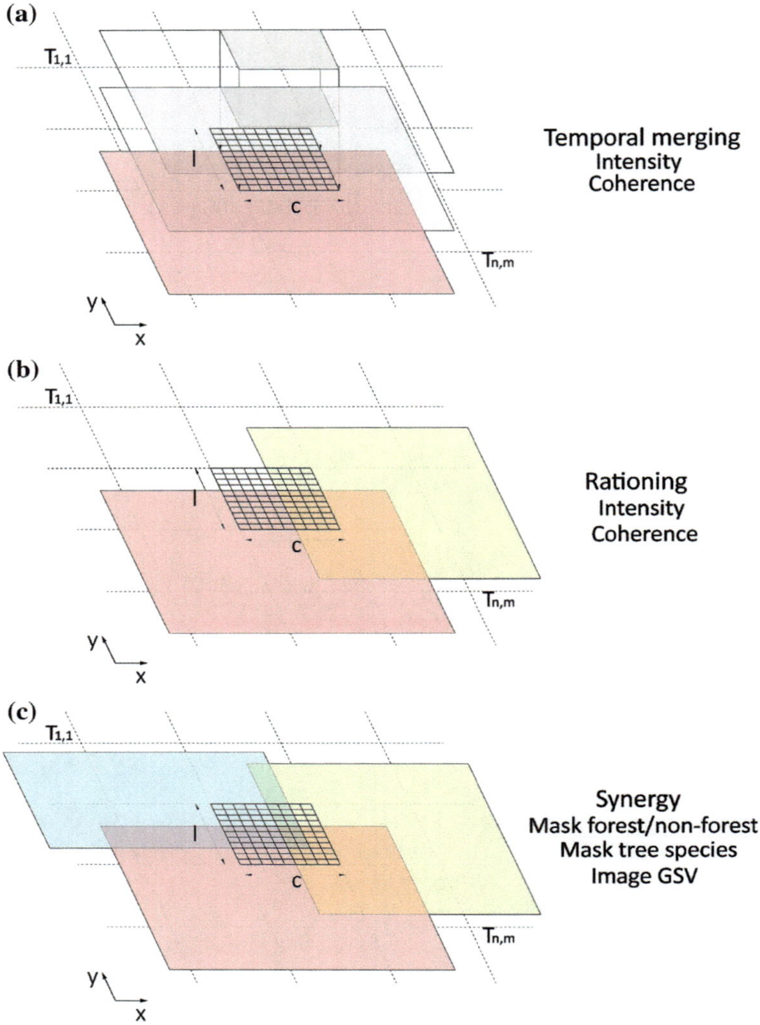

Fig. 4.10 Main processing steps in the fusion approach. The illustrated processes are: temporal merging fusion (**a**) ratio fusion (**b**) and synergy (**c**)

of correlation values, respectively. At the end of the process, the masks consisted of pixel-based data with a digital value of $DN = 1$ for forest or the species composition of interest and $DN = 0$ for the non-forested areas or for the tree species which needed to be masked.

As indicated in Fig. 4.10c, the synergy process considered the masks and *GSV* images. In the case of the forest/non-forest and tree species masks a pixel-based process that involved selecting pixels with the minimum correlation value (good separability) was employed for synergy. Pixel selection was performed iteratively over all pixel locations using two for loops. This approach was performed for all tiles ($T_{n,m}$) and repeated for each class in order to create distinct masks for each landcover class and tree species composition. At the end of the process, the landcover masks were combined to obtain two final masks: one containing only forested areas and the other only the species of interest. One consequence of the aforementioned synergy process is that the accuracy of the masks may vary spatially in accordance with the available input datasets. Thus, to inform the user of the varying spatial accuracies, correlation images corresponding to the selected minimum correlations were processed. Additionally, to simplify the interpretation of the correlation images, the correlation values were converted into a score, which is defined as follows:

$$S(R_{\text{norm}}(0)) = (1 - |R_{\text{norm}}(0)|) \cdot 100\,. \tag{4.18}$$

The score S was calculated as a function of the cross-correlation $R_{\text{norm}}(h)$ at lag $h = 0$. A high score indicates good separability, while a low score indicates poor separability. *GSV* images were combined on the basis of the principle of synergy, but in this case, minimum *RMSE* was the criterion for pixel selection. The outcomes of this process provided each *GSV* image tile with potentially the best estimates of *GSV*. Similar to the synergy of the masks, to keep track of the accuracy of the *GSV* images, the *RMSE* values were also selected with their corresponding *GSV* in order to create correlation flag images for the masks, which are *RMSE* flag images in this case. The fusion processes shown in Fig. 4.10 correspond to the definition of fusion given in Sect. 3.3. However, synergy in the abovementioned context differs from its theoretical definition. In this respect, it is convenient to emphases the principal difference between the two terms. In fusion, information content from multiple images is merged to obtain a fused image. Therefore, at least two images are required for this purpose. In synergy, information in the output data is obtained by the selection of the optimal data sources. In this situation, two images are required, but they are not necessarily overlaid. Thus, the fusion process may provide information of 'greater quality', while the synergy process does not necessarily improve the quality of information.

End Products

After the completion of the synergy process, the masks were applied to the *GSV*-combined information in order to create tiled products. Finaly, the tiled products were mosaicked to obtain the end products (Fig. 4.11).

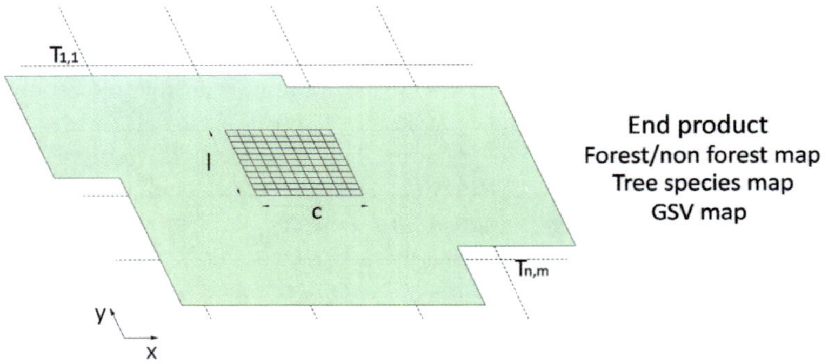

Fig. 4.11 End products obtained with the developped fusion approach

References

ASI (2007) Cosmo-SkyMed SAR products handbook. Technical report, Italian Space Agency (ASI)

Castel T, Beaudoin A, Stach N, Stussi N (2001) Sensitivity of space-borne SAR data to forest parameters over sloping terrain. Theory and experiment. Int J Remote Sens 22(12):2351–2376

Chamara S (2009) Thuringian forest inventory accuracy. Private communication

Eineder M, Fritz T, Mittermayer J, Roth A, Börner E, Breit H (2009) TerraSAR-X ground segment basic product specification document. Technical report, DLR, Munich, Germany

Eriksson LEB, Santoro M, Wiesmann A, Schmullius C (2003) Multitemporal JERS repeat-pass coherence for growing-stock volume estimation of Siberian forest. IEEE Trans Geosci Remote Sens 41(7):1561–1570

ESA (2007) Information on ALOS PALSAR products for ADEN users. Technical report, European Space Agency (ESA)

Fritz T, Duque S, Bräutigram B, Bueso Bello L (2012) TanDEM-X experimental product description. Technical report, DLR, Munich, Germany

Gatelli F, Guamieri AM, Parizzi F, Pasquali P, Prati C, Rocca F (1994) The wavenumber shift in SAR inferometry. IEEE Trans Geosci Remote Sens 32(4):855–865

Grassmann J (2012) Prüfung der Aktualität der Daten der Thüringer Forsteinrichtung. Bachelor thesis, Fachhochschule Erfurt

Hansson R (2001) Radar interferometry: data interpretation and error analysis. Baker and Taylor, New York

JAXA (2012) PALSAR user's guide. Technical report, Japan Space Systems

Lee J-S, Schuler DL, Ainsworth TL, Boerner WM (1999) POLSAR data compensation for terrain azimuth slope variation. In: IEEE 1999 international geoscience and remote sensing symposium. IGARSS'99 (Cat. No.99CH36293), IEEE, vol 5, pp 2437–2439

Lee J-S, Schuler DL, Ainsworth TL (2004) A review of polarization orientation estimation from polarimetric SAR data. In: Proceedings of the workshop on POLinSAR—applications of SAR polarimetry and polarimetric interferometry. Frascati, Italy, p 3.1, CD-ROM

Massonnet D, Souyris J-C (2008) Imaging with synthetic aperture radar. EPFL Press, Lausanne

Oliver CJ, Quegan S (2004) Understanding synthetic aperture radar images. SciTech Publishing, Boston

Pottier E, Ferro-Famil L, Allain S, Cloude S, Hajnsek I, Papathanassiou K, Moreira A, Williams M, Minchella A, Lavalle M, Desnos Y-L (2009) Overview of the PolSARpro V4.0 software. The open source toolbox for polarimetric and interferometric polarimetric SAR data processing. In: 2009 IEEE international geoscience and remote sensing symposium, p IV-936–IV-939

Pulliainen JT, Mikhela PJ, Hallikainen MT, Ikonen J-P (1996) Seasonal dynamics of C-band backscatter of boreal forests with applications to biomass and soil moisture estimation. IEEE Trans Geosci Remote Sens 34(3):758–770

Reinholz J (2011) Waldzustand 2011. Technical report, Freistaat Thüringen, Gotha, Germany

Santoro M, Askne J, Smith G, Fransson JES (2002) Stem volume retrieval in boreal forests from ERS-1/2 interferometry. Remote Sens Environ 81:19–35

Santoro M, Beer C, Cartus O, Schmullius C, Shvidenko A, McCallum I, Wegmüller U, Wiesmann A (2011) Retrieval of growing stock volume in boreal forest using hyper-temporal series of Envisat ASAR ScanSAR backscatter measurements. Remote Sens Environ 115(2):490–507

Shimada M, Isoguchi O, Tadono T, Isono K (2009) PALSAR radiometric and geometric calibration. IEEE Trans Geosci Remote Sens 47(12):3915–3932

Ulander LMH (1996) Radiometric slope correction of synthetic-aperture radar images. IEEE Trans Geosci Remote Sens 34(5):1115–1122

Wegmuller U (1998) Automated terrain corrected SAR geocoding. In: IEEE 1999 international geoscience and remote sensing symposium. IGARSS'99 (Cat. No.99CH36293), IEEE, vol 3, pp 1712–1714

Chapter 5
Results

The main purpose of this study is the estimation of forest *GSV* using a combination of SAR data. To achieve this objective, several processing stages were defined. On selecting a test site located in a hilly area, the first step involved investigating the variations in the SAR signal with respect to topography. With the available datasets, the second and third steps referred to the examination of the scattering and decorrelation mechanisms, respectively, which occur in the Thuringian Forest. On the basis of these investigations, the fourth step involved SAR-data modelling and *GSV* estimation. Finally, the last phase involved production of a *GSV* map with focus on the potential fusion of the SAR data.

The results of these different steps are presented in this chapter. The first section introduces the effects of topography. In particular, visual observations and scatterplots of the backscatter intensity and interferometric coherence are presented. The second section describes the investigation of the scattering mechanisms. Simple visualisations as well as description of the signatures of the SAR parameters and their sensitivity to *GSV* are provided in this section. The third section highlights the results of examination of the decorrelation mechanisms. Visualisations of coherence, characterisations of the signatures of the InSAR parameters and sensitivity of the coherence to *GSV* are described in this section. In addition, owing to the availability of PolInSAR data, some results corresponding to loci coherence are discussed. In the fourth section, the modelling results are presented with the results of the parametric and non-parametric approaches introduced in Sect. 4.4. The last section of this chapter presents the results of the SAR data fusion. With a view to providing a *GSV* map product for the Thuringian Forest, the first section (5.5.1) describes the histogram analysis performed for the derivation of a forest/non-forest and species map. Then, the second section (5.5.2) suggests some radar data configurations for retrieving forest parameters. Finally, the last section (5.5.4) illustrates examples of the products derived by applying the fusion approach and present the validation results.

For the sake of clarity and to limit the size of this document, only selected examples are presented for the first three sections. Appendix B can be referred to for the plots and figures for further results. It should be noted that the satellite data only

N. Ackermann, *Growing Stock Volume Estimation in Temperate Forested Areas Using a Fusion Approach with SAR Satellites Imagery*,
Springer Theses, DOI 10.1007/978-3-319-13138-2_5

covered some parts of the test site. To locate the data over the site and facilitate an understanding of the results, the acquisition frames are summarised in Appendix A.2. When necessary, the reader may refer to this Appendix regarding the spatial coverage of the examined dataset. Finally, for the last section, although fusion and synergy were referred in the developed fusion methodology (Sect. 4.4) to different processes, both involve a combination of the data (see Sect. 3.3). For the sake of clarity, if not explicitly mentioned, the terms 'fusion', 'merge', 'combination', 'synergy' and 'integration' in the last section indicate the combination of the data.

5.1 Topography Analysis

5.1.1 Visual Observations

5.1.1.1 Scattering Mechanisms

The investigated SAR data were visualised in order to distinguish potential trends in the topography. Figure 5.1 depicts visualisations of the forested areas around the Schönbrunn Lake (see Appendix A.1). In the first row, ALOS PALSAR (a) HH and (b) HV as well as TSX (c) HH and (d) HV non-normalised intensities are depicted. The corresponding normalised data which were processed according to Eq. (4.1) are shown in the second row. It should be noted that the grey values were stretched

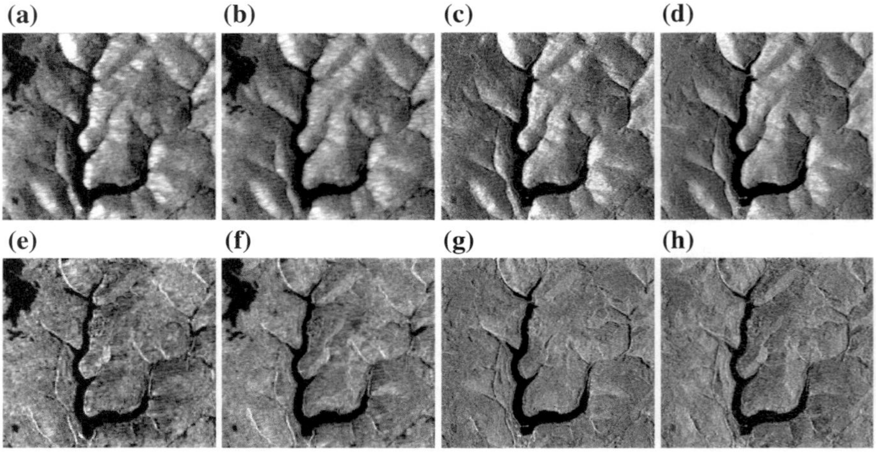

Fig. 5.1 Visualisation of topographic effects on the backscatter intensity before (*first row*) and after (*second row*) topographic normalisation. The SAR data are composed of ALOS PALSAR FBD 39° Asc. **a**, **e** HH and **b**, **f** HV (frame FBD7) as well as TSX SM 37° Asc. **c**, **g** HH and **d**, **h** HV backscatter intensity (frame SM6). The PALSAR and TSX scenes were acquired on 10th June 2010 and 1st September 2010, respectively. The forested areas are located around the Schönbrunn Lake (see the aerial photography in Appendix A.1)

between each set of the normalised and non-normalised data in order to enable interpretations of topographic normalisation. In addition, it should be recalled that the SAR acquisitions were performed in the ascending pass with right looking and at a direction of approximately 350°.

The topography is distinct in the non-normalised SAR data (first row) in Fig. 5.1, particularly with the bright areas facing the SAR sensors. These bright regions are attributed to the layover effect and the larger energy received by the radar on slopes facing the sensor compared to back slopes. Further discussions about this observation are given in Sect. 5.1.2. From the normalised and non-normalised data, it is observed that the topographic-normalisation approach performed well. However, small topographic variations appear to remain in the normalised data. These variations are particularly well-observed for L-band and X-band HV polarisation, which depict brighter grey values on the slopes facing away from the radar rather than on slopes facing the SAR spaceborne platform. This observation is discussed by examining the scattering mechanisms in Sect. 5.1.2.

5.1.1.2 Interferometric Coherence

Similar to backscatter intensity, the interferometric coherence was examined over topographic areas. As shown later in Sect. 5.1.3, PALSAR, CSK and TSX were significantly affected by temporal decorrelations, which generally masked the topographic effects. In the case of TDX, the single-pass interferometric acquisitions were not affected by temporal decorrelations and enabled clear observations of the topographic trends. Thus, TDX was chosen here for the visualisation of topographic variations in the interferometric coherence. Figure 5.2 presents the TDX coherence over the same region as that shown in Fig. 5.1. The coherence (a) before and (b)

(a) (b)

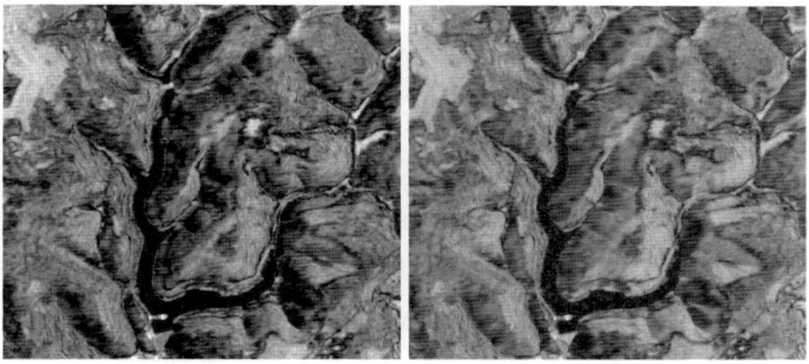

Fig. 5.2 Visualisation of topographic effects on interferometric coherence before (**a**) and after (**b**) slope adaptive common band filtering. The InSAR data are composed of TDX SM 38° HH Asc. $B_n = 258$ m coherence (frame TDX3) acquired on 30th August 2011. The forested areas are located around the Schönbrunn Lake (see the aerial photography in Appendix A.1)

after the application of slope adaptive common band filtering is depicted. It should be noted that for comparison purposes, both images were stretched using the same statistics. In addition, it should be recalled that TDX acquisitions were performed in the ascending pass with right looking and at an orientation of approximately 350°.

In Fig. 5.2a, it can be observed that the coherence image clearly shows darker grey values, i.e. lower phase correlation, for slopes facing the radar (negative slopes) compared to slopes facing away from the radar (positive slopes). This result may be explained by two dominant spatial decorrelation mechanisms, namely spectral-shift and volume decorrelation. Spectral-shift decorrelation appears because of the slight difference in the incidence angle of the two interferometric acquisitions. Owing to their geometry, the negative slopes would increase the spectral-shift, thus increasing the spatial decorrelation (see Sect. 3.2). Volume decorrelation in X-band may be greater for canopies located on slopes facing the radar compared to canopies facing away from the spaceborne platform. In such configurations, the steep local incidence angle enhances the penetration of microwaves into the forest and therefore increases returns from the canopy. It should be noted that the spectral-shift spatial decorrelation may be corrected by filtering the common signal, as introduced in Sects. 3.2 and 4.3. However, volume decorrelation cannot be corrected with such filters and may require other techniques such as the RVoG model, which is discussed at the end of Sect. 5.1.3. in comparing the coherence images shown in Fig. 5.2a with and b without slope adaptive common band filtering, it can be seen that the coherence generally increases after applying filtering corrections. However, as shown in Fig. 5.2b, topographic variations continue to occur in the filtered InSAR data. These observations are subject to further examination and are further addressed in Sect. 5.1.3.

5.1.2 Topography and Scattering Mechanisms

5.1.2.1 Scattering Mechanisms

Topography may be described using two parameters, inclination and orientation of the terrain. These two parameters were investigated together with the SAR backscattering intensity. The first examinations were performed by comparing the variations in the signal with the orientation of the slopes for normalised and non-normalised data. This was illustrated by polarplot representations. The slope orientations, which ranged between 0° and 360°, were intuitively depicted on the circumference of the plot, while the backscatter intensity, with 0 dB as the maximum amplitude at the centre of the plot, was depicted on the polarplot radius. To facilitate the apprehension of the potential symmetries in azimuth and range directions, the azimuth direction of the sensor was aligned with the vertical axis of the plot. The representation of the slope gradients was performed using three different classes: one ranging between 0° and 10°, second between 10° and 20° and the last one over 20°. Figure 5.3 illustrates an example of polarplots derived from ALOS PALSAR FBD and PLR data over spruce forest stands. It should be noted that geometrical distortions such as layover, foreshortening

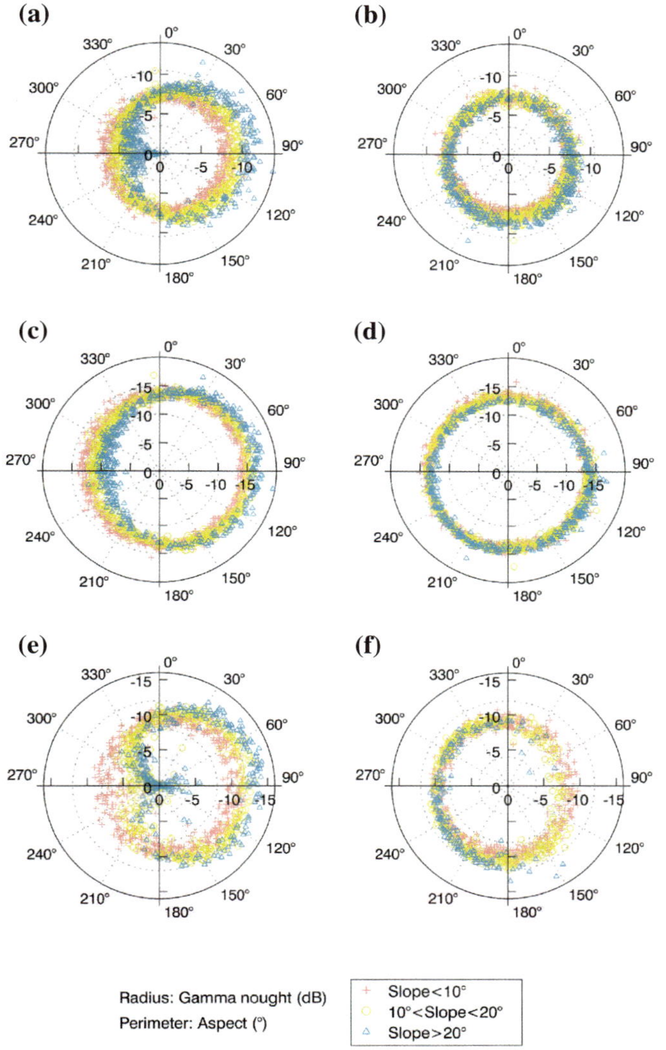

Fig. 5.3 Topographic effects on backscatter intensity before (*left column*) and after (*right column*) topographic normalisation. The graphs consist of polarplots with the aspect angle on the circumference and the backscatter intensity (γ^0) on the radius of the plots. The sensor orientation is aligned at 0° on the ordinate. The classes represent the inclination of three slopes, namely below 10° (*pink*), between 10° and 20° (*yellow*) and above 20° (*blue*). The SAR data are composed of ALOS PALSAR FBD 39° Asc. **a** HH and **b** HV (frame FBD7) as well as **c** ALOS PALSAR PLR 21° Asc. VV (frame PLR3) backscatter intensity. The PALSAR FBD and PLR scenes were acquired on 10th June 2010 and 12th April 2009, respectively. The forested areas represent Norway Spruce

and shadowing were masked in the investigated data. This process led to missing information, especially for negative steep slopes (blue points) measured by PLR data, which were acquired with a steep incidence angle (see intensity analysis Sect. 5.2). FBD HH, HV and PLR VV data are shown together for this example in order to provide information about both VV polarisation (PLR data) and limited geometrical distortion effect for HH and HV datasets (FBD data). For the interpretations, it is important to note the different incidence angles between PLR (21°) and FBD (39°) data. Additionally, both datasets were acquired in the ascending direction. The plots (a), (c) and (e) depict the non-normalised backscatter intensities, and the graphs (b), (d) and (f) illustrate the corresponding normalised amplitudes.

Figure 5.3 indicates that the PALSAR data before normalisation depict significant variations, while these variations are significantly reduced after normalisation. Then, by viewing only the non-normalised plots and considering the right-looking acquisition mode of the sensor, it can be seen that the signal's amplitude increases for negative slopes, while it decreases for positive slopes. Furthermore, comparing the different polarisations, with the highest peaks reaching 0 dB for slopes facing the radar, this phenomenon appears to be more significant in HH and VV polarisations than in the HV channel. The higher backscatter intensity for negative slopes relative to positive slopes can be explained using two different phenomena. The first and most significant one involves varying the pixel sizes on the basis of the gradient and orientation of the slopes. This effect is compensated for on the normalised data by the pixel area factor introduced in Eq. (4.1). The second phenomenon is related to backscattering properties, which increase backscatter with incoming incident waves aligning orthogonal to the terrain surface, according to Fresnel formulas (Eqs. 3.26 and 3.27). This effect is corrected in the normalised data by considering a local incidence angle. Finally, in addition to slopes in the range direction, slopes in the sensor flight direction, namely azimuth slopes, may also be examined. The observations of the polarplots depicting non-normalised data (left column plots) show that the signal on azimuth forested slopes attains the same level of intensity as that in flat forested areas represented by pink crosses. Thus, in this case, the azimuth slopes may be interpreted by the radar signal as virtually flat forested areas.

The comparison of the PALSAR intensity before and after topographic normalisation in Fig. 5.3 clearly showed improvements in the pixel area and local incidence angle. However, as suggested by the visualisations in Sect. 5.1.1, a more detailed examination of the corrected data using scatterplots showed that systematic variations in the signal with topography remained. Figure 5.4 depicts an illustration of the obtained scatterplots. The aspect angle (β) is presented on the abscissa and the amplitude on the ordinate. For the sake of clarity, the aspect angle was set between $-180°$ and $180°$ and the azimuth direction at 0°. Therefore, the negative slopes (facing the radar) would be depicted between $-180°$ and 0°, while the positive slopes (facing away from the radar) would be depicted between 0° and 180°. Similar to the case of polarplots, the three slope classes are described using the same three colours pink, yellow and blue for 0°–10°, 10°–20° and above 20°, respectively. This example is based on the same datasets as those used for Fig. 5.3, namely PALSAR FBD (a) HH and (b) HV 39° and (c) PALSAR PLR VV 21°.

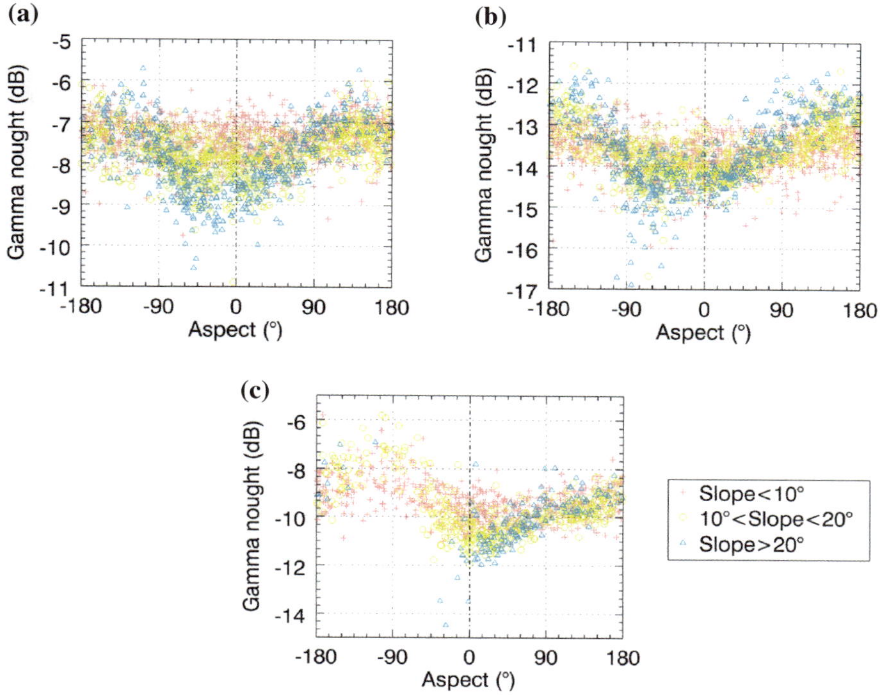

Fig. 5.4 Topographic effects on the normalised backscatter intensity. The scatterplots depict the aspect angle on the abscissa between −180° and 180°, with azimuth slopes centred at 0° and backscatter intensity (γ^0) on the ordinate. Negative slopes are described between −180° and 0° and positive slopes between 0° and 180°. The classes represent the inclination of three slopes, namely below 10° (*pink*), between 10° and 20° (*yellow*) and above 20° (*blue*). The SAR data are composed of ALOS PALSAR FBD 39° Asc. **a** HH and **b** HV (frame FBD7) as well as **c** ALOS PALSAR PLR 21° Asc. VV (frame PLR3) backscatter intensity. The PALSAR FBD and PLR scenes were acquired on 10th June 2010 and 12th April 2009, respectively. The forested areas represent Norway Spruce

Figure 5.4 indicates that the variations in the backscatter intensity with slopes and orientations remain in the normalised PALSAR data. These variations appear to be generally polarisation-dependent, but the different graphs indicate a common trend for HH, HV and VV polarisations. The latter is characterised by a clear decrease in amplitude for all polarisations on the azimuth slopes oriented in the moving direction of the radar ($\beta = -90°$ to 90°). This specific amplitude reduction may be difficult to interpret because it appears only on azimuth slopes exposed in the flying direction of the radar (asymmetry of the signal for azimuth slopes). A possible reason for this general decrease is the different growth conditions of the trees owing to the different slope expositions (see Sect. 3.1). Factors such as water stress, light availability, wind and topography influence the growing conditions. The latter two factors lead especially to eccentric stem cross-sections which may influence the backscatter intensity. This result should be examined in more detail in subsequent studies.

The orientations outside the aforementioned range, namely from −180° to −90° and 90° to 180°, depict polarisation-dependent phenomena. These orientations can be analysed separately for each polarisation. First, considering HH polarisation in Fig. 5.4a, it can be observed that the amplitude of the signal for forests located on flat areas is relatively constant (pink crosses). It can also be seen that the HH amplitude over hilly terrains does not exceed the amplitude measured over flat areas, except at two different orientations which show two intensity peaks, namely those around 150° and −135°. These maxima may be related to double bounce scattering. Assuming the penetration of the canopy in L-band frequency, the tree trunks act as corner reflectors under certain conditions. In the case of the flat area, the dihedral cannot be highlighted because the forest conditions remain constant. However, in the case of the sloping terrain, the two aforementioned orientations may provide conditions which produce dihedrals. For these conditions, it should be noted that while the ground is inclined, the trees maintain their vertical extension because they usually grow in the vertical direction. Consequently, the trunk and the ground surface in sloping terrains are not perpendicular to each other; the angle between the trunk and ground is between 90° and 180° for slopes facing the radar and between 0° and 90° for slopes facing away from the radar. This physical difference between flat and sloping terrains and in particular the specific oblique angle formed between the trunk and the inclined surface may explain the position of the two peaks observed in Fig. 5.4a. In addition to the above FBD and PLR HH polarisation observations, PALSAR FBS HH data were also investigated with aspect and slope angles in order to compare the effect of topography between frozen and non-frozen conditions. The frozen conditions given by FBS acquisitions undertaken under temperatures below 0 °C showed equivalent but more amplified signal variations when compared to the non-frozen conditions. This result may be explained by the greater penetration of microwaves when the canopy and ground conditions reach the freezing point. Thus, returns from the ground level are enhanced, and dihedrals for the positive and negatives slopes oriented at 150° and −130°, respectively, are amplified.

Observing the HV polarisation in Fig. 5.4b, it is found that the trends exhibited by this channel are close to HH polarisation. In fact, maximum intensities are shown around 135° and −135°, which can be referred to as dihedrals (similar to HH polarisation). However, in contrast to HH polarisation, the intensity of the steep slopes in HV between 90° and 180° and between −90° and −180° appeared to generally remain relatively higher than the flat terrain intensity. The latter range corresponds mainly to the positive-range slopes as well as azimuth slopes oriented in the direction opposite to the sensor's flight direction. Typically, the maximum amplitude in HV polarisation should be observed exactly on slopes facing away from the radar ($\beta = +90°$). The path length travelled by the microwaves in the canopies is the longest for forests located on positive-range slopes. The volume viewed by the radar increases, thus increasing the HV backscatter intensity. With respect to the decrease in the intensity described above, this surprising result is attributed to the physical variations (e.g. stem cross-section) in the forest with different slope expositions.

Finally, the VV polarisation shown in Fig. 5.4c indicates a different trend for the HH and HV channels. In the present case, the intensity is maximum only for slopes facing the radar ($\beta = -45°$ to $-135°$). In theory, microwaves in VV polarisation are extremely sensitive to the terrain surface, especially in Bragg-scattering conditions. Considering that the vertically polarised waves penetrate the forest canopies with a slope facing the radar, it can be conceived that the response of the terrain surface to VV polarisation should increase in such conditions. Thus, the observations made in Fig. 5.4c appear to be realistic.

Analogous to the PALSAR backscatter intensity, TSX, TDX and CSK, amplitudes were investigated. Examinations using X-band sensors showed results which were generally equivalent to the L-band results presented above. A few divergences were observed, especially with VV polarisation, because of the opaque canopy seen at the X-band frequency. In this case, the intensity fluctuations were found to be similar to those corresponding to HH polarisation. Some further processed results for the X-band sensors are provided in Appendix B.1. In addition to the spruce species, beech forests were investigated. The results for beech were found to be similar to those for spruce (refer to Appendix B.1). It should be noted that pines were not considered for this analysis. As discussed in Sect. 4.3 and showed in Fig. A.6, Appendix A.1, this species was located mainly in flat areas.

5.1.2.2 Optical Crown Depth

The analysis described above highlighted the various contributions of forest canopies to the radar signal when dealing with rough terrain conditions. In particular, it showed the importance of considering forest canopies in the topographic-normalisation approach. As introduced in Eq. (4.2), the correction of backscatter intensity for forest canopies over a sloping terrain may be optimised using an exponential term n. The n exponent may take into account the cases where the terrain is covered by vertically distributed targets.

The estimation of the n-coefficient was performed in accordance with the empirical approach described in Sect. 4.3.4. The results showed some improvements of the normalisation of the signal over the sloping terrain. However, the improvements were insignificant and mainly concerned positive slopes ranging between 80° and 150°. The reason is that in the case of positive slopes, microwaves are mainly returned from the canopies in the form of volume scattering, while in the case of negative slopes, ground-level contributions such as surface and double bounce scattering may be predominant contributors to the signal. Because Castel's approach was based on the WCM, its assumptions only consider the forest canopy, and potential returns from the ground level are neglected. Thus, volume contributions can be corrected well, but surface and dihedrals appearing for slopes facing the radar can be hardly normalised. Owing to the insignificant improvements obtained by using the n-coefficient and the non-physical approach used to estimate the n-value, $n = 1$ was set for the normalisation, and the original amplitude variations related to topography were maintained. It should be noted that the application of the n-coefficient may also further alter the signal rather than improving it. Forest canopy parameters such as the coefficient of extinction may vary within the forest, particularly under various growth or weather conditions. Thus, the application of the n-coefficient may introduce new sources of errors when it is not considered individually for each forest stand.

5.1.3 Topography and Decorrelation Mechanisms

5.1.3.1 Decorrelation Mechanisms

The investigations of the coherence over topographic areas showed two opposing trends, which are discussed in Fig. 5.6. To support the interpretations, before presenting these results, the main decorrelation mechanisms in forested areas can be recalled in brief. According to Eq. (3.35), the decorrelation components which may be considered for the interpretations are mainly noise, temporal, volume and spectral-shift decorrelations. Decorrelation due to thermal noise was presented in Sect. 4.3, and for the investigated sensors, it was found to be insignificant in forested areas (see Fig. 4.7 and Table 4.4). Temporal decorrelation may be related to the revisit time of the repeat-pass systems and their carrier frequency. PALSAR and CSK may be subject to temporal decorrelation, while TDX is free of temporal decorrelation because interferometric acquisition for TDX is performed in a single-pass. Two important parameters are significant with respect to volume decorrelation: canopy crown height and canopy extinction (Cloude and Papathanassiou 2003). In hilly areas, the projected crown height varies with slope gradient and orientation, leading to different optical paths through the canopy and eventually different volume decorrelations. In this regard, a slope facing the radar presents a shorter optical path, i.e. less volume decorrelation, than a slope facing away from the sensor. The canopy extinction involves crown structural properties and environmental conditions such as air temperature. The greater are the elements composing the canopy, the greater is the extinction. The air temperature influences the dielectric permittivity of the trees and thus affects the canopies' extinction properties. Typically, frozen conditions tend to decrease the trees' dielectric constant, thereby increasing the transparency of the canopy to SAR microwaves. Spectral-shift decorrelation, which was introduced in the theory Sect. 3.2 with range (γ_{rg}) and azimuth (γ_{az}) frequency shift components, is related mainly to the hilly terrain and the slightly varying incidence angle between the two acquisitions comprising the interferometric pairs. The processing of interferometric coherence involved spectral-shift corrections using slope-adaptive common band filtering. However, the filtering may not completely remove spectral-shift spatial decorrelation, especially for steep slopes facing the radar (see Sect. 3.2). Thus, as shown in the visualisations in Sect. 5.1.1, spectral-shift decorrelation due to the topography may remain. To estimate the contribution of spectral-shift decorrelation, a theoretical analysis of the loss of coherence due to spectral-shift decorrelation, based on Eq. (3.38), may be performed. For each interferometric configuration studied in Fig. 5.6, Fig. 5.5 presents spatial correlation as a function of slope gradient.

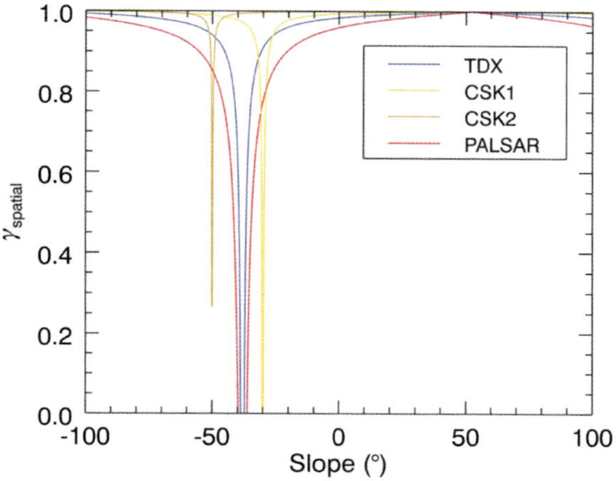

Fig. 5.5 Modelling of spectral-shift decorrelation as a function of terrain slope (°). The functions describe four different acquisition configurations, namely TDX SM 38° $B_n = 258$ m (*blue*), CSK SM Himage 30° $B_n = 185$ m (*yellow*), CSK SM Himage 50° $B_n = 59$ m (*brown*) and ALOS PALSAR FBS 39° $B_n = 475$ m (*red*)

Comparing the slopes facing the radar (negative values) and those away from the radar (positive values), it can be generally seen that the spectral-shift decorrelation is significant in the first situation, while it is negligible in the second case. On examination of negative slopes for the PALSAR FBS sensor, it is observed that spectral-shift decorrelation begins to be determinant at slopes inclined at −25° and is maximal at about −38°. In fact, the latter angle coincides with the radar incidence angle. For CSK 30°, CSK 50° and TDX 38°, spectral-shift decorrelation becomes significant on slopes inclined at approximately −25°, −49° and −35°, respectively. As discussed in Sect. 4.1, most slopes in the Thuringian Forest generally lie between 0° and 35°. Thus, in the case of CSK 30° and PALSAR FBS 39° interferometric configurations to a lesser extent, spectral-shift decorrelation needs to be considered in the evaluation of coherence over steep slopes facing the radar.

As introduced in Sect. 3.2, it should be noted that spectral-shift decorrelation may not only be expressed from a spectral perspective but also from a geometrical point of view. In this regard, it should be first noted that according to Eq. (3.37), the spectral-shift is intrinsically related to the perpendicular baseline. Proportionally, the greater the perpendicular baseline, the greater the spectral shift and therefore the spatial decorrelation. This relationship recalls the definition of the critical baseline given in Eq. (3.39). The influence of the perpendicular baseline may be observed in Fig. 5.5 using the blind angle, i.e. where slopes totally decorrelate the signal. In fact, an increase in the normal baseline expands the blind angle. The blind angles in Fig. 5.5 show that they remain narrow for all sensors. The reason is that the available sensor configurations showed relatively small perpendicular baselines compared to their respective systems' critical baselines (see Sect. 4.2). Also, in terms of geometry, it can be noted that negative slopes occurring above the blind angle refer to layovers, while positive slopes with an angle greater than the sensor depression angle corresponds to shadow areas. In this respect, as shown in Fig. 5.5, spatial decorrelation due to shadowing has less importance.

After recalling the various decorrelation mechanisms, the observations made from the available interferometric data can be presented. PALSAR, CSK and TDX coherence were examined using scatterplots, with the terrain aspect angle processed on the abscissa and the interferometric coherence on the ordinate. To facilitate the interpretations, the target direction of the sensor (azimuth direction) was set at 0°, and the aspect angle was processed between −180° and 0° and between 0° and 180° for negative and positive slopes, respectively. The same three slope classes, namely 0°–10°, 10°–20° and above 20° were indicated in pine, yellow and blue, respectively. In general, owing to temporal decorrelation, the investigated interferometric data did not show any clear trend. However, specific configurations presented some tendencies, which appeared to highlight different decorrelation mechanisms. Figure 5.6 presents some examples of these results with (a) CSK 30°, (b) TDX 38°, (c) CSK 50° and (d) ALOS PALSAR 39° coherences all measured in HH polarisation in the right-looking ascending mode and with an orientation of about 350°. It should be noted that the coherence data was corrected for the spectral shift (slope adaptive common band filtering). It should also be noted that the forest stands comprised only of spruce species, as the number of forest stands for beech and pine were limited for this kind of analysis. Finally, it may be remarked that coherence bias should be insignificant as the high resolution of sensors provided a sufficient number of samples.

Two different trends should be noted from the scatterplots in Fig. 5.6. The first one can be seen on the graphs (a) and (b) which depict steep slopes (blue) with a clear drop in the interferometric phase coherence between −160° and 0°, while between 0° and 180°, the coherence appears to be relatively constant. The second trend can be observed on plots (c) and (d) which present an increase in coherence between −180° and −60° for steep terrain and a decrease in coherence between 0° and 180°.

In the case of plot (a), the decrease in coherence between −160° and 0° is considered to be mainly caused by the spectral-shift spatial decorrelation because the slopes concerned by the loss of coherence are above 20°, and reviewing the decorrelation mechanisms, it is observed that slopes above 25° for CSK 30° are subject to spectral-shift decorrelation. Owing to the temporal decorrelation occurring with X-band 1-day repeat-pass acquisitions, the coherence of CSK on plot (a) is also distinct, with a significant dispersion. In plot (b) of Fig. 5.6, the clear decrease in the coherence between −160° and 0° is not due to spectral-shift decorrelation but volume decorrelation, according to the analysis made from Fig. 5.5. However, this result should be considered carefully because with steep slopes facing the radar, it is still expected to have some spectral-shift decorrelation contributions. This is particularly the case in forested areas as the scatterers are mostly unstable, which usually leads to uncorrelated signals and reduced performances of the spectral-shift filter. To better interpret the decrease in coherence for TDX negative slopes, the coherence in forested areas was compared with that in open areas on the basis of the same slope gradient and orientation (see Appendix B.1). Although slopes in the case of open areas were limited to 20°, the signal was relatively constant with the different orientations particularly for an aspect angle range of −160° to −60°. Thus, the decrease in coherence observed in Fig. 5.6b for forests over slopes facing the SAR sensor may actually be caused by the contribution of volume decorrelation.

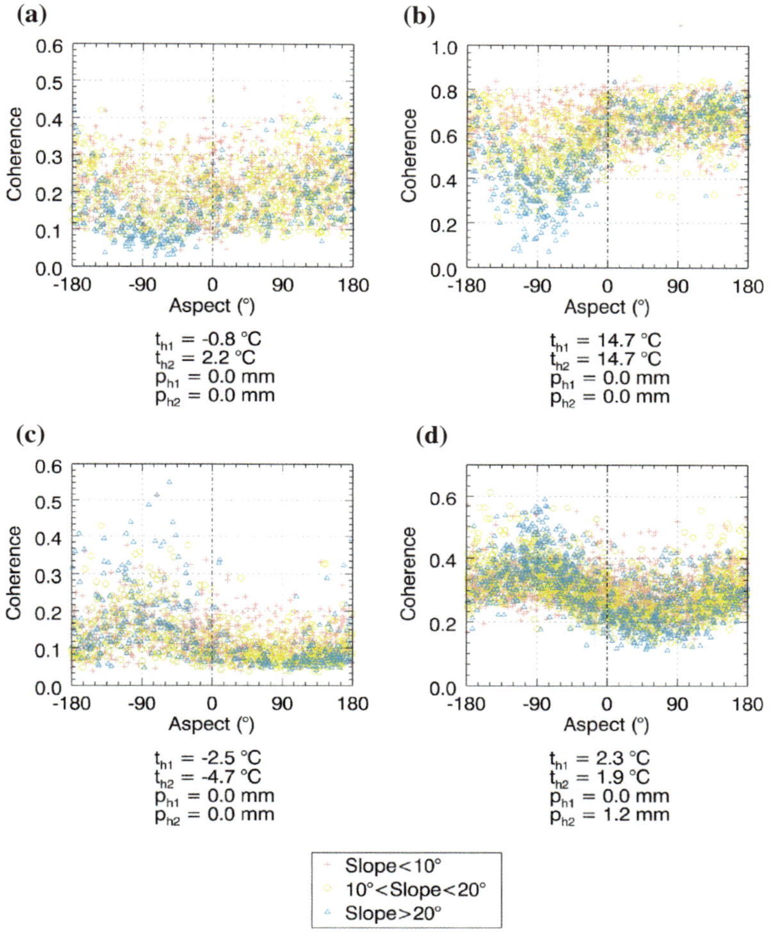

Fig. 5.6 Topographic effects on interferometric coherence corrected for spectral shift (adaptive slope common band filtering). The scatterplots depict the aspect angle on the abscissa between $-180°$ and $180°$ with the azimuth slopes centred at $0°$ and the interferometric coherence (γ) on the ordinate. Negative slopes are described for the range of $-180°$ to $0°$ and positive slopes between $0°$ and $180°$. The classes represent the inclinations of the three slopes, namely below $10°$ (*pink*), between $10°$ and $20°$ (*yellow*) and above $20°$ (*blue*). The InSAR data are composed of **a** CSK SM Himage HH 30° Asc. $B_n = 185$ m (frame CSK1), **b** TDX SM HH 38° Asc. $B_n = 258$ m (frame TDX3), **c** CSK SM Himage HH 50° Asc. $B_n = 59$ m (frame CSK1) and **d** ALOS PALSAR FBS HH 39° Asc. $B_n = 475$ m (frame FBS7) coherence acquired on 22nd/23th March 2011, 30th August 2011, 19th/20th February 2011 and 4th March/19 April 2008, respectively. The forested areas represent Norway Spruce

Similar to plots (a) and (b), plot (c) depicts X-band phase coherence measurements. However, the observations are clearly different in this case. This can be explained by the variation of the air temperature. In the acquisition shown in the graph (c), the air temperature remained below −2.5 °C, which produced frozen conditions on the canopies and the ground. In such a situation, the opacity of the forest canopies decreases and in turn enhances the contribution of returns from the ground for slopes facing the radar ($\beta = -160°$ to $0°$). It should be noted that the acquisition was measured at an incidence angle of about 50° and with a perpendicular baseline of 59 m. Therefore, the spatial decorrelation due to spectral shift and layover was not significant in such a configuration (see Fig. 5.5). It should also be noted that the dispersion of the data is extremely high. This significant variation can be explained by temporal decorrelation as well as the high versatility of the temperature and forest conditions (i.e. snow). The PALSAR coherence depicted in plot (d) presented results similar to those depicted by the CSK coherence acquisition shown in graph (c). Under equivalent temperature conditions, PALSAR microwaves transmitted in L-band frequency are prone to better penetration through the canopy than CSK X-band microwaves. Hence, as shown in Fig. 5.6d, the contribution from the ground increases the coherency of the signal for slopes facing the radar in L-band configuration. The slight decrease in coherence in the range 0°–100° (facing away from the radar) may be interpreted as volume decorrelation. It should be noted that the description of spectral-shift shown in Fig. 5.5 indicated potential spectral-shift decorrelations over negative slopes for the PALSAR FBS configuration. However, while the coherence obtained from this configuration decreases with spectral-shift decorrelation, the measured coherence in Fig. 5.6 depicted an increase in coherence for the slopes facing the radar. Thus, it can be assumed that spatial decorrelation due to spectral-shift was of minor importance in this case of PALSAR configuration.

The decrease in the coherence of TDX observed in Fig. 5.6b is potentially expressing volume decorrelation. Thus, the decrease in coherence could be used to estimate parameters such as microwave penetration depth in the canopy. Below, a broad evaluation of the penetration using the RVoG model is shown. The RVoG model was evaluated using the equations described in Sect. 3.2 and the interferometric parameters of the TDX acquisition. The interferometric coherence could be modelled as a function of vertical height reflectivity. Figure 5.7 presents the obtained results and summarises the parameters applied to the modelling. The volume height is depicted on the abscissa and the volume decorrelation on the ordinate. It should be noted that for the given modelling scenario, the ground to volume ratio was chosen to be near zero ($m_G/m_V = 0.1$). It is fair to assume that the ground contribution is insignificant in X-band frequency and with unfrozen conditions. Thus, the model in the given scenario comprises only a single scattering random volume layer, namely the forest canopy. It should also be noted that according to the noise evaluation undertaken in Sect. 4.3, TDX noise decorrelation should reach approximately $\gamma_{SNR} = 0.9$. This factor was deduced in the RVoG model by multiplying it with the modelled volume decorrelation.

The different curves presented in Fig. 5.7 express coherence as a function of volume height for various extinction coefficients σ. These coefficients are usually forest- and species-specific; they were unknown for Norway Spruces in the Thuringian Forest. Thus, σ was relaxed in the model to a range of values between 0 and 2 m^{-1}. The lowest functions describe low extinction coefficients, while the top functions

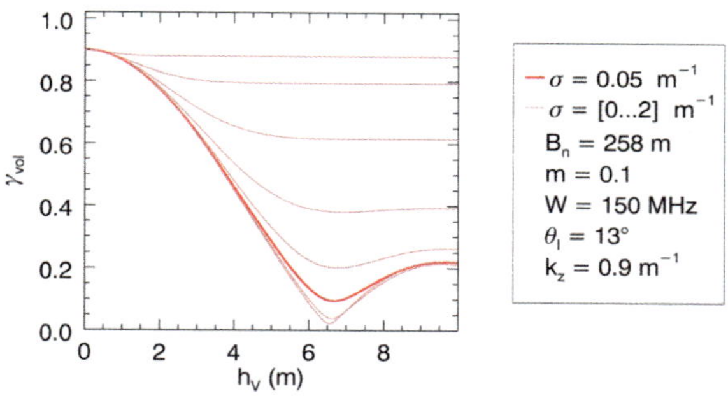

Fig. 5.7 Modelled volume decorrelation as a function of volume height for various extinction coefficients σ. The curves refer to the RVoG model with TDX SM 13° $B_n = 258$ m configuration. In the present case, the local incidence angle (θ_l) is considered

depict high extinction coefficients. By observing the curves in Fig. 5.7, it is noted that coherence decreases with volume height until a height of about 6.5 m is attained. Assuming for this example a coefficient of extinction of $0.05\,\mathrm{m}^{-1}$, the volume height corresponding to a coherence of 0.2 is $h_V = 5.5$ m. After projecting this height in slant range with the cosine of local incidence angle ($\theta_l = 13°$), a volume height h_V of approximately 5.4 m is obtained. This value would correspond to the penetration through the canopy layer for the specific configuration shown in Fig. 5.6b, namely TDX coherence acquisition at 38° incidence angle and with slopes facing the sensor. It should be noted that the presented RVoG modelling only assumed volume decorrelation with noise decorrelation correction. The decrease in coherence observed in Fig. 5.7 may also potentially include the remaining spectral-shift correlation, as discussed beforehand in Fig. 5.6. Thus, the vertical height derived in this example should only be considered to be the first approximation of the penetration depth. It may also be noted that the forest should also satisfy the principal assumption of the model, i.e. the canopy can be considered as a random volume. Comprising the trunk and branches, the spruce canopies may not fit this assumption, particularly because the branches present a systematic horizontal orientation (refer to Sect. 4.1).

5.1.4 Summary

The influence of topography on the X-band and L-band backscatter intensity and interferometric coherence over forested areas was investigated in this section. Visual comparisons of the amplitude and phase coherence both with and without topographic corrections was first performed. Then, a more detailed examination was undertaken using polarplots and scatterplots.

For backscatter intensity, topographic normalisation including the pixel area and local incidence angle performed well. However, fluctuations were still observed. The fluctuations were found to be related to specific scattering mechanisms depending on the SAR systems' polarisation. In particular, VV polarisation in L-band enhanced surface scattering for negative (facing the radar) slopes, while HV polarisation presented for the same system enhanced volume contribution in mostly positive (facing away from the radar) slopes. Unexpected effects such as asymmetries in the azimuth direction were also observed, which were not clearly understood. It was suggested that various tree physiologies with different slope expositions may have affected the scattering mechanisms and may therefore have explained the asymmetric azimuth observations. In this respect, it would be interesting to examine the degree of trunks eccentricities due to slopes and wind and investigate the overall contribution of this effect to the backscatter intensity in future studies. Results in X-band were similar to those of the L-band sensor, except for VV polarisation, which did not depict any surface scattering owing to the high frequency. To reduce the effect of topography in forested areas, a correction based on the optical crown depth coefficient n was also tested. Because the application of an n-value different from 1 resulted only in minor improvements, $n = 1$ was set for normalisation and to consequently maintain the original amplitude variations associated with topography.

The InSAR coherence was studied using PALSAR L-band 46-day and CSK X-band 1-day repeat-pass systems as well as a TDX X-band single pass instrument. Among the available interferometric data, four coherence pairs were selected to illustrate different decorrelation contributions. With CSK 30° under unfrozen conditions, it was first observed that coherence decreased with negative slopes. This observation was explained by the spatial decorrelation induced by the spectral shift between the two interferometric acquisitions. With the same SAR system with a 50° incidence angle under frozen conditions, it was then observed that on this occasion, a decrease in coherence occurred for positive slopes and an increase in coherence was observed for negative slopes. This result could be explained by the frozen conditions of the ground and canopy, which increased the penetration of the microwaves, thus increasing the volume decorrelation for positive slopes and surface contribution for negative slopes. Under unfrozen conditions, the ALOS PALSAR observations presented similar results because the low frequency of the L-band system may penetrate the forest and is sensitive to the volume of the canopies. Thus, the negative slopes indicated that the microwaves reached the ground, thus suggesting high coherence, while positive slopes indicated that the microwaves returned from canopies, thus enhancing volume decorrelation. Finally, a significant decrease in the coherence for slopes facing the radar could be highlighted in the case of the TDX 38° InSAR system configuration. This decrease in coherence was found to be related mainly to volume decorrelation because temporal decorrelation did not affect the TDX system, spatial decorrelation due to spectral shift was found of minor importance and open areas over steep negative slopes depicted high coherence.

In summary, the correction of radar amplitude signals over hilly forested areas is complex and requires deep understanding of the different interacting parameters. To completely understand the problem, simulations using different scenarios may be envisaged. This topic, which was recently addressed in Villard et al. (2010), is beyond the scope of this study and is required to be examined in future studies. Spatial, volume and temporal decorrelations could be distinguished for interferometric coherence. Although spectral-shift deccorelation was corrected with a slope adaptive common band filter, spectral-shift decorrelation appeared to remain in the data, especially in steep sloping areas. Therefore, the results presented in this study, such as the characterisation of volume decorrelation for TDX negative steep slopes, should be carefully considered. Temporal decorrelation was found to be significant with the two repeat-pass systems investigated in this study, namely CSK and ALOS PALSAR.

5.2 Scattering Mechanisms Analysis

5.2.1 Visual Observations

Before proceeding with an in-depth analysis of the backscatter intensity, it may be useful to visualise some of the available intensity data. In Fig. 5.8, an example of (b) TDX and (c) PALSAR amplitude is presented together with (a) a RE

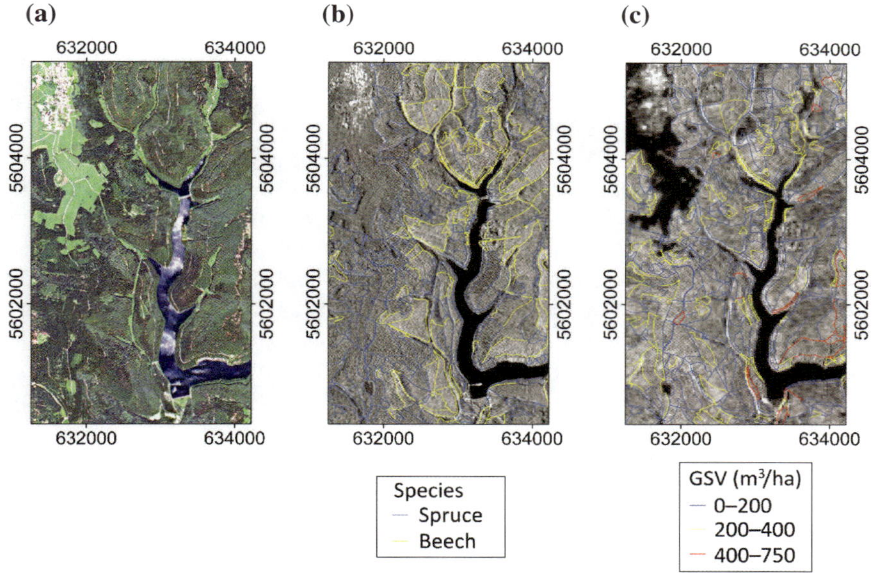

Fig. 5.8 Visualisation of backscatter intensity (γ^0) scaled in dB for (**b**) TDX SM HH 38° Asc. (frame TDX3) and **c** ALOS PALSAR FBD HH 39° Asc. (frame FBD7) data. The PALSAR and TDX scenes were acquired on 10th June 2010 and 30th August 2011, respectively. A RE optical image (RGB) acquired on 13th June 2009 is provided in (**a**) for comparison among different landcovers. Tree species compositions and *GSV* information obtained from the forest inventory are overlaid onto TDX and PALSAR images, respectively. Spruces are depicted in *blue* and beeches in *yellow*. The *GSV* are presented in *blue*, *yellow* and *red* for the ranges 0–200, 200–400 and 400–750m^3 ha^{-1}, respectively. The coordinates are provided in the reference system UTM zone 32

scene for comparison purposes. The area covered is the Schönbrunn Lake region introduced in Sect. 5.1 and Appendix A.1. For this example, the forest inventory data is superimposed on the different images in order to highlight the information corresponding to the forested areas. In the case of TDX, the inventory depicts the Norway Spruce and European Beech species in blue and yellow, respectively. For the PALSAR image, three different ranges of *GSV* were displayed, namely 0–200, 200–400 and 400–700 $m^3\,ha^{-1}$ in yellow, blue and red, respectively.

Before describing the radar images, the elements observed over the selected area in the optical RE scene may be listed (Fig. 5.8a). Referring to the general landcover types, it is observed that the forests surround a lake at the centre right of the area. The lake is shown in the RE scene with white patches. The patches represent the reflection of clouds on smooth water. On the upper left side of the image, a small village known as Frauenwald appears, which is depicted in white owing to the high reflectance of asphalt and other man-made surfaces. Some open areas close to the village are shown in light green, indicating large reflection of the grass in the visible (more particularly in the green) spectrum.

Now, considering the SAR images (Fig. 5.8b, c), on comparing the boundaries of the forest inventory vector data and different areas in the radar images, such as forests surrounding the lake, it is noted that the forest stands are in good agreement with the forest borders observed in the satellite images. On closer examination of the TDX image in Fig. 5.8b, it can be observed that there is a net contrast between the Norway Spruce and European Beech. In particular, beech appears brighter than spruce in the image, reflecting its higher intensity. By focusing on the different landcovers shown in the scene, it is observed that open areas display the same grey tone as the Norway Spruce. The scatterers of these two different areas appear to have similar characteristics in X-band. Water, which reflects the microwaves in a specular manner, shows the lowest intensity, while the highest amplitudes appear to occur in urban areas, where double reflections are dominant. The PALSAR images indicate that the forest appears to be bright in comparison to other land surfaces. The forest expresses the high intensities inferred from the forest canopies in L-band. In comparison to TDX HH polarisation, the PALSAR HH scene illustrates that the distinction between the Norway Spruce and European Beech is limited in L-band. Moreover, considering the *GSV* ranges depicted by the inventory data, it is observed that the differentiation of different biomass levels using the PALSAR amplitude does not appear to be possible. This topic has been one of the main points of focus of the scattering mechanisms investigation and is presented in more detail later in this section. In addition to forested areas, it can be observed that the signal is weak over open areas as well as over water. In both cases, the roughness level appears to be insignificant compared to the L-band wavelengths, thus allowing specular reflections of the electromagnetic waves. In contrast to open areas and water, the urban zone shows bright pixels suggesting a high backscatter produced mainly by double reflections.

5.2.2 SAR Parameter Signatures

5.2.2.1 Multitemporal

A multitemporal analysis provides a good overview of the investigated signals and allows the first global examination of potential phenologies. In this study, this analysis was chosen to introduce some of the factors affecting the radar backscatter over the Thuringian Forest. First, the time-series observations for the X-band systems are presented. Then, the results of the L-band temporal analysis are presented. It should be noted that the results shown below represent one frame per dataset, which can be viewed in Appendix A.2.

X-Band

TSX and CSK were configured in order to acquire time-series datasets (see Sect. 4.2). Figure 5.9 depicts the latter data with the acquisition date on the abscissa, backscatter intensity on the left ordinate and weather information on the right ordinate.

In the example shown in Fig. 5.9a, the TSX time-series were acquired in the HS mode with HH and VV polarisations, ascending direction and at an incidence angle of 37°. The time window of the acquisitions extends from August 2009 to November 2009, and the frame location is given in the Schmiedefeld region (see frame HS1), which consists mainly of spruce and beech species. The TSX temporal graph indicates that the amplitudes in HH and VV polarisations are comparable for both spruce and beech species. The average intensities of the signals for the two species are about −5.5 and −7.0 dB. The error bars show ±0.7 and ±0.9 dB for spruce and beech, respectively. According to these observed amplitudes, the separation between these two species may be envisaged. Similar results are obtained by Ortiz et al. (2012). The higher backscatter for beech compared to spruce results mainly from the difference in the structural foliage characteristics of these species. In particular, the sizes of the leaves and needles as well as the gaps produced by the foliage of the two species play an important role in X-band. From this perspective, if the foliage composition and structure are important, there should also be some seasonal variations, particularly during leaf fall. On examining spruce and beech signals over different months, it is observed that the separation between the two species slightly improves during the leaf-off season (October–November) in comparison to the case during the leaf-on season (June–September). However, the signal intensity of beech decreases during the leaf-on season, while it was expected to increase with the apparition of the leaves. This consistent decrease in backscatter may be explained by the higher absorption of radar microwaves by the leaves. Difference in species characteristics is not the only cause of the variations shown by the different acquisition dates. In fact, they can also be explained by external factors such as weather conditions. A comparison of spruce and beech amplitude signals with the processed weather data indicates some relations. In particular, the precipitation histograms appear to follow the amplitude curves,

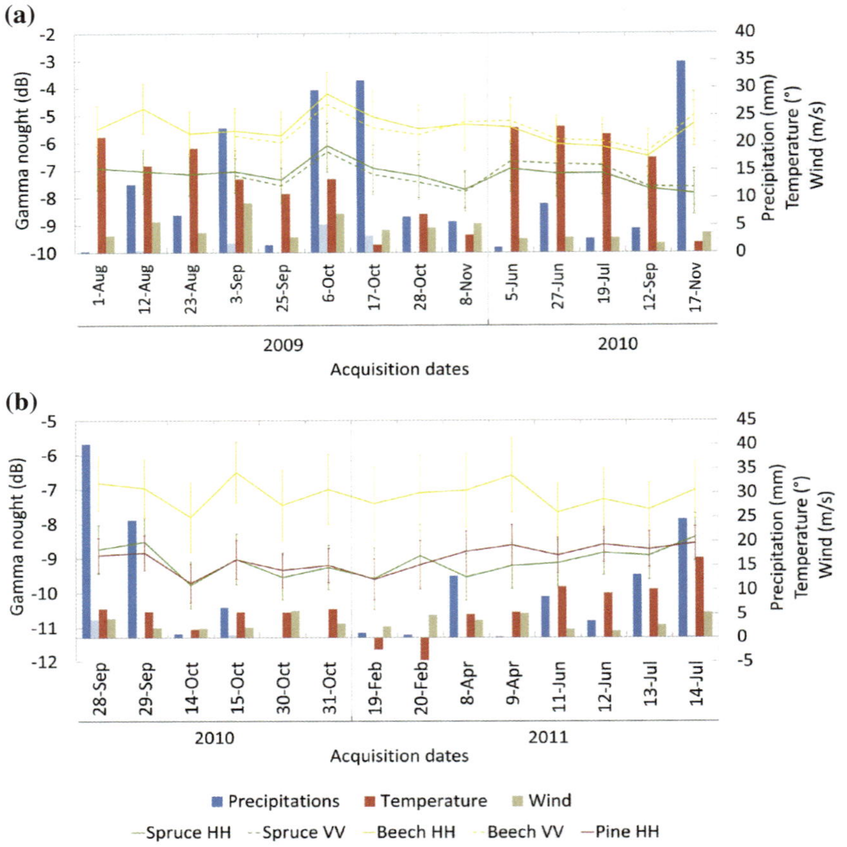

Fig. 5.9 Time-series analysis with **a** TSX HS HH 37° Asc. (frame HS1) and **b** CSK SM Himage HH 50° Asc. (frame CSK1) backscatter intensity (γ^0). HH polarisation is represented by a *continuous line*, while VV polarisation is shown by a *detached line*. Beech, spruce and pine are shown in *yellow*, *green* and *red*, respectively. Weather information is depicted in *dark blue*, *bright blue*, *red* and *grey* histograms for daily precipitation (p_d), hourly precipitation (p_h), daily air temperature (t_d) and daily wind (w_d), respectively. Here **a** represents data obtained by processing 40 spruce and 99 beech stands and **b** represents data obtained by processing 1,545 spruce, 314 beech and 136 pine stands

particularly on 6th October 2009, which shows a pick of intensity agreeing with the 30-mm and especially the 5-mm precipitation which occurred 4 days (dark blue histogram) and 4 h (light blue histogram) before the given acquisition, respectively. This observation agrees well with the results of other studies (see Chap. 2).

Figure 5.9b depicts the temporal datasets processed for CSK. These were acquired between September 2010 and July 2011 in the SM Himage mode, ascending pass, HH polarisation and at an incidence angle of 50°. Owing to the large extension

(40 km × 40 km) of CSK's frame and its location centred in the Frauenwald region (see frame CSK1, Appendix A.2), the three dominant species of the site, i.e. spruce, beech and pine, were covered. It should be noted that CSK was principally acquired to investigate X-band short temporal baseline interferometric coherence (see Sect. 5.3 coherence analysis). Hence, CSK acquisitions depicted in Fig. 5.9b consists of image pairs with a 1-day repeat-pass. The CSK temporal plot indicates amplitudes of about 1.5 dB, which are lower than the TSX backscatter amplitude. In addition, the separation between spruce and beech at an amplitude difference of approximately 2–2.5 dB appears to be more pronounced in the case of CSK. The reasons for this difference can be explained not only by the different incidence angles but also the different acquisition time selected for the two sensors. Some discussions on these topics are given later in this section by the description of the incidence angle and pass direction results. The red curve representing the amplitude of pine agrees well with the green signal corresponding to spruce. However, it should be noted that a standard deviation of 0.5 dB for pine is clearly smaller than that for spruce. This difference may be explained by the smaller number of pine stands available ($n = 136$) in comparison to spruce stands ($n = 1{,}545$). Similar to TSX time-series observations, the separation between beech and conifer species is enhanced during the leaf-off period (October–April) in comparison to the leaf-on period (June–September). However, no relationship between backscatter intensity and precipitation was observed. Also, the CSK scenes acquired under frozen conditions (19th February 2011 and 20th February 2011) did not highlight any obvious amplitude variation.

L-Band

A large amount of overlying PALSAR acquisitions provided a good basis for L-band frequency multitemporal analysis. Figure 5.10 depicts an example of the multitemporal PALSAR (a) HH and (b) HV dataset processed in the same manner as the TSX and CSK time-series presented previously.

PALSAR HH data shown in Fig. 5.10a were acquired between 13th November 2006 and 9th January 2011 in the FBD and FBS modes with an ascending-pass direction and at an incidence angle of 39°. The FBD acquisitions were measured during the vegetative period, while the FBS ones were captured during winter when the metabolisms of plants and trees are mostly constant. To observe the backscatter signal over a large number of acquisitions, the overlay of two different frames was considered (frames 7 and 8 defined in Appendix A.2). By observing the curves in Fig. 5.10a, some general remarks can be made. First, in contrast to the X-band sensors TSX and CSK, spruce and pine show higher absolute amplitudes than beech for the PALSAR L-band instrument. More precisely, a difference of 1 and 1.5 dB separates the average signals of beech and the two conifers taken together. Also, a comparison of pine and spruce suggests that the absolute amplitudes of these species are extremely close. However, pines still show a slightly (~0.3 dB) lower intensity compared to spruce. The standard deviation of the three species indicates a spread extending between 0.5 and 1.5 dB, with the smallest dispersion values for

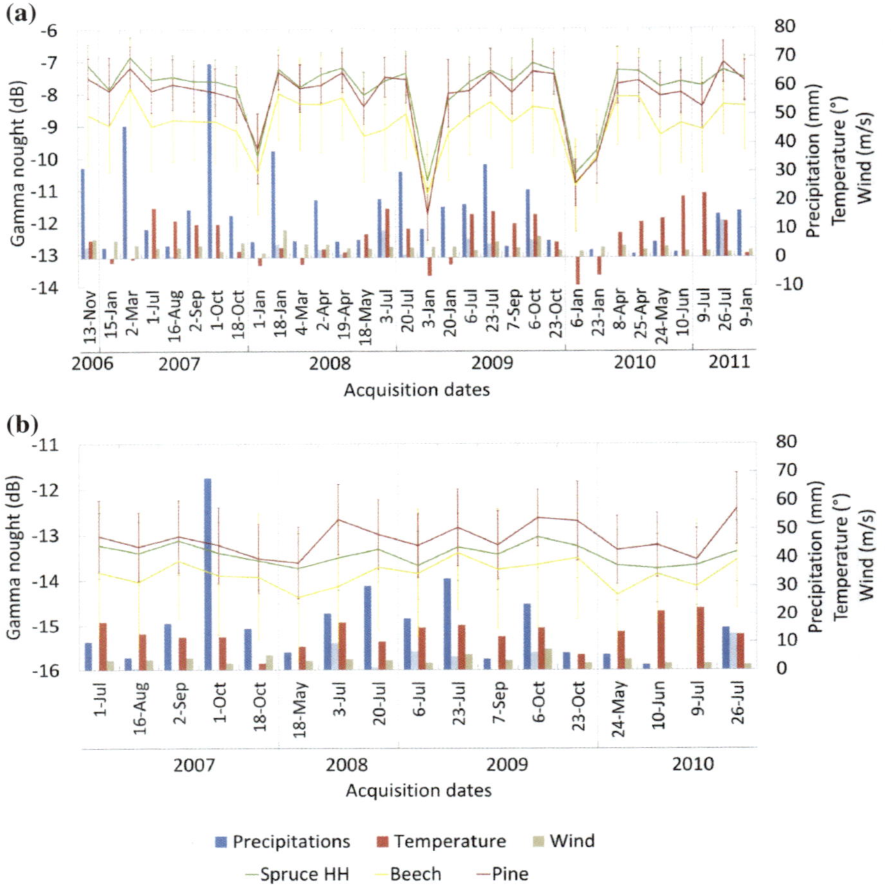

Fig. 5.10 Time-series analysis with **a** ALOS PALSAR FBD FBS 39° Asc. HH (frames FBD7,8 and FBS7,8) and **b** ALOS PALSAR FBD 39° Asc. HV (frames FBD7,8) backscatter intensity (γ^0). Beech, spruce and pine are in *yellow*, *green* and *red*, respectively. Weather information is depicted by *dark blue*, *bright blue*, *red* and *grey* histograms for daily precipitation (p_d), hourly precipitation (p_h), daily air temperature (t_d) and daily wind (w_d), respectively. Here **a** represents data obtained from 590 spruce, 29 beech and 30 pine stands and **b** represents data obtained from 616 spruce, 31 beech and 30 pine stands

the conifers. The size of conifer branches may generally better correspond to the size of the L-band microwaves than the broadleaves branches, which explains why the energy returned from conifers is higher than that returned by beech. However, it should be noted that many other factors need to be considered in order to understand the observed amplitude differences between the species. The interaction of radar microwaves may differ with the canopy properties of the species (i.e. size, shape, orientation, dielectric constant of the branches) as well as structures of the trees (i.e. density, understorey). The microwaves may also penetrate more deeply into beech forests than into spruce or pine canopies, thus increasing the ground contribution of beech and decreasing backscatter. To clarify these observations, other results need to be introduced. Hence, further discussions concerning these results are given later in this chapter (see Sect. 5.2.3). As mentioned previously, the PALSAR time-series data include not only FBD but also FBS acquisitions, which enables comparisons between winter and summer acquisitions. The FBS measurements in Fig. 5.10 indicate a significant decrease in the intensity during winter. This decrease, which exceeds 4 dB at its highest (3rd January 2009), appears to be closely related to the temperature depicted by the red histograms. In fact, when the temperature reaches the freezing point, the dielectric constant (relative permittivity) of the trees reduces, and all their metabolic activities (e.g. sap flow) are terminated (see Sect. 3.1), thus enabling better penetration of the microwaves through the canopy under such conditions and a higher contribution from the ground, which generally conducts in L-band to poor backscatter intensity. It should be noted that the backscatter intensities of spruce, beech and pine are significantly closer for acquisitions performed under frozen conditions than for those performed under non-frozen conditions. This observation supports the interpretations which suggest that the microwaves penetrate through the canopy and reach the ground. However, it should be noted that snow generally covers the Thuringian Forest during winter, and this may have affected the backscatter intensity. It should also be noted that this result agrees well with the findings of other authors (see Sect. 2.1). A comparison of the precipitations or wind to backscatter intensity indicated no significant trends.

The PALSAR HV multitemporal data are presented in Fig. 5.10b. These data were acquired with the FBD HH data shown in Fig. 5.10a simultaneously. Therefore, the acquisition configurations and selected frame remain the same as those for the FBD HH dataset. PALSAR HV time-series data acquisition started on 1st July 2007 and ended on 26th July 2010. The graph indicates that in terms of backscatter intensity, the average HV signal amplitudes for spruce, pine and beech are −13.2, −13.4 and −14.0 dB, respectively. These backscatter amplitudes are relatively 5 dB lower than the ones described for HH polarisation. Also, a comparison of the backscatters of the species indicates that similar to HH returns, spruce and pine show higher amplitudes than that of beech. In this case, the difference between the average signal amplitudes of beech and the two conifers taken together is only 0.5–1 dB. Spruce and pine also depict different backscattering coefficients. However, in contrast to HH polarisation, it can be observed that on this occasion, pine's amplitude remains 0.3 dB higher than the spruce backscatter amplitude. Finally, the dispersion shows that the largest spread is given by beech (±1.5 dB), followed by spruce (±0.6 dB) and pine (±0.5 dB). With respect to the HH results presented in Fig. 5.10a, the observations made in Fig. 5.10b cannot be explained unless several components are considered. This is described and further discussed during the comparison of the PALSAR sensor and *GSV* (Sect. 5.2.3). Comparing the weather data with the different acquisitions, no specific trends could be found. As mentioned above, for HH polarisation, the L-band frequencies appear to be less sensitive to precipitation events than the X-band frequencies.

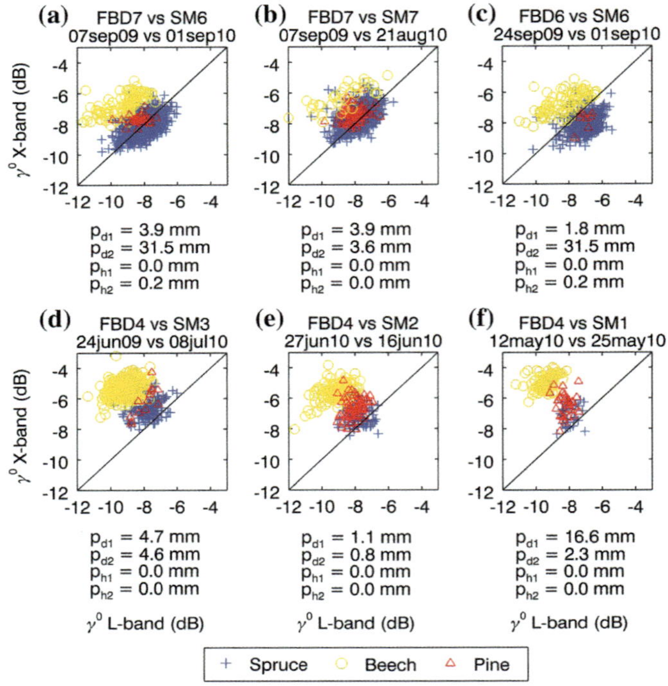

Fig. 5.11 Frequency analysis with ALOS PALSAR FBD HH 39° Asc. (frames SM1–3 and SM6,7) versus TSX SM HH 32°–38° Asc. (frames SM1–3 and SM6,7) backscatter intensity (γ^0). *Blue*, *yellow* and *red* indicate spruce, beech and pine, respectively

5.2.2.2 Multifrequency

ALOS PALSAR and TSX instruments transmit microwaves at different frequencies. In this regard, several investigations were performed in order to compare their respective signals. Figure 5.11 depicts 2D scatterplots with L-band and X-band measurements on the abscissa and ordinate, respectively. The L-band and X-band data shown in Fig. 5.11 were performed in the ascending mode, HH polarisation and at an incidence angle ranging between 32° and 39°. It should be noted that the acquisition dates between each compared dataset were selected to be as close as possible and from the same season so that no phenological phenomena affected the examinations.

The scatterplots indicate that broadleaves can be well-separated from conifers. A virtual oblique line may be traced between the clouds of points representing these two tree categories. This observation suggests that the ratio L-band/X-band may be useful for separating broadleaves and conifers. In terms of amplitude, it can be observed that the clouds of points for the three species are generally shifted above the 1:1 line. In other words, the backscatter intensity over the Thuringian Forest is generally higher at X-band frequencies than at L-band frequencies. This result may be explained by the structural composition of the canopies and more specifically by the number and size of the scatterers. Considering that X-band scatterers mainly return from twigs and small branches while L-band scatterers mostly return from large branches, the X-band microwaves may have proportionally more scatterers than the L-band electromagnetic waves. It should be noted that the scatterplot '(c) 24sep09 versus 01sep10' is an exception to this observation. In fact, clouds of points of spruce and pine are shown to be slightly below the 1:1 line. On comparing this plot with the first one '(a) 07sep09–01sep10', it can be observed that the L-band acquisition dates are different, while the X-band dates are the same. Consequently, the apparition of spruce and pine stands below the 1:1 line is explained by the variation in PALSAR measurements. The reasons for this amplitude difference ($\Delta\gamma^0 \approx 1$ dB) in L-band data could not be clearly identified. The weather conditions with no hourly and less than 4-mm daily precipitation appear to be insignificant. Intuitively, the canopies' structure cannot be related to this particular observation because all the three tree species show a 1-dB amplitude shift, and the consideration of the same stands between subplots (a) and (c) revealed equivalent results.

Similar to Figs. 5.11 and 5.12 presents the scatterplots processed for the HV polarisation. It should be noted that the selected acquisition dates are the same as those chosen for the HH polarisation in Fig. 5.11.

Similar to the HH polarisation in Fig. 5.11, comparing the Fig. 5.12 amplitudes for the X-band and L-band systems, the forest backscatter intensity for the HV polarisation appears to be higher in X-band than in L-band frequency. The subplot (c) 24sep09 versus 01sep10, where the strength of the backscatter is higher in L-band than in X-band again, appears as an exception. The explanations given above for HH polarisation should be applicable to the case of HV polarisation. In contrast to HH, the enhanced separation of the conifers and broadleaves using an X-band/L-band ratio does not appear to be possible with HV polarisation anymore. Indeed, neither X-band nor L-band systems in HV allow a clear separation of the investigated tree species. These observations suggest that scatterers such as needles, leaves and branches possess similar characteristics in terms of structural composition. Finally, the clouds of points in the subplots of Fig. 5.12 indicate that positively oriented ellipse regions at approximately 45° can be conceived. The orientation of the ellipses indicates the tendency that the forest backscatter increasing in X-band causes an increase in backscatter in L-band and vice versa. In this respect, a parallel can be drawn between the amounts of vegetation considering HV polarisation. A higher volume of small twigs and small branches at the top of the canopy indicates a proportionally higher volume of branches inside the canopy.

5.2.2.3 Multi-angle

As presented in Sects. 3.2 and 5.1, radar acquisitions with varying incidence angles may show different scattering mechanisms. Hence, the radiometry of the SAR data may be affected. In this framework, the various incidence angle configurations of the

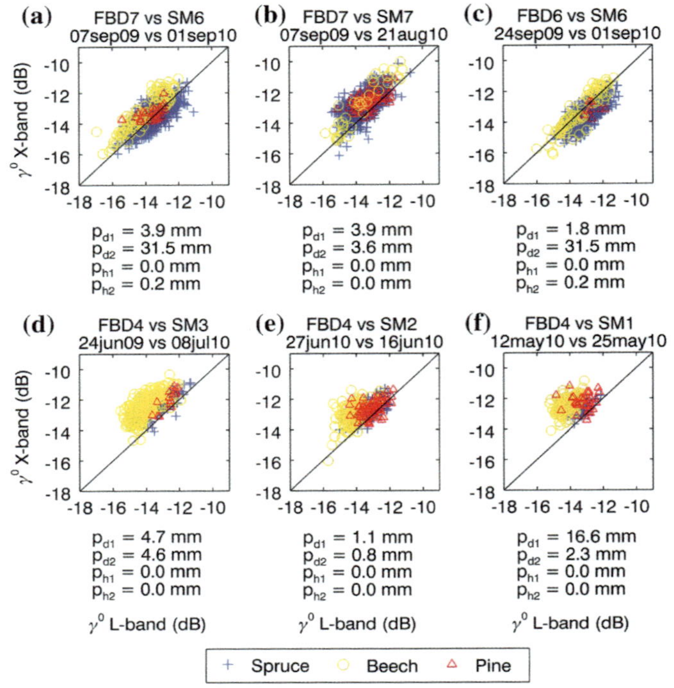

Fig. 5.12 Frequency analysis with ALOS PALSAR FBD HV 39° Asc. (frames FBD4,6,7) versus TSX SM HV 32°–38° Asc. (frames SM1–3 and SM6,7) backscatter intensity (γ^0). *Blue*, *yellow* and *red* indicate spruce, beech and pine, respectively

available TSX, CSK and ALOS PALSAR data were analysed. In particular, whisker boxplots were used to observe potential amplitude variations associated with the different incidence angles. It should be noted that according to the investigations performed in Sect. 5.1, the analysis performed here focused on flat or moderate slope areas. In fact, as it was presented in Sect. 5.1, the topography introduced amplitude variations, which appeared to be related to not only the radar scattering mechanisms but also the physiology of the forest. Hence, to distinguish the scattering mechanisms related to the incidence angle from the steep terrain effects, sloping criteria were defined for the incidence angle analysis. It should also be noted that when possible (i.e. when the number of available ROIs was sufficient), the investigations focused on open areas as well as spruce, beech and pine forest stands. The open areas were considered to reveal potentially different scattering mechanisms occurring in non-forested areas relative to forested areas. For the sake of synthesis, only the results of the forested regions are presented below focusing on spruce stands. The reader is referred to Appendix B.2 to refer to further processed graphs.

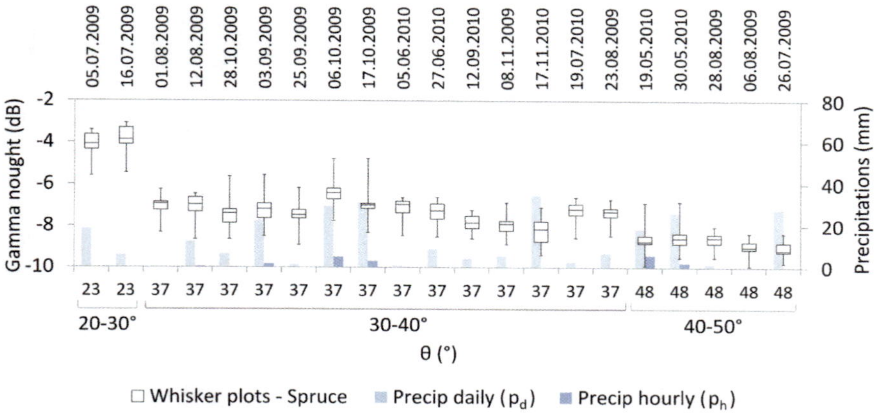

Fig. 5.13 Incidence angle analysis with TSX HS HH 23°–48° Asc. backscatter intensity (γ^0) (frame HS1). The *boxplots* represent spruce species with the sample minimum, lower quartile (Q1), median (Q2), upper quartile (Q3) and sample maximum. Weather information is depicted in *bright blue* and *dark blue* for daily precipitation (p_d) and hourly precipitation (p_h), respectively. A slope of <10° was selected, leading to 11 remaining spruce forest stands

X-Band

Figure 5.13 depicts the boxplots computed for the TSX HH polarisation data over spruce stands. All the depicted data were acquired in the ascending mode. Each boxplot represents one radar acquisition for which the incidence angle and acquisition date are described on the primary and secondary abscissae, respectively. It should be noted that for the sake of clarity, three different incidence angle ranges, namely 20°–30°, 30°–40° and 40°–50°, are considered on the primary abscissa. The backscatter intensity in dB and weather information, such as daily and hourly precipitations in mm, are represented on the primary and secondary ordinates, respectively. According to the methodology presented in Sect. 4.4, the intersecting area of the investigated TSX scenes was selected and the common nodata values were masked so that the same forest stands could be compared for the different acquisitions. Additionally, as mentioned above, forest stands with limited topography were selected. In the case of TSX, owing to the small size of the HS frames, the number of forest stands was limited. Thus, only slopes below 10° were selected. It should be noted that the incidence angle considered here is the one measured at the mid-swath of the SAR data. It should also be noted that the number of beech and pine stands was not sufficient to compute reliable statistics. Therefore, these classes were not investigated in this study.

The boxplots in Fig. 5.13 indicate that the amplitude decreases with increasing incidence angle. More specifically, the median of the first two acquisitions lies at around −4 dB with an incidence angle of 23°, while that of the other acquisitions fluctuates between −7 and −8 dB and −9 and −9.5 dB at incidence angles of 37° and 48°, respectively. According to theoretical models for forest backscatter (see Sect. 2.1), the total backscatter of the forest should decrease with increasing incidence angle. This trend is mainly explained by the high surface scattering contribution occurring at steep angles and the relatively lower contribution of the volume at large incidence angles. Thus, it is not surprising to see an intense backscatter at 23° and lower amplitude signals at 37° and 48°. However, the variation between 23° and 37° is very significant (~3 dB). No particular reasons could explain this large difference. As reported by some authors (Rauste 1990; Sun et al. 1991; Westman and Paris 1987; Moghaddam and Saatchi 1993), in addition to volume, double bounces may increase under large incidence angle acquisition conditions. In the present case, the contributions of double bounce do not appear to be significant in the large incidence angle configuration because the amplitude shown by the acquisitions performed at an incidence angle of 48° remain low compared to the other incidence angles. This can be justified by the frequency used by TSX. Indeed, at X-band frequencies and in such large incidence angle configurations, the canopy is not sufficiently transparent to allow electromagnetic waves backscattered from the ground level. As mentioned earlier, precipitations may increase the backscatter intensity. The consideration of precipitations in Fig. 5.13 did not show any clear trend with the backscatter intensity. However, it was noticed that in comparison to dry weather conditions, the dynamic range increased when hourly precipitations occurred.

Similar to Fig. 5.13 for TSX, Fig. 5.14 depicts the boxplots processed over spruce stands for CSK HH polarisation. The configuration is the same as TSX, except that CSK acquisitions were performed during night at 5 a.m. UTC, while the plots for TSX represent daily measurements with images acquired at 5 p.m. UTC. For comparison

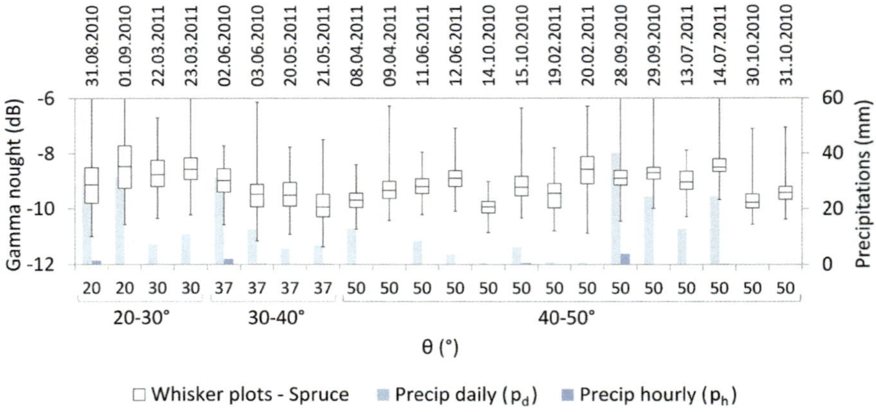

Fig. 5.14 Incidence angle analysis with CSK SM Himage HH 20°–50° Asc. backscatter intensity (γ^0) (frame CSK1). The *boxplots* represent spruce species with the sample minimum, lower quartile (Q1), median (Q2), upper quartile (Q3) and sample maximum. Weather information is depicted in *bright blue* and *dark blue* for daily (p_d) and hourly (p_h) precipitations, respectively. A slope of <5° was selected, leading to 165 remaining spruce forest stands

purposes, the boxplots computed over open areas, beech stands and pine stands are provided in Appendix B.2.

In contrast to the boxplots illustrating TSX in Fig. 5.13, the graphs depicted in Fig. 5.14 for CSK do not present any clear trend of the backscatter intensity with the incidence angle. However, it should be noted that the dynamic ranges appear larger in the case of CSK, suggesting that water droplets referring to the typical early morning humidity have affected the CSK data. It should also be noted that the CSK dimensions of SM Himage scenes are 40 km × 45 km, while those of TSX HS images are 10 km × 5 km, which implies that the number of stands available is much higher for CSK than that for TSX. In addition, the incidence angle variation in the range direction within the scenes is also larger for CSK ($\Delta\theta_{\mathrm{CSK}} = \pm 4°$) than that for TSX ($\Delta\theta_{\mathrm{TSX}} = \pm 2°$).

L-Band

Considering the current ALOS PALSAR data, the incidence angle was investigated by comparing PLR and FBD acquisitions, which were measured in the ascending direction at about 22° and 34°, respectively. The large amount of available PALSAR data allowed many observations, including those of spruce, beech and pine forest stands, as well as open areas. Figure 5.15 illustrates the processed boxplots for the backscatter in HH polarisation over spruce forest stands.

The tendency of the L-band frequency system seemed to be opposite to that of the X-band frequency system. In fact, as shown in Fig. 5.15, the backscatter intensity

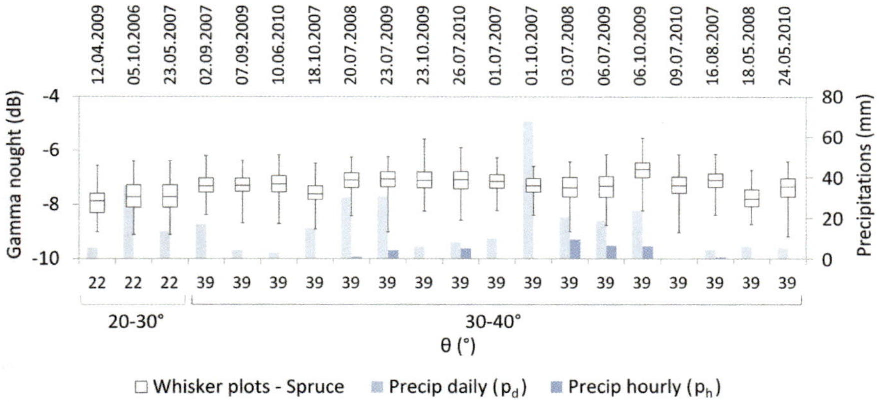

Fig. 5.15 Incidence angle analysis with ALOS PALSAR FBD HH 39° Asc. and PLR HH 22° Asc. backscatter intensity (γ^0) (frames FBD7,8 and PLR3). The *boxplots* represent spruce species with the sample minimum, lower quartile (Q1), median (Q2), upper quartile (Q3), and sample maximum. Weather information is depicted in *bright blue* and *dark blue* for daily (p_{d}) and hourly (p_{h}) precipitations, respectively. A slope of <5° was selected, leading to 85 remaining spruce forest stands

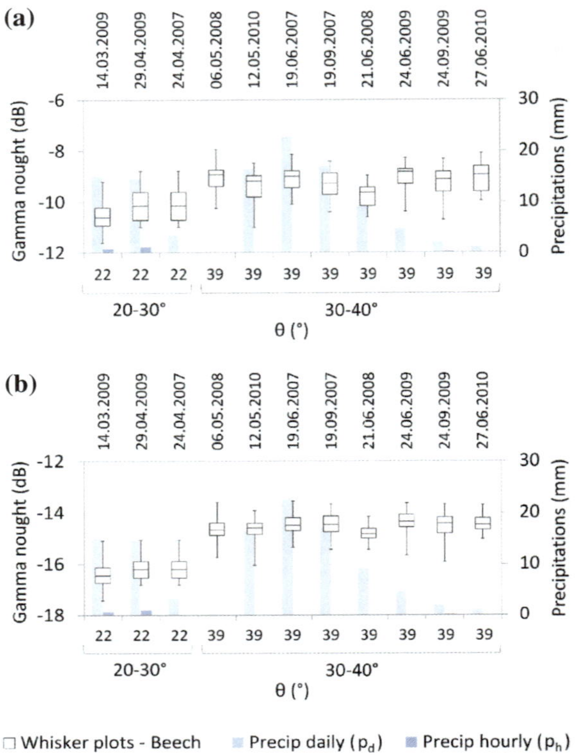

Fig. 5.16 Incidence angle analysis with **a** ALOS PALSAR FBD HH Asc. and PLR HH Asc. as well as with **b** ALOS PALSAR FBD HV Asc. and PLR HV Asc. backscatter intensity (γ^0) (frames FBD2 and PLR1). The *boxplots* represent beech species with the sample minimum, lower quartile (Q1), median (Q2), upper quartile (Q3) and sample maximum. Weather information is depicted in *bright blue* and *dark blue* for daily precipitation (p_d) and hourly precipitation (p_h), respectively. A slope of $<10°$ was selected, leading to 49 remaining beech forest stands

slightly (~0.5 dB) increased with increasing incidence angle. The descriptive statistics show the median to be slightly above −8 dB for PLR data ($\theta = 22°$) and slightly under −7 dB for FBD acquisitions ($\theta = 39°$). As mentioned earlier, the theoretical models in previous studies describing forest backscatter with varying incidence angles showed decreasing intensity with increasing incidence angles (see Sect. 2.1). Therefore, the obtained observations contradict the reported models. To understand this contradictory result, beech forest, pine stands and open areas were examined in addition to the analysis of spruce in Fig. 5.15. Figure 5.16 depicts the boxplots generated for PALSAR data over beech forest stands in (a) HH and (b) HV polarisations. Owing to the location of the beech stands in mainly topographic areas, slopes below 10° were selected, ensuring a sufficient number of stands while removing areas with high topography.

As shown in Fig. 5.16, the observation of the backscatter intensity for different incidence angles over beech forest stands confirmed the results addressed above for spruce, but on this occasion, the observed difference in amplitude between steep and shallow incidence angles appeared to be more significant. In fact, the beech backscatter amplitude in Fig. 5.16a varies by about 2 dB between 22° and 34°, while the spruce amplitude in Fig. 5.15 varies by only 0.5 dB in the same incidence angle range. Carrying along the investigations, the HV polarisation presented for beech has the same trends as those presented by HH with a more pronounced difference between steep and shallow incidence angles. The observations for pine and open areas revealed that the results obtained for pines were equivalent to those obtained for spruce, namely an increase in the backscatter for higher incidence angles; however, in the case of open areas, the signal in HH and HV appeared to be relatively constant with the incidence angle (see Appendix B.2).

With reference to these observations, it appears difficult to clearly explain the forest scattering mechanisms occurring at various incidence angles. In a typical situation, surface scattering should be dominant at a steep incidence angle (22°), volume scattering at large incidence angles ($\theta > 39°$) and the backscatter intensity should decrease with increasing incidence angle. In the present case, the backscatter increases with increasing incidence angle. Thus, it is presumed that at L-band frequencies, the Thuringian Forest does not fit the theoretical models shown in the literature for forest backscatter (see Chap. 2). At this stage of the investigations, only assumptions can be addressed. For instance, it may be assumed that dihedrals enhance the backscatter at large incidence angles. It may also be assumed that several forest properties such as canopy extinction or forest horizontal structure may have affected the signal in different ways for various incidence angles. Considering the more marked variation in intensity for beech compared to spruce or pine, it may be assumed that beech presents a greater penetration of the microwaves in comparison to conifers, thus decreasing the backscatter intensity with greater ground contribution for steep incidence angles and increasing the backscatter with more apparent volume for large incidence angles. With reference to forest structure, these results may be explained by the differences in height and basal area between the conifers (spruce and pine) and beech. By comparing the species in Sect. 4.3, mature beech were characterised by low density (i.e. basal area) and large height, while mature conifers showed high density and small height (see Fig. 4.4). Finally, the slightly more pronounced variation in HV compared to HH may be explained in a similar manner. HV may highlight a larger volume than HH for a large incidence angle and present a smaller sensitivity to ground reflections at steep angles. Some of these elements require a more in-depth examination, which is presented later in this section.

So far, the discussions about the incidence angle focused on the radiometry of the SAR data. However, as introduced in Sects. 3.2 and 5.1, in addition to radiometry, the side-looking acquisition of radar data may also affect the geometry of the images. In this regard, a visualisation of the SAR data with different incidence angles was performed. In Fig. 5.17, the backscatter intensity images for PALSAR (a) PLR HH and (b) FBD HH data are depicted. The first image was acquired on 12th April 2009 in the PLR mode at an incidence angle of 22°, while the second one was acquired on the 10th June 2010 in the FBD mode. Both images were measured in the ascending pass direction and were linearly stretched between −20 and 0 dB to enable comparisons.

In the PALSAR PLR data ($\theta = 22°$) shown in Fig. 5.17a, geometric distortions such as layover and foreshortening can be well distinguished in bright stretched areas near the Leibis-Lichte lake. However, in the case of the FBD data ($\theta = 39°$) (Fig. 5.17b), these effects are greatly reduced owing to a shallower incidence angle.

(a) (b)

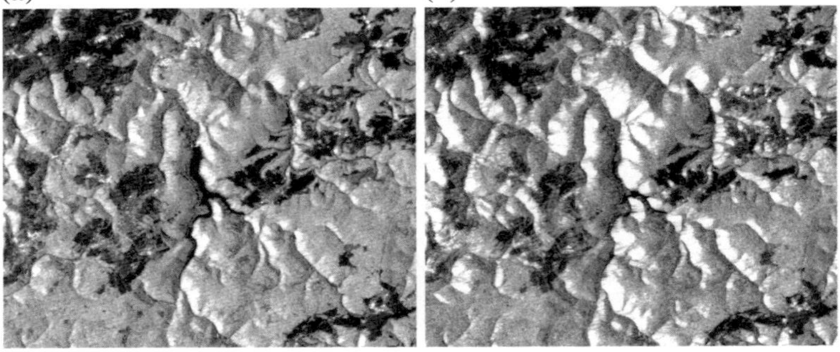

Fig. 5.17 Comparison of layover and foreshortening between **a** ALOS PALSAR PLR HH 22° Asc. (frame PLR3) and **b** ALOS PALSAR FBD HH 39° Asc. backscatter intensity (γ^0) (frame FBD7). The PALSAR PLR and FBD scenes were acquired on 12th April 2009 and 10th June 2010, respectively. Linear stretching was applied between −20 and 0 dB for comparison purposes. The area covered is the Leibis-Lichte Lake region

5.2.2.4 Multipolarisation

An additional characteristic of the SAR systems is their ability to simultaneously acquire images in different polarisations. In this respect, the present section focuses on the comparison of the different polarisations measured in the dualpol or quadpol acquisition mode. The discussions first concentrate on the X-band sensors and then shifts to the acquisitions performed with the L-band instruments.

X-Band

The TSX system performed several acquisitions in dual polarisation (dualpol) over the Thuringian test site. The acquisitions were measured in the HHVV, HHHV and VVVH polarisation modes (see Sect. 4.2). Figure 5.18 first compares the HHVV mode by representing the data in 2D scatterplots with HH polarisation on the abscissa and VV polarisation on the ordinate. The data shown in this figure were measured in the descending pass and at a large incidence angle in the range 40°–50°. Further results for the other pass directions and incidence angle ranges are provided in Appendix B.2. It should also be noted that the tree species compositions spruce (blue), beech (yellow) and pine (red) were considered for the polarisation analysis. However, they were not always available together in the same frame, which explains the apparition of only one or two species on each plot.

The scatterplots in Fig. 5.18 show that the clouds of points approximately follow the 1:1 line, indicating that the amplitude for HH and VV polarisations is rather comparable. Upon closer examination, it is observed that there is a positive (i.e. plots (a) (k) or (l)) and in some other cases negative (i.e. plots (c)–(f), or (j))

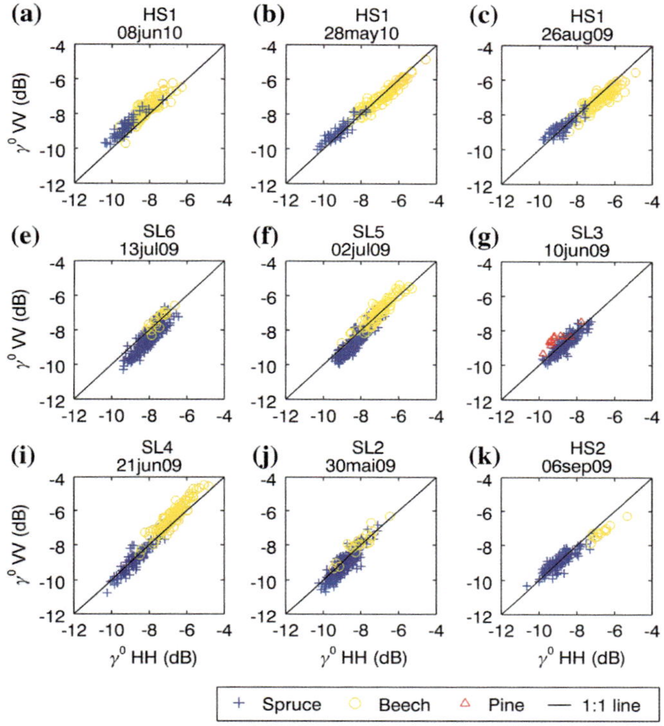

Fig. 5.18 Polarisation analysis with **a–d**, **k**, **l** TSX HS HH 40°–50° Desc. versus TSX HS VV 40°–50° Desc. backscatter intensity (γ^0) (frames HS1–3 and HS5) as well as with **e–j** TSX SL HH 40°–50° Desc. versus TSX SL VV 40°–50° Desc. backscatter intensity (γ^0) (frames SL1–6). *Blue*, *yellow* and *red* represent spruce, beech and pine, respectively

difference of 0.5 dB between the HH and VV backscatter intensity for the tree species under investigation. Comparing the available radar acquisitions, these amplitude differences appeared to be independent of the acquisition dates (Fig. 5.18) and radar acquisition configurations (refer to Appendix B.2) because both positive and negative differences were observed for several epochs and acquisition parameters. A single frame of the HS data showed a positive (i.e. plots (a), (b)) and a slightly negative (i.e. plot (c)) amplitude difference. This suggests that the 0.5 dB variation is also not related to the structure of the forest. Distinguishing the scatterplots for spruce, beech and pine by blue, yellow and red, respectively, it can be observed that the conifers can be separated to broadleaves. Indeed, spruce and pine amplitudes extend between approximately −8 and −10 dB, while beech's intensity ranges between −6 and −8 dB. As discussed above during the description of the time-series datasets, the foliage composition and branching structures of the investigated species differ, especially between the broadleaf and the two considered conifers. At X-band frequencies, these differences result in an increase of approximately 2 dB for beech compared to spruce and pine. It should be noted that in some cases, such as on the scatterplot (a) or (f), spruce and beech separation is limited because in the first case, the amplitude of beech decreases to -9 dB, while in the second case spruce's amplitude increases to -7 dB. This variation in intensity for beech in the first case and spruce in the second may be related to the structure of the forest. Indeed, as described in Sect. 2.1, in terms of structure, the forest may vary locally from one location to another depending on the defined logging strategies. The forest structure may also be determined by the environmental conditions of the forest such as slope exposition or groundwater availability. In this regard, there may be diverse forest structures which may explain why the dynamic range shown by spruce and beech in the plots (a) and (f), respectively, is higher. Additionally, owing to the different microwave absorption or reflection capabilities of the canopies, the various complex permittivity values given for the three different tree species in different periods of the year may produce variability in the signal, thus contributing to the observed variability.

After examining HHVV polarisations, the HH linear polarisation can be compared to the HV channel in order to interpret the differences in scattering mechanisms for the like- and cross-polarisations. Figure 5.19 depicts the scatterplots computed over spruce, beech and pine forests with HH channel on the abscissa and HV polarisation on the ordinate. The presented plots refer to the TSX SM data acquired at incidence angles of 30°–40° in the descending pass direction. Further results of X-band cross-polarisation comparisons are provided in Appendix B.2.

An overview of the plots in Fig. 5.19 shows the difference in the trends between HH and HV polarisations. As depicted on plots (b) and (d)–(h), broadleaves may be

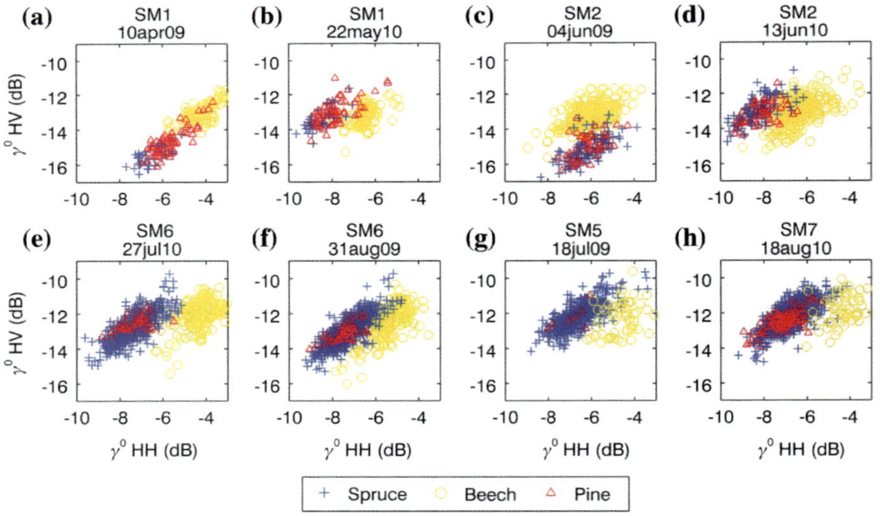

Fig. 5.19 Polarisation analysis with TSX SM HH 30°–40° Desc. versus TSX SM HV 30°–40° Desc. backscatter intensity (γ^0) (frames SM1,2 and SM5–7). *Blue*, *yellow* and *red* represent spruce, beech and pine, respectively

distinguished from conifers with HH but not with HV. The like-polarisation results were already discussed during the interpretation of Fig. 5.18. It may be observed that comparing the HH amplitudes of the plots in this figure with the ones in Fig. 5.19, the intensity shown by Fig. 5.19 is generally 1–2 dB higher. This is explained by the steeper incidence angle configuration used for the data presented in Fig. 5.19. The cross-polarisation results suggest that conifers and broadleaves have the same physical behaviour because the amplitudes shown by the three species are generally located within the same range of −12 to −14 dB. In fact, both conifer and broadleaf canopies are composed of twigs and small branches, which may be assimilated into small thin cylinders. These cylinders contribute to the same amount of depolarisation of the X-band EM waves for conifers and broadleaves and explain why the backscatter signal cannot be separated for these two classes in such a polarisation configuration (Ferrazzoli and Guerriero 1995).

Although a general tendency could be observed, it should be noted that some exceptions to the above discussed plots were noticed. Graph (a) was processed on the same forested areas as plot (b), and scatterplot (c) represents the same forest stands as plot (d). For conifers, plots (a) and (c) clearly show an amplitude shift of 2 dB in HV polarisation. Indeed, the HV backscatter intensity ranges between −14 and −16 dB in plots (a) and (c), while the intensity extends between −12 and 14 dB in plots (b) and (d). Neither the weather conditions nor structure of the forest could explain these differences because the precipitations were low and the same forest stands were selected on plots (a) and (b) as well as on graphs (c) and (d).

L-Band

As introduced in Sect. 4.2, ALOS PALSAR acquired full polarimetric radar data over the Thuringian Forest. The data acquired in the four linear polarisations (HH, VV, HV, VH) are compared below. Figure 5.20 depicts the processed scatterplots with the backscatter intensity in HH and HV polarisations on the abscissa and ordinate, respectively. The scatterplots show the PALSAR FBD data acquired on frame FBD3, as defined in Appendix A.2. Here this frame was selected as an example because it provides sufficient forest stands for the three principal tree species compositions, i.e. spruce (blue), beech (yellow) and pine (red). Further scatterplots for the PALSAR FBD frames are provided in Appendix B.2.

The plots in Fig. 5.20 indicate that there are generally two distinct clouds of points, one representing broadleaves with beech and the other representing conifers with spruce and pines. The average absolute amplitude of beech is −9 to −14 dB for HH and HV polarisations, respectively, while the combined average of pine and spruce is about −7 and −13 dB for HH and HV channels, respectively. With regard to the dynamic range on the abscissa, namely HH polarisation, beech extends between −8 and −12 dB, while the combined range of spruce and pine is approximately from −6 to −8 dB. The dynamic range on the ordinate, namely HV polarisation, indicates that the amplitude of beech varies between −12 and −16 dB, while that of spruce and pine combined varies between −11 and −15 dB. Thus, the dynamic ranges are

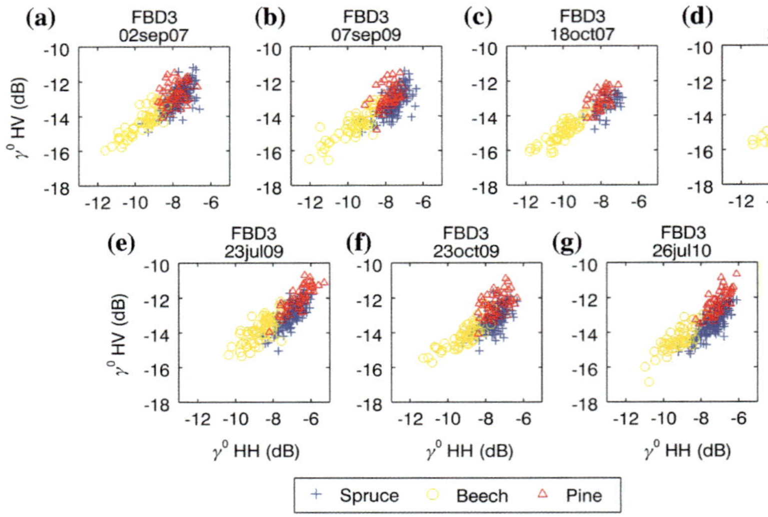

Fig. 5.20 Polarisation analysis with PALSAR FBD HH 39° Asc. versus PALSAR FBD HV 39° Asc. backscatter intensity (γ^0) (frame FBD3). *Blue*, *yellow* and *red* represent spruce, beech and pine, respectively

larger for beech than that for spruce and pine and in HV polarisation than in HH polarisation. With respect to the above mentioned amplitude values, the separation of broadleaves and conifers in L-band is more suitable with HH than with HV.

Considering the PALSAR PLR data, it is possible to compare the two copolarisations, namely HH and VV channels. Figure 5.21 depicts the processed scatterplots on the abscissa and ordinate with the amplitudes in HH and VV polarisations, respectively. The number of spruce (blue), beech (yellow) and pine (red) forest stands each was dominant on a specific PLR frame. Thus, the backscatter intensity for each tree species composition was presented separately in Fig. 5.21. In all cases, slopes below 5° were selected. Also, owing to full polarimetry, the backscatter intensity was compensated for azimuth slopes (see Sects. 3.2 and 4.3).

For the three tree species, the comparison of HH and VV polarisations in Fig. 5.21 clearly shows that HH presents a higher amplitude than VV. This observation may be explained by three different phenomena. First, the attenuation of the VV signal in the forest canopy is higher than that for the HH component. The scatterers comprising the canopy include the trunk, which is mainly vertically distributed, leading to a greater absorption of the vertically transmitted waves. Second, for the waves which potentially reach the ground, the ground-trunk or trunk-ground double reflections are higher in HH than in VV. In fact, the magnitude of the Fresnel coefficients always remains higher for HH than for VV (Cloude 2009). Owing to the higher attenuation of the vertically polarised waves in the canopy, the VV double reflections are also intrinsically reduced relative to HH. Finally, the lower VV amplitude relative to the HH backscatter may be explained by the Brewster angle effect (see Sect. 3.2).

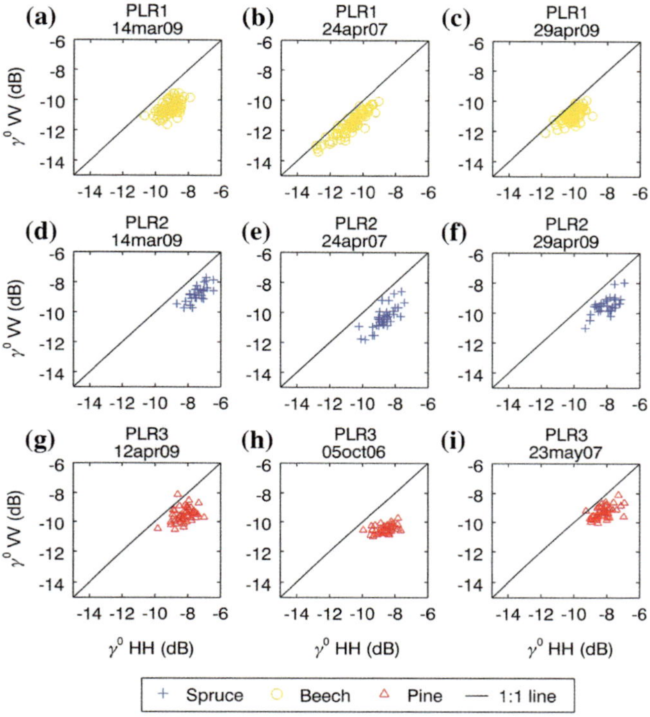

Fig. 5.21 Polarisation analysis with PALSAR PLR HH 22° Asc. versus PALSAR PLR VV Asc. 22° backscatter intensity (γ^0) (frames PLR1–3). *Blue*, *yellow* and *red* represent spruce, beech and pine, respectively

The vertically polarised incident waves transmitted at this specific angle tend to be absorbed by the ground, thus resulting in such a situation for a minor fraction of energy scattered back towards the radar.

The plots depicted in Fig. 5.21 were selected mainly for their large number of forest stands. The reader is referred to Appendix B.2 for the consideration of the remaining L-band HH versus VV plots. Also, the scatterplots comparing HH and HV polarisations with PLR datasets were not considered here because the results of the comparison were similar to those presented in Fig. 5.20. Comparisons of cross-polarisations were also not depicted because the results of HV polarisation were equivalent to those of VH (see Sect. 3.2.3).

5.2.2.5 Pass Direction

As presented in Sect. 4.2, TSX and CSK scenes were acquired in both descending and ascending modes. The descending-pass scenes were measured at different times of

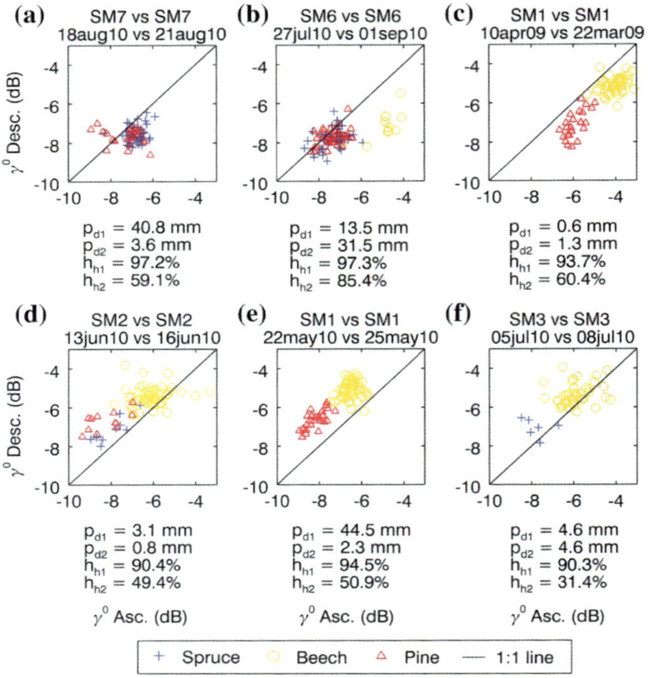

Fig. 5.22 Pass direction analysis with TSX SM HH 32°–39° Desc. versus TSX SM HH 32°–39° Asc. backscatter intensity (γ^0) (frames SM1–3 and SM6,7). *Blue*, *yellow* and *red* represent spruce, beech and pine, respectively. Slopes of $<5°$ were selected for spruce and pine and slopes of $<10°$ were chosen for beech

day. For TSX, the ascending and descending passes were measured at around 5 p.m. and 5 a.m. UTC, respectively, while for CSK, the ascending and descending passes were acquired at around 5 a.m. and 5 p.m. UTC, respectively. The different pass directions were compared in order to detect possible systematic variations. Figure 5.22 depicts 2D scatterplots with early morning (5 a.m. UTC) and late afternoon (5 p.m. UTC) acquisitions on the abscissa and ordinate, respectively. The depicted plots present TSX SM data measured at an incidence angle of around 35°. The air humidity information summing 4 daily measurements is given in the different subplots and is noted as h_{d1} and h_{d2} for the early morning (abscissa) and late afternoon (ordinate) acquisitions, respectively. When possible, the three dominant species, namely spruce (blue), beech (yellow) and pine (red), from the test site were considered. It should be noted that only forest stands with a slope below 5° were considered for spruce and pine species. In fact, owing to the topography, the positive and negative slopes are not equivalent for a right-looking ascending or descending pass direction. Beech species were mostly located on topographic areas. In the latter case, to maintain a sufficient number of available stands while limiting the effect of a steep terrain, slopes below 10° were selected. The discussions below describe the TSX data.

As presented in Fig. 5.22, the backscatter signal is slightly unequal in the ascending and descending passes. More precisely, a difference of about 0.5 to -1 dB and 0.5–2 dB separate the descending and ascending backscatter intensity for conifers and beech, respectively. This difference appears to be inconstant because on plots (a)–(c), the clouds of points are below the 1:1 line, and on plots (d)–(f), they appear above it. The air humidity shown by h_{d1} and h_{d2} values does not appear to explain these results; for instance, on plots (a) and (b), a similar trend of backscatter intensity is observed, and plot (a) shows high humidity only in the morning, while plot (b) shows high humidity in the morning and in the afternoon. In fact, the analysis of the pass direction expected to show systematic variations owing to the dew effect, which occurs early in the morning and increases the backscatter intensity owing to the apparition of condensed water particles on the branches of trees. Thus, owing to the dew effect, the backscatter intensity should have always been higher in the morning acquisitions. However, other dominant factors such as the difference between the radar observations of the forest structure in the ascending and descending modes and different weather conditions during the acquisitions may have masked this effect. Therefore, it is not possible to highlight any trend, and it remains difficult to draw any clear conclusion.

5.2.3 Sensitivity of Radar Backscatter Intensity to Forest GSV

After comparing the different SAR acquisition configurations, a closer examination of the relationship between the SAR parameters and forest properties was performed. The investigations were conducted by first comparing the SAR parameters with general forest attributes such as stand height or *DBH* and then analyzing the relationship between the SAR signal and *GSV*. The *GSV* synthesises the different stand attributes together through an allometric equation (see Sect. 3.1 theory). With *GSV* being the primary parameter used by foresters and considering the relatively high correlation between the forest attributes and *GSV* (see Sect. 4.3), the following discussions describe the relationship between *GSV* and SAR signal intensity.

As discussed in Sect. 2.1, many studies have examined the relationship between radar backscatter and *GSV* or biomass. Most of the results presented in these studies showed high sensitivity to *GSV* levels in cross-polarisation and when using low-frequency radar systems (i.e. L-band, P-band). According to these results, some investigations were performed to determine whether similar observations could be obtained in the Thuringian Forest. Surprisingly, the obtained results were shown to be different from the ones presented in previous studies. In this context, deeper analyses were conducted to clarify the unexpected results. The following discussions focus on these investigations. First, the results of the PALSAR backscatter intensity for the different species are presented. Then, some comparisons considering parameters such as stem density and weather conditions are made. Finally, further analyzes based on polarimetric decompositions are described.

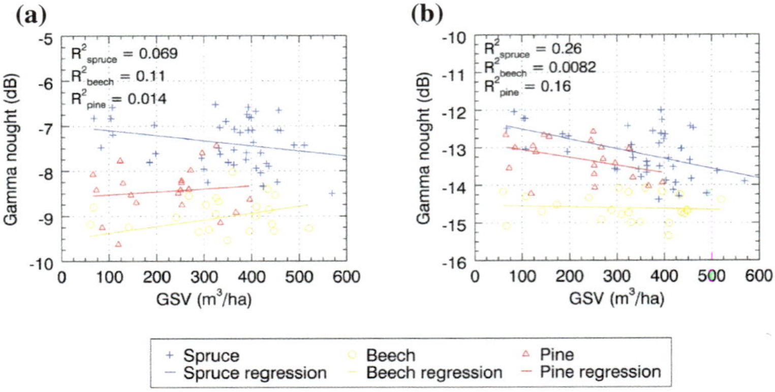

Fig. 5.23 *GSV* sensitivity analysis with ALOS PALSAR FBD 39° Asc. **a** HH and **b** HV backscatter intensity (γ^0) (frame FBD2). The PALSAR scene was acquired on 19th September 2007. *Blue*, *yellow* and *red* represent spruce, beech and pine, respectively. Slopes of <5° were selected for spruce, beech and pine

5.2.3.1 Species Comparisons

One of the main objectives of this thesis is to study the relationship between backscatter intensity and *GSV* and highlight the scattering mechanisms occurring in the Thuringian Forest. In this framework, the backscatter for the different tree species compositions was compared with *GSV*. Figure 5.23 depicts an example of the scatterplot obtained by plotting the PALSAR backscatter intensity on the ordinate versus *GSV* on the abscissa for spruce, beech and pine. It should be noted that to limit the topographic effects discussed in Sect. 5.1, slopes below 5° were selected for spruce, beech and pines stands. The scatterplots presented in Fig. 5.23 are derived from a PALSAR FBD scene acquired on 19th September 2007 at an incidence angle of 39°. It is noteworthy that in this example, the range of *GSV* did not fully cover the first 100 $m^3\,ha^{-1}$. This range is specifically discussed later in this section. The results presented in these graphs are representative of the investigations performed on the entire PALSAR FBD datasets, which include different acquisition dates as well as different acquisition frames. Please refer to Appendix B.2 for further scatterplots.

The plots in Fig. 5.23 indicate three different elements. First, the sensitivity of spruce, pine and beech backscatter to *GSV* appears to be extremely poor. Second, the theoretical increase in backscatter intensity with *GSV* cannot be observed. Finally, the trends given by the tree species composition appear to differ. These first observations deserve some attention because they differ from those which were found in previous studies. Upon closer examination, particularly for the R^2 values, the highest sensitivity is shown by spruce in HV polarisation with $R^2 = 0.26$, followed by pine with $R^2 = 0.16$ also in HV polarisation and beech with $R^2 = 0.11$ in HH polarisation. Although the r-square is low, the minor correlation is shown to be rather positive for beech and negative for spruce (refer also to Appendix B.2).

The correlation concerning pines was also shown to be negative, but this was not always the case, especially in HH polarisation. Comparing the polarisations, it should be noted that spruce presented slightly higher sensitivity to *GSV* in HV than in HH polarisation. Conversely, the sensitivity of beech appeared to be higher in HH than that in HV polarisation. However, these correlations remained very low. In terms of backscatter intensity, first considering the *GSV* range 0–200 $m^3\,ha^{-1}$ in the scatterplots from Fig. 5.23, it is observed that there is a difference of about 2.5 and 2.0 dB in HH and HV polarisations between spruce and beech, respectively. Then, at higher *GSV* levels ($GSV > 200\,m^3\,ha^{-1}$), the amplitude difference in the low *GSV* range appears to decrease until it reaches about 1.5 dB in HH and less than 1 dB in HV. This is observed on most of the scatterplots given in Appendix B.2. Considering the dispersion of the data for each species, the intensity varies within approximately 1–1.5 dB over different *GSV* ranges. This variation is fairly acceptable for a radar system. In addition to HH and HV polarisations, the VV and VH intensities obtained using the PALSAR PLR data could be plotted against *GSV*. In these cases, low correlations were also obtained and similar trends could be observed. It should be noted that the azimuth-slope correction performed on PLR data did not show any significant improvement in the correlations.

5.2.3.2 Temperature

Following the above description of Fig. 5.23, further investigations were performed in order to find some complementary information which may help to understand the obtained results. The focus lay on the analyses of the weather conditions, particularly air temperature, because the multitemporal plots in Sect. 5.2.2 indicated a drop in the amplitude under frozen conditions. Similar to Fig. 5.23, scatterplots comparing backscatter intensity and *GSV* were processed using the PALSAR FBS data acquired during winter and spring. Figure 5.24 depicts the results obtained for four FBS scenes, which were all acquired at different air temperatures. The four images were measured with an incidence angle of 39°, and in all cases, only forest stands with slopes below 5° were selected. Also, the overlapping area of the four acquisition frames was processed so that the same stands could be compared from one acquisition to another. Figure 5.24 consists of an example showing the influence of air temperature on the radar signal. Further FBS acquisitions processed with their respective measured air temperature are referenced in Appendix B.2.

As shown in Fig. 5.24, the amplitude decreases when the air temperature decreases, particularly when the air temperature reaches negative values. This is observed on the subplots presented in Appendix B.2. As mentioned during the interpretation of the time-series data, this decrease in amplitude may be explained by the reduced sap flow during the forest dormancy period (see Sect. 3.1) and the decrease in the relative permittivity (dielectric constant) of the canopy and ground owing to the frozen forest and ground conditions, respectively. In other words, the forest canopies should be more transparent to radar microwaves when the air temperature reaches the freezing point.

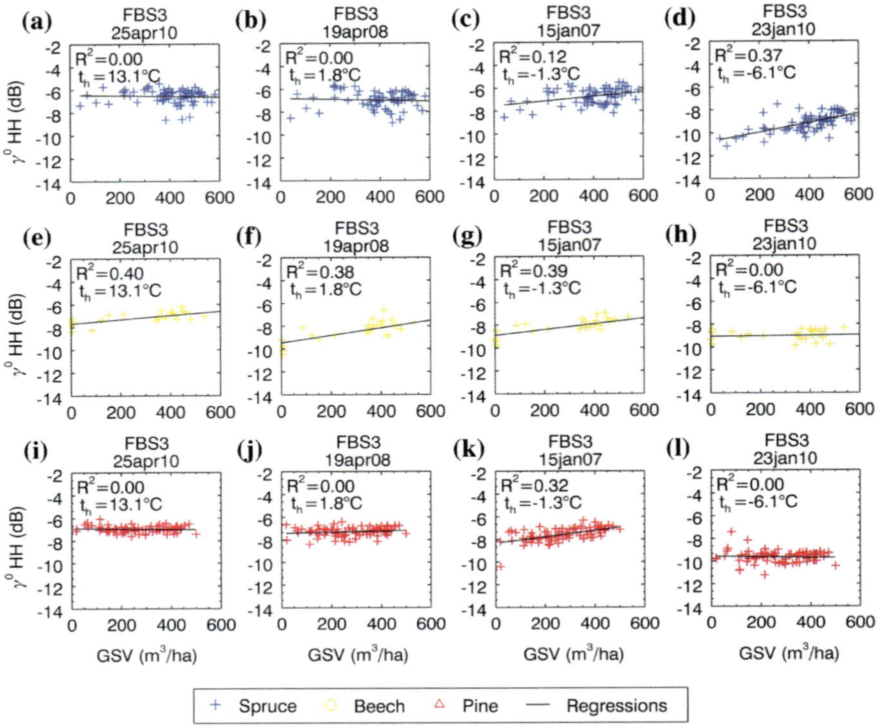

Fig. 5.24 *GSV* sensitivity analysis for different air temperatures with ALOS PALSAR FBS HH 39° Asc. backscatter intensity (γ^0) (frame FBS3). *Blue*, *yellow* and *red* represent spruce, beech and pine, respectively. Slopes <5° were selected for spruce, beech and pine

Comparing the absolute intensity values for the different tree species between the frozen (first subplots column) and non-frozen (last subplots column) acquisitions, the conifers appeared to be affected by the temperature to a greater extent than the broadleaves. In fact, in the example shown in Fig. 5.24, the backscatter intensity drops by about 2 dB for conifers and about 1 dB for the broadleaves represented by beech. Furthermore, young regenerative trees in the range 0–200 $m^3\,ha^{-1}$ appear to be affected by the decrease in temperature first. Then, under frozen weather conditions with air temperatures below −5 °C, mature trees represented by a *GSV* of greater than 400 $m^3\,ha^{-1}$ are also significantly affected by the temperatures below 0 °C and resulting frozen conditions.

Upon distinguishing between the different tree species, subplots (a)–(d) indicate that the relationship of spruce with *GSV* changes from a rather negative trend under non-frozen conditions to a positive trend under frozen conditions. More specifically, there is a larger amplitude variation for young spruces relative to mature spruces between non-frozen and frozen conditions. Thus, assuming that the ground-level signal is relatively weak (minor double bounce contributions) compared to the forest signal, this observation may suggest that in non-frozen weather conditions, there is greater penetration of the L-band microwaves through mature spruce canopies compared to the young regenerative spruce canopies, while under frozen conditions, this tendency is the opposite, i.e. greater penetration for young spruces compared to mature spruces. It should be noted that penetration refers not only to the penetration via the holes resulting from canopy openness but also via tree components (see the scattering pathways in Sect. 3.2). In Fig. 5.24, comparing the different acquisitions for beech (subplots (e)–(h)), it should be noted that the positive correlation is slightly more pronounced under frozen conditions compared to that under non-frozen conditions. Thus, with the same weak ground-level signal assumption as that made above for spruces, it can be deduced that under non-frozen conditions, the radar microwaves penetrate to a greater extent through young beech canopies than through mature beeches, and the penetration difference is more significant under frozen conditions. In extremely frozen conditions (subplot (h)), the backscatter signal for both young and mature beech species is equivalent. This observation suggests that under these conditions, the scatterers returning from the ground are dominant for both young and mature beech forests. Finally, in Fig. 5.24, comparing the frozen and non-frozen conditions for the pine signal (subplots (i)–(l)), it is observed that the overall intensity decreases from -7 to -9.5 dB for young regenerative pines as well as mature pines. Also, similar to the other species, the air temperature first affects young regenerative pines (subplot (k)) and then mature pines under frozen states (subplot (l)). In such weather conditions, the ground signal appears to be dominant, as *GSV* and backscatter intensity depict no correlation.

Summarizing the effect of temperature on forest backscatter, it can be deduced that young and mature trees react differently to the transition of air temperature; in particular, young regenerative forests tend to be affected by frozen conditions before mature forests. Additionally, the different tree species, each presenting specific dielectric properties, also show their own sensitivity to temperature changes. It should be recalled that snow occurs during winter in the Thuringian Forest. The snow might have contributed to an increase in the dispersion of the signal, especially when it melted, but it did not appear to be a determinant factor for the observations made in Fig. 5.24.

5.2.3.3 Horizontal Structure

As presented in the literature review of this thesis, many studies have examined the relationship between biomass or *GSV* and radar backscatter intensity. Most of these studies concerned forested areas with low *GSV*, typically ranging between 0 and 200 $m^3\,ha^{-1}$. In comparison to other temperate forests in Europe and around the world, the Thuringian Forest possesses large volumes of wood per hectare, with more than 600 $m^3\,ha^{-1}$ in some cases. Thus, the Thuringian Forest may show different structural characteristics in comparison to other forests. The above mentioned investigations examined the relationship between the principal species composition of the test site and *GSV*. However, as noted on the investigated plots, the full range of

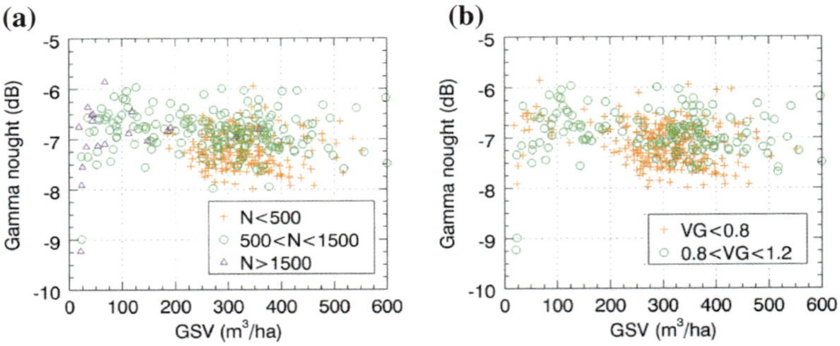

Fig. 5.25 *GSV* sensitivity analysis for different spruce **a** stem densities (N) and **b** relative stocking (VG) with ALOS PALSAR FBS HH 39° Asc. backscatter intensity (γ^0) (frame FBS7). The PALSAR scene was acquired on 19th April 2008. The density classes in **a** comprise $N < 500$, $500 < N < 1{,}500$ and $N > 1{,}500$ trees.ha^{-1} in *orange*, *green* and *purple*, respectively. The relative stocking classes in **b** distinguish $VG < 0.8$ and $0.8 < VG < 1.2$ in *orange* and *green* respectively. Slopes of <5° were also selected

GSV was not covered, especially between 0 and 100 m^3 ha^{-1}, as it was not possible to obtain a large number of stands for all the species together in a unique scene. In this context, the radar data presenting a particularly large number of stands in the low *GSV* range were selected and further investigated. Figure 5.25 shows two plots which illustrate the obtained results. The graphs consist of a 2D scatterplot, with *GSV* depicted on the abscissa and PALSAR backscatter intensity on the ordinate. The two plots represent the same spruce stands and radar acquisition, but plot (a) depicts different classes of stem density (N), while plot (b) describes different classes of relative stocking (*VG*). These two parameters are described below. The FBS data acquired in spring at temperatures above the freezing point were used because the number of spruce stands available was higher in this mode than in FBD. Also, similar to previous investigations, only spruce forest stands with slopes below 5° were selected to reduce the topographic effects.

Observing the cloud of points in Fig. 5.25 without considering the classes, it is seen that the backscatter intensity for spruce can be split into two different groups. The first one can be delimited between 0 and 200 m^3 ha^{-1}. The second one may be observed between 200 and 600 m^3 ha^{-1}.

In the first case, the backscatter appears to increase with *GSV* until a saturation point is reached at around 100 m^3 ha^{-1}. This tendency corroborates with the commonly accepted idea that radar backscatter increases with increasing biomass and the saturation levels, as reported in the literature (see Sect. 2.1). However, it should be noted that this trend could not be clearly observed on all acquisitions which presented a sufficient number of young regenerative spruce stands. The reason is that backscatter intensity depends not only on the amount of vegetation but also on the structure of vegetation (i.e. the forest). In terms of stem density (plot (a), Fig. 5.25), the young spruces consisted mainly of dense forests with more than 500 trees ha^{-1} (green and purple points). Thus, the few young regenerative spruces of the Thuringian

Forest with sparse stands may explain why the increase in backscatter with *GSV* is not clearly shown in all acquisitions, particularly in the acquisitions represented by a frame with less forest stands than those in frame FBS7. In addition to stem density, the relative stocking parameter also affects the observed backscatter amplitudes (plot (b), Fig. 5.25). In fact, the full stocked forest stands involve a better repartition of the trees over the stands relative to understocked stands, implying that for understocked stands, the radar backscatter decreases, which is caused by the higher contribution of the ground surface to the signal in this case. In Fig. 5.25, relative stocking with the understocked class depicted in purple explains why the amplitude is found to be less than −7 dB for some of the stands between 0 and 100 $m^3\,ha^{-1}$.

Then, the second cloud of points ranging between 200 and 600 $m^3\,ha^{-1}$ indicates that the radar signal for this range extends between −6 and −8 dB. On the assumption that the radar signal is saturated, one of the reasons for this large variation is the forest structure. Indeed, considering the density and relative stocking classes described in Fig. 5.25, it can be seen that the purple points presented in graph (a) and (b) are slightly shifted by 0.5–1 dB relative to the green points. In fact, while the young spruces have a higher volume, they compete with each other, and upon maturity, this leads to only a few spruces remaining with more free space between them. In terms of structure, these mature spruces are sparse and understocked, which certainly enhances the penetration of the L-band radar microwaves through the canopy and the reflected signal from the ground, thus explaining why their backscatter intensity is generally lower than young or mature dense spruces. Focusing now only on the green points depicted in both graphs, it is observed that after the saturation point at about −6.5 dB at 100 $m^3\,ha^{-1}$, the signal remains approximately constant at this amplitude, even when the *GSV* increases. In this case, in terms of (horizontal) structure, the forest conditions may not have been altered relative to young regenerative spruces, thus leading to a relatively constant radar signal.

More specifically, the relationship between *GSV* with radar backscatter may be linked to a combination of two properties, namely amount and structure of vegetation. The amount of vegetation may be described by *GSV*, while the horizontal structure component may be given by relative stocking and stem density. Both these parameters should affect the amount of ground seen by the radar. The above observations showed that a high backscatter generally involved fully stocked or overstocked and relatively dense stands, while a low backscatter generally implied understocked and relatively sparse stands. However, it was still difficult to determine which of these two parameters, namely *GSV* or structure of the trees, was the most significant in the total radar backscatter signal.

To provide a visual impression of the above results, Table 5.1 presents several spruce forest stands which were depicted in Fig. 5.25 at a particularly low or high intensity. According to the above discussions, these stands were assigned to different classes including a combination of young or mature, understocked or fully stocked and sparse or dense spruce stands. Overall, four different types of forest structure were distinguished.

Briefly, the first two forest types shown in Table 5.1 depict young spruces with a distribution of sparse and dense trees, respectively. The next two types present

Table 5.1 Visualisation of the types of forest structure with ALOS PALSAR FBS HH 39° Asc. backscatter intensity (γ^0) (frame FBS7)

Designation	Orthophotos	PALSARHH backscatter	
Young Understocked Sparse	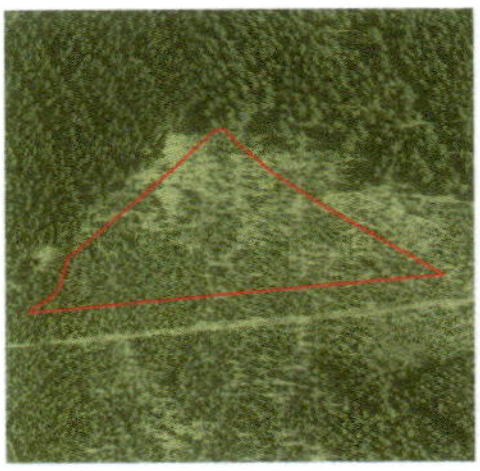	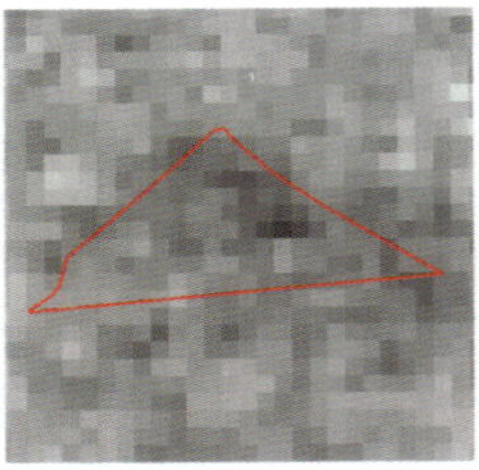	$GSV = 21\,\mathrm{m^3\,ha^{-1}}$ $VG = 1.0$ $N = 1975$ trees $\mathrm{ha^{-1}}$ $\mu = -9.2\,\mathrm{dB}$ $\sigma = \pm 1.4\,\mathrm{dB}$
Young Full stocked Dense		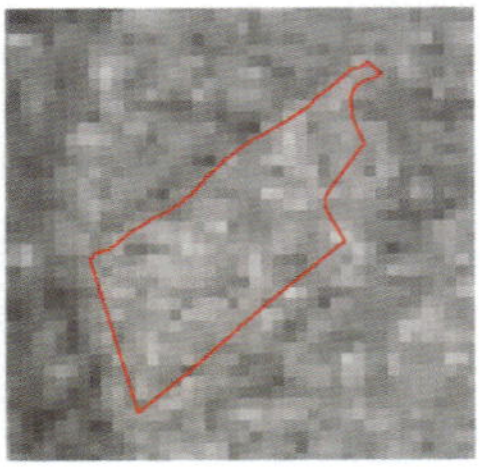	$GSV = 107\,\mathrm{m^3\,ha^{-1}}$ $VG = 0.9$ $N = 1241$ treesha^{-1} $\mu = -6.2\,\mathrm{dB}$ $\sigma = \pm 1.4\,\mathrm{dB}$
Mature Understocked Sparse	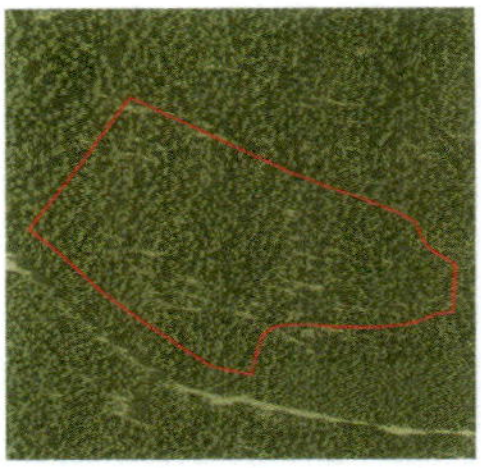	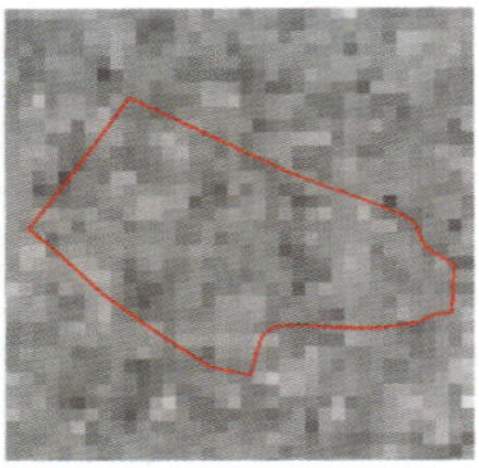	$GSV = 250\,\mathrm{m^3\,ha^{-1}}$ $VG = 0.6$ $N = 264$ trees $\mathrm{ha^{-1}}$ $\mu = -7.7\,\mathrm{dB}$ $\sigma = \pm 1.4\,\mathrm{dB}$
Mature Full stocked Dense	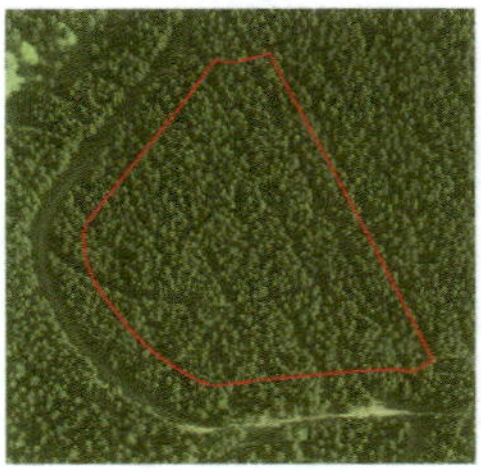	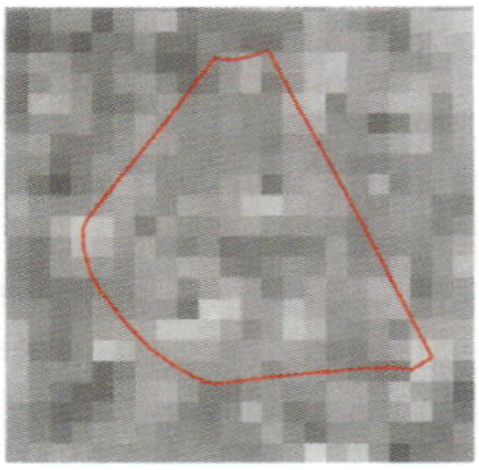	$GSV = 337\,\mathrm{m^3\,ha^{-1}}$ $VG = 1.0$ $N = 875$ trees $\mathrm{ha^{-1}}$ $\mu = -6.2\,\mathrm{dB}$ $\sigma = \pm 1.4\,\mathrm{dB}$

The PALSAR scene was acquired on 19th April 2008. Digital orthophotos (RGB) are provided for comparison with the radar data. Stands' borders are delimited with red lines over the images and statistics related to each stand are provided in the last column of the table. *VG* denotes relative stocking and *N* density attributes. The stands consist of spruce tree species two mature forest structures, one including sparse and another dense spruce forests. Comparing the orthophotos with the PALSAR backscatter intensity, it is observed that the dense forests show rather bright pixels, indicating that the intensity in these cases is relatively high. In fact, in both cases, the average stand intensity is -6.2 ± 1.4 dB. Unlike the dense forests, the sparse forests show low intensity or very low intensity in the case of the young sparse forest. In fact, the amplitude reaches -9 dB in the latter case. These observations corroborate well with the observations made in Fig. 5.25. It should be noted that owing to forest-management activities, young regenerative sparse spruces (i.e. case 1) were uncommon in the Thuringian Forest. Thus, few stands with such low backscatter were depicted in the different data under investigation. Also, it can be seen that the inventory do not always match to the observations performed on the orthophotos. In this example, the orthophoto for the designation 'young understocked sparse' (Fig. 5.25a) clearly show an undestocked forest while the inventory reported full-stocked forest with '$VG = 1$'.

The results presented in Fig. 5.25 and visualised in Table 5.1 were discussed in a few studies (see Sect. 2.1). For instance, Rauste et al. (1994) identified that spruce species from the Black Forest (Germany) showed an increase in backscatter up to a *GSV* level of 100–150 $m^3\,ha^{-1}$ followed by a decrease in backscatter. However, the author did not explain this observation. Some first explanations were given recently by Woodhouse (2006) and Brolly and Woodhouse (2012). In Brolly and Woodhouse (2012), the forest was represented as a 'matchstick', namely a collection of identical cylinders with no canopy. Removing the random volume scattering aspect of the forest, the authors showed that the relationship between radar backscattering signal with biomass is not only due to the volume contribution but also to the forest structure.

5.2.3.4 Double Bounce–Polarimetry

The results presented above illustrate the importance of the horizontal structure of the Thuringian Forest in studying the relationship between *GSV* and radar backscatter intensity. In particular, it was shown that the penetration of the L-band microwaves was enhanced in cases where the forest was sparser. In such circumstances, the radiations generally reach the ground surface and scatters specularly away from the radar sensor, thus leading to a reduced measured signal. However, the path followed by the waves may differ from this original scheme. In fact, under specific conditions, the radar electromagnetic waves are reflected on the ground surface and may then be reflected again off the tree trunks and vice versa. This phenomenon, which was described in Sect. 3.2 and mentioned above, is commonly known as double bounce or double reflection. The various scattering contributions were investigated in this study using polarimetry techniques. Coherent decomposition of the backscatter was performed on the basis of the Yamaguchi algorithm. The decomposition led to the separation of the signal into three different scattering mechanisms, namely surface, volume, and double bounce. Figure 5.26 depicts an example of the results obtained for spruce, beech and pine using PALSAR PLR data. The figure shows subplots with scattering power on the ordinate and *GSV* on the abscissa. The subplots were sorted by columns

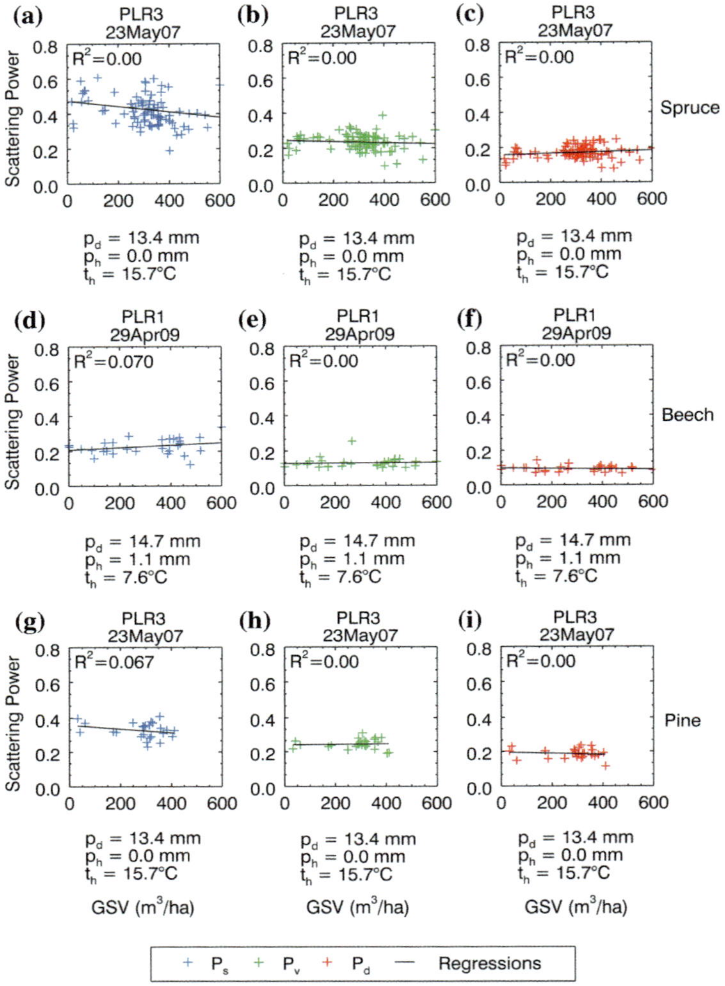

Fig. 5.26 *GSV* sensitivity analysis for different scattering mechanisms with ALOS PALSAR PLR HH 22° Asc. power (frames PLR1,3). On the basis of the Yamaguchi target decomposition algorithm, the power was decomposed in the surface (P_S), volume (P_V) and double bounce (P_d) scattering mechanism. P_S, P_V and P_d are shown as *blue*, *green* and *red*, respectively. The tree species spruce, beech and pine are depicted in the first, second and third *subplot lines*, respectively. Slopes of $<5°$ were selected for the three tree species compositions. Azimuth-slope corrections were included in the polarimetric decomposition

with surface, volume and double bounce on the first, second and third columns, respectively, and by lines with spruce, beech and pine species on the first, second and third lines, respectively. Forest stands with slopes below 5° were selected for all the cases. Also, the three decompositions were corrected for azimuth slopes. For the sake of clarity, the subplots depicted only one species and scattering mechanism.

Before interpreting the plots given in Fig. 5.26, it may be recalled that the PLR data was measured at a steep incidence angle ($\theta = 22°$). This implies that the results presented in Fig. 5.26 may not be directly comparable to the ones shown above, as FBD data was acquired with an incidence angle of 39°, and the scattering mechanisms with varying incidence angles may not be comparable (see incidence angle analysis in Sect. 5.2.2). First, observing the relative power of the three scattering mechanisms for each species in Fig. 5.26, it is noted that the surface dominates ($P_s = 0.2 - 0.5$), then volume ($P_v = 0.1 - 0.3$) and finally double bounce ($P_d = 0.1 - 0.2$). Thereafter, comparing the power between the different species, it is observed that pine and particularly spruce exhibit greater surface scattering power ($P_s = 0.4 - 0.5$) than beech ($P_s = 0.2$). According to the L-band frequency and steep incidence angle of the PALSAR PLR data, the high power for surface scattering relative to volume scattering may be conceivable. However, the greater surface contribution for spruce and pine relative to beech would signify greater penetration for spruce and pine compared to beech and would therefore contradict the results shown above in this section. These results can be interpreted in two different manners. Either the physical model is not suitable for the conifers or broadleaves of the Thuringian Forest, or higher surface scattering occurs for spruce and pine compared to beech. The Yamaguchi decomposition is a physically based scattering model which assumes oriented dipoles scatterers. On the basis of the VV/HH amplitude ratio, the orientation of the dipoles in the model is defined as random, horizontal or vertical (see Sect. 3.2). In the case of the Thuringian Forest, the VV/HH amplitude ratios of spruce, beech and pine were between −2 and 0 dB (see Fig. 5.21), signifying that the Thuringian Forest canopy principally consisted of randomly oriented dipoles. However, this assumption may not be suitable for the Thuringian Forest at L-band frequency, especially for spruce and pine, which essentially show horizontally structured branches along the trunk (see Sect. 4.1 and the study performed by Neumann et al. 2012). To gain an insight into the model conformity, the Yamaguchi decomposition was investigated on different landcovers such as forest, crop, urban and open areas, i.e. ROIs defined in Sect. 4.3.

An examination of the decomposed signal showed a surface scattering of about 0.2 for crop and open areas. It is the highest power between surface, volume and double bounce. Thus, comparing the surface scattering power between forested and open areas in Fig. 5.27, the surface scattering for forests were found to be greater than that for open area. This result is not realistic and shows that the assumptions made in the Yamaguchi model do not fit the Thuringian Forest and is therefore not applicable in this case. Assuming that the model is correct, the high surface scattering for spruce relative to beech may suggest that the forest canopy is scattering in the optical regime with scatterers of about 10 times the size of the wavelength (see Sect. 3.2). In other words, the forest canopy would be seen by the radar as a surface. In fact, this behaviour would be hard to envisage as the dominant scatterers in such a situation would require a size of at least $10 \times 0.23 = 2.3$ m, which is not conceivable for a forest. Thus, the results of the Yamaguchi decomposition must be carefully considered because (1) the incidence angle of the PLR data differs from the FBD data and (2) the model may not have been suitable for the different species structures of the Thuringian Forest. The obtained results should be examined in more detail in subsequent studies.

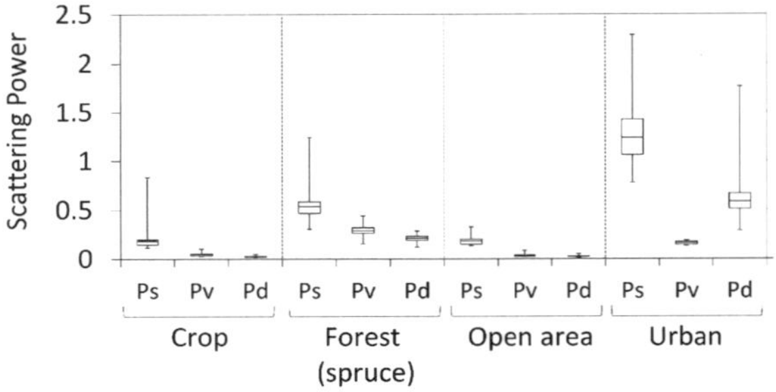

Fig. 5.27 Comparison of surface (P_s), volume (P_v) and double bounce (P_d) scattering mechanisms in various landcovers. The scattering mechanisms were decomposed on the basis of Yamaguchi target decomposition algorithm

5.2.4 Summary

The scattering mechanisms occurring in the Thuringian Forest were investigated using X-band and L-band radar data. After initial visualisation, comparisons of several radar parameters were established, including time-series, frequency, incidence angle, polarisation and pass direction. In the second step, some examinations regarding the relationship between radar backscatter and *GSV* were performed. The visual observations and investigations of the various parameters showed that conifer and broadleaves tree species present different signal magnitudes, thus allowing their discrimination. To this end, the leaf-off period, co-pol ratio X-band/L-band, positive temperature and large incidence angle appeared to be optimal. However, the separation of conifer species such as spruce and pine appeared to be limited using only radar backscatter intensity. The relationship between backscatter intensity and *GSV* was found to be poor for the three investigated species, even with L-band data. In addition, the trends depicted by the different species were shown to be different. In fact, the spruce backscatter signal tends to decrease, while that of beech tends to increase with increasing *GSV*. To understand the poor correlation and these different trends, further investigations were performed. Winter scenes were first considered because the temperature was shown to decrease the L-band SAR signal under frozen conditions in the time-series analysis. Comparing the backscatter intensity with *GSV*, it was observed that young regenerative forests responded to the frozen ground and canopy conditions resulting from the temperature decreasing to subzero values earlier than mature forests. The drop in the amplitude was also found to be species specific. From these observations, it was suggested that the penetration is more limited for spruce and pine compared to beech species and for young regenerative forests compared to mature forests. After studying the effect of temperature on backscatter intensity, the horizontal structure of the forest was examined. To this end, stem density and relative stocking parameters were selected in order to distinguish various stands' structures. The results for sparse and understocked spruce forests showed that the backscatter intensity increased with increasing *GSV* between 0 and 100 $m^3\,ha^{-1}$, which corresponded to the saturation point. Above this level, the backscatter decreased, particularly owing to the open canopy and the greater contribution of the ground. In the case of dense and fully stocked spruce forests, the amplitude remained relatively constant for the entire range of *GSV*. Owing to the management practices performed in the Thuringian Forest, the forest mainly consisted of dense stands. Thus, with reference to this particular forest structure, the obtained results supported the idea derived from the temperature analysis that in the Thuringian Forest the penetration for young spruce is generally more limited than the penetration for mature spruce forests. Benefiting from the ALOS PALSAR polarimetric datasets, PolSAR techniques were used to gain an insight into the scattering mechanisms occurring in the Thuringian Forest. After applying azimuth-slope corrections, Yamaguchi decomposition was performed. The results obtained from the polarimetric decomposition denoted the difficulty involved in understanding the scattering mechanisms that occur in the Thuringian Forest. In fact, the dominant contribution of surface scattering compared to double bounce and volume was first observed for the three investigated species. This result was explained by the steep incidence angle of the PLR data. Then, the surface scattering was found to be higher for spruce and pine compared to beech, suggesting less penetration for the broadleaves than for the conifers. As this result had not been expected, two possible interpretations were proposed: (1) the model is not suitable for the Thuringian Forest or (2) the model fits the forest and surface scattering occurs, which suggests that the forest canopy components are large enough to be seen by the radar as a surface. The first interpretation appeared to be more reasonable than the second one because comparisons of surface scattering between forested and open areas showed higher surface scattering contributions in the case of forested areas, and at L-band frequencies, for the canopy to act as a surface, it must contain scatterers which are at least 2.3 m long and large, which is also not realistic. Therefore, although it could not be fully confirmed, the different investigations led to the conclusion that the penetration through the Thuringian Forest canopies was more limited for dense forests compared to sparse forest and more so for spruce and pine relative to beech. To gain a better understanding of the scattering mechanisms, especially penetration in the forest, further analyses such as the use of phase information are expected to be useful (see Sect. 5.3). Texture parameters such as Haralick parameters (Haralick et al. 1973; Champion et al. 2011) or lacunarity (Allain and Cloitre 1991; Sun and Ranson 1998) may also bring additional information, especially concerning forest stand density. Finally, for the scatterers returning from the ground, the possible effects related to the understory and soil moisture were assumed to be of minor importance (see Sect. 4.1) and were therefore neglected.

5.3 Interferometric Phase and Coherence Analysis

5.3.1 Visual Observations

Similar to the backscatter intensity presented in the last section, a simple visualisation of the interferometric data provides a rich source of information. In Fig. 5.28, the

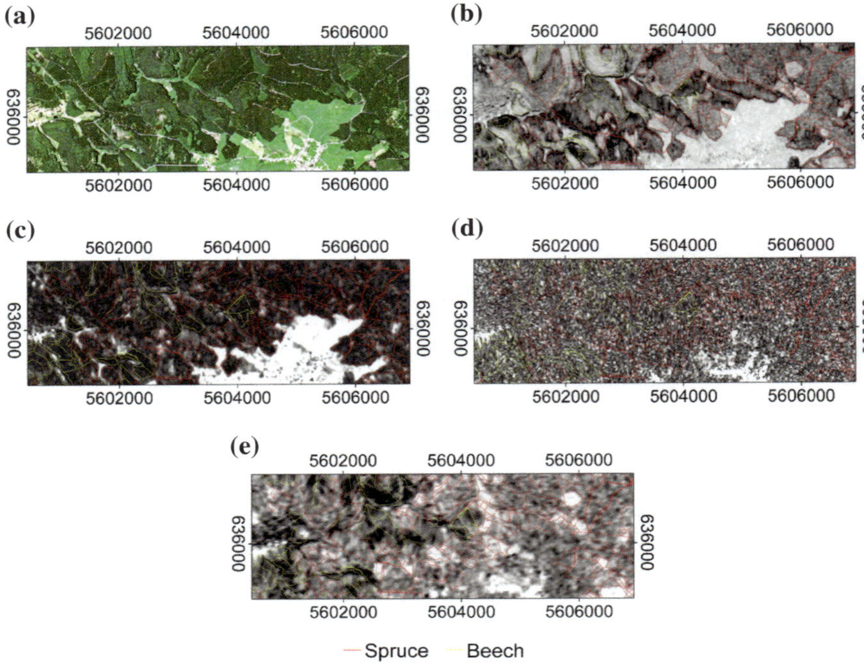

Fig. 5.28 Visualisation of interferometric coherence (γ) for (**b**) TDX SM HH 38° Asc. $B_n = 258$ m, **c** CSK SM Himage HH 50° Asc. $B_n = 296$ m, **d** TSX HS HH 23° Asc. $B_n = 209$ m and **e** ALOS PALSAR FBD HH 39° Asc. $B_n = 379$ m InSAR data (frames TDX3, CSK1, HS1 and FBD7). The TDX, CSK, TSX and PALSAR interferometric scenes were acquired on 30th August 2011, 30th/31st October 2010, 5th/16th July 2009 and 7th September/23rd October 2009, respectively. A RE optical image (RGB) acquired on 13th June 2009 is provided in (**a**) for comparison of the different landcovers. The coherence images are overlaid by tree species compositions obtained from the forest inventory. Spruce and beech are depicted in *red* and *yellow*, respectively. The coordinates are provided in the reference system UTM zone 32

interferometric coherence for TDX, CSK, TSX and PALSAR are compared using an optical image.

Some general comments can be made regarding the images depicted in Fig. 5.28. First, the microwave frequencies appear to play an important role. Comparing the L-band coherence from PALSAR with the X-band coherences from TDX, TSX and CSK, it is observed that even with a repeat pass of 46 days, the level of coherence for PALSAR remains relatively high compared to that for TDX, which is acquired in a single pass. In other words, phase is better correlated for low-frequency systems (L-band, P-band) than that in the case of high frequency instruments (X-band, C-band). This observation corroborates the InSAR theory presented in Sect. 3.2. Second, examining the repeat-passes of the X-band coherences, it is observed that the coherence increases when the time between the first and second acquisitions is reduced. In this respect, as expected, TDX shows the highest coherence with a single-pass acquisition, while TSX shows the lowest coherence with an 11-day repeat-pass.

Examining additional details regarding the different landcovers, it is observed that urban areas show very high coherence and are easily differentiated from other landcovers in the case of TSX and PALSAR. This indicates good stability of the scatterers in these areas. Open areas (grass) can also be easily distinguished with coherence in TDX and particularly in the CSK interferometric data, where the contrast between open areas and forest is pronounced. The high contrast between forests and open areas in CSK coherence reveal unstable scatterers such as leaves, needles, twigs and small branches of the trees, which are very sensitive to weather variations, particularly wind and precipitations. For the forests, when comparing the compositions of beech and spruce in the TDX data, the coherence for the broadleaves appears to be higher than that for the conifers, while the opposite is observed in the PALSAR interferometric data. For these two sensors, an intraspecies comparison depicts variations in coherence. These observations may reflect some different decorrelation mechanisms associated with the different canopy structures and development stages of the trees as well as the topographic effects described in Sect. 5.1. Some further analyses in this regard are presented in Sect. 5.3.3.

5.3.2 InSAR Parameter Signatures

Although some of the coherence properties of the test site could be introduced by a simple visualisation of the interferometric data in Fig. 5.28, factors such as weather conditions or the acquisition configuration of the sensor need to be considered for an optimal interpretation of interferometric coherence. These factors are discussed in further detail in this section.

5.3.2.1 Comparison of Interferometric Systems

To compare the magnitude of phase coherence derived from the different interferometric configurations in this study, diagrams including simple whisker boxplots and weather information were produced. Similar to the multi-angle backscatter analysis in Sect. 5.2, the boxplots were performed with the 25th and 75th percentile, each representing the lower and upper quartiles of the data. Indications of the median and the smallest and largest observations were also computed. Figures 5.29 and 5.30 present the plots for two different landcover surfaces, namely (a) open areas and (b) forested areas. Owing to the high number of available spruce stands over each acquisition relative to the other species, spruce was selected for representing the forested areas. Open areas are expected to highlight temporal decorrelation mechanisms, while in addition to the temporal component, spruce areas are expected to indicate the contribution of volume decorrelation caused by the vertical extension of the canopies. The different sensors and the perpendicular baseline of each interferometric acquisition are depicted on the primary (bottom) abscissa, and the respective acquisition dates are shown on the secondary (top) abscissa. For better readability, the acquisition

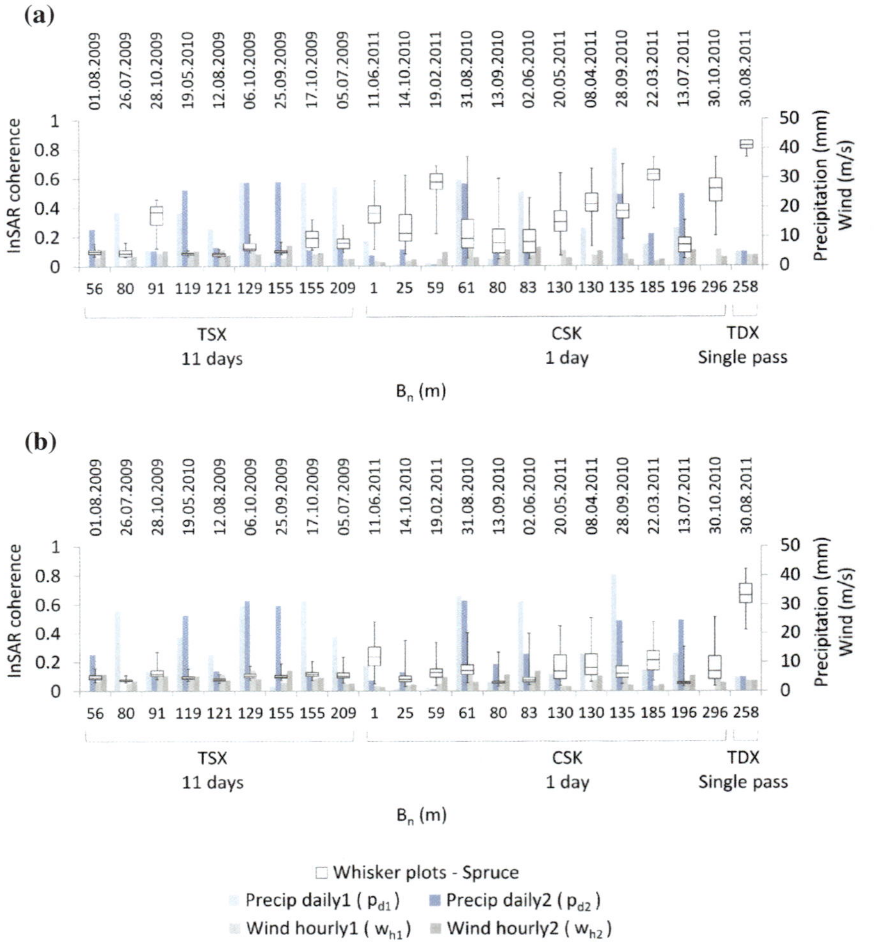

Fig. 5.29 Comparisons of interferometric coherence (γ) between **a** open areas and **b** forested areas with TSX HS HH 23°–48° Asc., CSK SM Himage HH 20°–50° Asc./Desc. and TDX SM HH 38° Asc. (frames HS1, CSK1 and SM3). The *boxplots* represent spruce with the sample minimum, lower quartile (Q1), median (Q2), upper quartile (Q3) and sample maximum. Weather information is depicted in *bright blue*, *dark blue*, *bright grey* and *dark grey* for daily precipitations acquisition 1 ($p_{\mathrm{d}1}$), daily precipitations acquisition 2 ($p_{\mathrm{d}2}$), hourly wind acquisition 1 ($w_{\mathrm{h}1}$) and hourly wind acquisition 2 ($w_{\mathrm{h}2}$), respectively. Slopes of <5° were selected to limit topographic effects

dates refer only to the first acquisition of the coherence pairs. The interferometric coherence and weather information are displayed on the primary (left) and secondary (right) ordinates, respectively. The weather parameters shown on the graphs reflect the amount of precipitation and wind speed 4 days and 4 h before the radar acquisitions, respectively (see Sect. 4.3). The plots are sorted by increasing the baseline

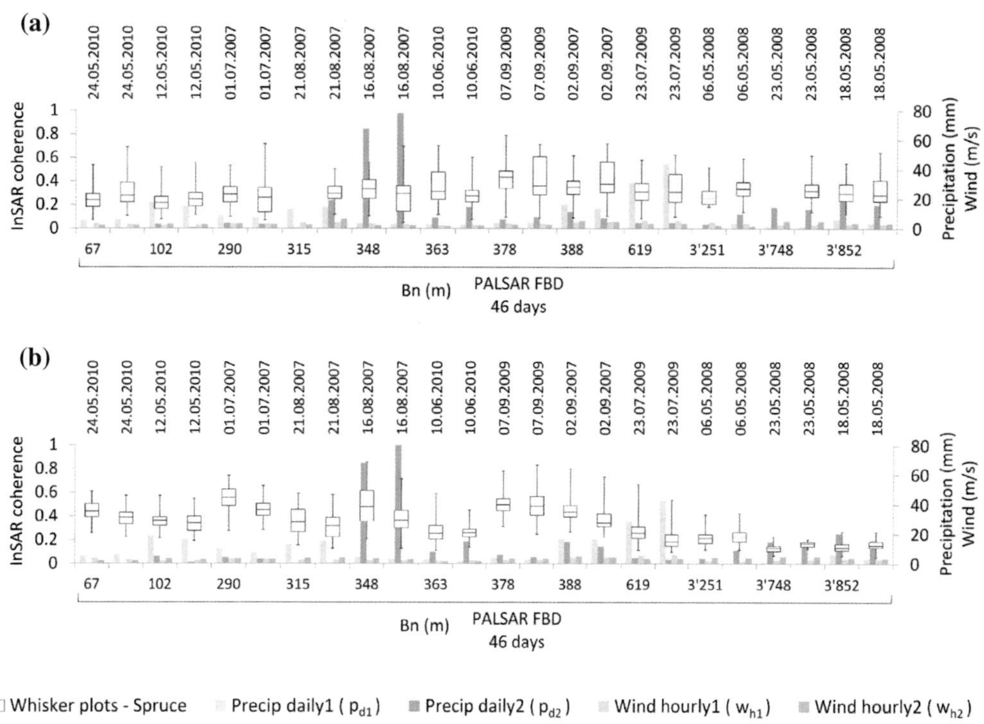

Fig. 5.30 Comparisons of the interferometric coherence (γ) between **a** open areas and **b** forested areas with ALOS PALSAR FBD HH 39° Asc. (frames FBD1–8). The *boxplots* represent spruce species with the sample minimum, lower quartile (Q1), median (Q2), upper quartile (Q3) and sample maximum. Weather information is depicted in *bright blue*, *dark blue*, *bright grey* and *dark grey* for daily precipitations acquisition 1 (p_{d1}), daily precipitations acquisition 2 (p_{d2}), hourly wind acquisition 1 (w_{h1}) and hourly wind acquisition 2 (w_{h2}), respectively. Slopes of <5° were selected to limit topographic effects

so that the potential influence of this parameter may be considered. Also, to limit spatial decorrelation due to topography (see Sect. 5.1), open areas and spruce stands with slopes below 5° were selected. Finally, for the interpretations, the compared data may represent different frames, i.e. potentially different forest stands, and for the X-band sensors, the acquisitions may also depict various incidence angles.

First, by examining the boxplots of the different sensors on the diagram for open areas (plot (a) from Fig. 5.29), it is generally observed that with an 11-day repeat-pass, the signal is mostly decorrelated for TSX. Coherence values generally remain below 0.2, except for one specific interferometric acquisition (28th October 2009) where coherence reaches 0.4. In this case, few daily precipitations ($p_d = 4$ mm) and small hourly wind speeds ($w_h = 4\,m\,s^{-1}$) were measured for both radar images. This precipitation and wind speed did not appear to be sufficiently significant to affect the coherence of the phase. The boxplots given for CSK on an open area show that with a 1-day temporal baseline, the dispersion of the interferometric coherence generally increases to 0.6 in some cases. However, the temporal decorrelation for CSK appears to still be significant, and in addition to TSX, the coherence of CSK pairs with values below 0.2 can be observed. Generally, it is remarked that pairs with high coherences ($\gamma = 0.5 - 0.6$) such as 19th/20th February 2011, 22nd/23rd March 2011 and 30th/31st October 2010 showed low values of daily precipitations ($p_d < 11$ mm) and wind speed ($w_d < 4.8$ mm). The acquisition pair of 22/23nd March 2011 represented daily precipitations ($p_{d1} = 7.3$ mm, $p_{d2} = 11.0$ mm) but no hourly precipitations ($p_{h1} = 0.0$ mm, $p_{h2} = 0.0$ mm). In other words, precipitations occurred during the 4 last days before this radar acquisition, but no precipitations were measured during the 4 last hours, which explains why the coherence remains to be as high as the data recorded on 19th February 2011. The CSK interferometric acquisitions showing low coherences ($\gamma = 0.2$) were all affected by consequent daily and in some cases, hourly precipitations (31st August 2010, 2nd June 2010, 28th November 2010). For the TDX interferometric acquisition, the coherence is very high, with the median around 0.81, and the dispersion together with the 1st and 3rd quantiles ranging between 0.80 and 0.83 is extremely small. As for single-pass data, the comparison of TDX coherence with TSX and CSK coherence highlights the temporal decorrelation occurring in repeat-pass systems. The PALSAR data in Fig. 5.30a show that the L-band 46-day repeat-pass coherence over open areas generally ranges between 0.2 and 0.4, with some maximas reaching 0.7. The relationship of the weather data with L-band coherence was less evident than that with X-band coherence. In fact, high precipitations or winds did not always appear to induce decorrelation. A comparison of summer (FBD) and winter (FBS) acquisitions also did not show any significant trend. The boxplots processed for the FBS and PLR acquisitions are provided in Appendix B.3.

The boxplots processed for spruce stands (plots (b) from Fig. 5.29) indicate that for the TSX interferometric scenes, the signal is totally decorrelated, even for the scene acquired on 28th October 2009, where the weather conditions were shown to be favorable to less temporal decorrelation over open areas. A comparison of the CSK and TDX forest boxplots with the ones presented for open areas in Fig. 5.29a, b, respectively, indicates a decrease in coherence for the spruce stands. The decreased coherence appears to be more significant for CSK ($\Delta\gamma = 0.2 - 0.4$) than for TDX ($\Delta\gamma = 0.1$). In fact, as TDX interferometric pairs are measured in single pass and the topographic effects are potentially removed with slope selection, volume decorrelation is the only contribution to the TDX decorrelation mechanisms, while CSK acquisitions with a 1-day repeat-pass show volume decorrelation as well as temporal decorrelation. If volume decorrelation occurs at X-band frequency, a decrease in InSAR coherence should be observed with an increase in perpendicular baseline. However, in the present case, the temporal decorrelation in TSX and CSK appeared to be too significant to highlight the sensitivity of forest canopies to normal baselines. Considering PALSAR boxplots in Fig. 5.30b, higher variations can be observed relative to the boxplots given for open area in Fig. 5.30a. These variations are not easily interpreted from the FBD interferometric acquisitions. A significant decorrelation can be observed after a normal baseline of 3,251 m. In fact, this distance represents about 40 % of the critical baseline, which is generally sufficient for high decorrelation of the signal. In contrast to the X-band sensors, PALSAR coherence over forested areas in several cases (i.e. 24th May 2010, 12th May 2010, 1st July 2007, 16th August 2007) appeared to be higher than that over open areas. In particular, the coherence for forest appeared to be higher than that for open area during summer acquisitions (FBD) with a low perpendicular baseline ($B_n < 300$ m). Three reasons mainly explain this observation. First, the scattering properties of open areas may have been affected to a greater extent by weather conditions such as rainy precipitations than forested areas. This could be the case if the soils were water-logged or show high humidity in one of the acquisition pairs. Thus, high soil moisture tends to decrease the coherence for open areas. Second, with a smaller perpendicular baseline, volume decorrelation tends to be more limited for forests. Thus, a small normal baseline preserves high coherence for forested areas. Third, the noise depicted by forest is negligible compared to open areas (see Sect. 4.3). Thus, the coherence for open areas is affected to a greater extent by noise decorrelation than forests.

Although it is not always convenient to determine the relation between weather information and interferometric coherence, especially at L-band frequency, a single InSAR coherence scene may locally capture some meteorological phenomena, such as extreme rainfall. In Fig. 5.31, PALSAR interferometric coherence is shown with a precipitation map. This latter represents the summation of precipitations over 4 days occurring before the acquisition undertaken on 23rd July 2009.

A dark stripe corresponding to low coherence can be easily distinguished in Fig. 5.31. This coherence scene is the result of two radar acquisitions taken on 23rd July 2009 and 7th September 2009. Investigations showed that the high decorrelation is a source of temporal decorrelation which referred to a convective precipitation event that occurred during the first acquisition on 23rd July 2009. Indeed, this date corresponds to a specific configuration of cold front air masses over Europe, which led to violent storms and hail precipitations. According to the processed weather data, the weather stations measured precipitations of more than 40 mm during the 4 days before this date in the North of the Thuringian Forest, while in comparison, only 20 mm precipitations were measured in the South during the same period. Also, only on 23rd July 2009, about 20 mm of precipitation was measured in the North and

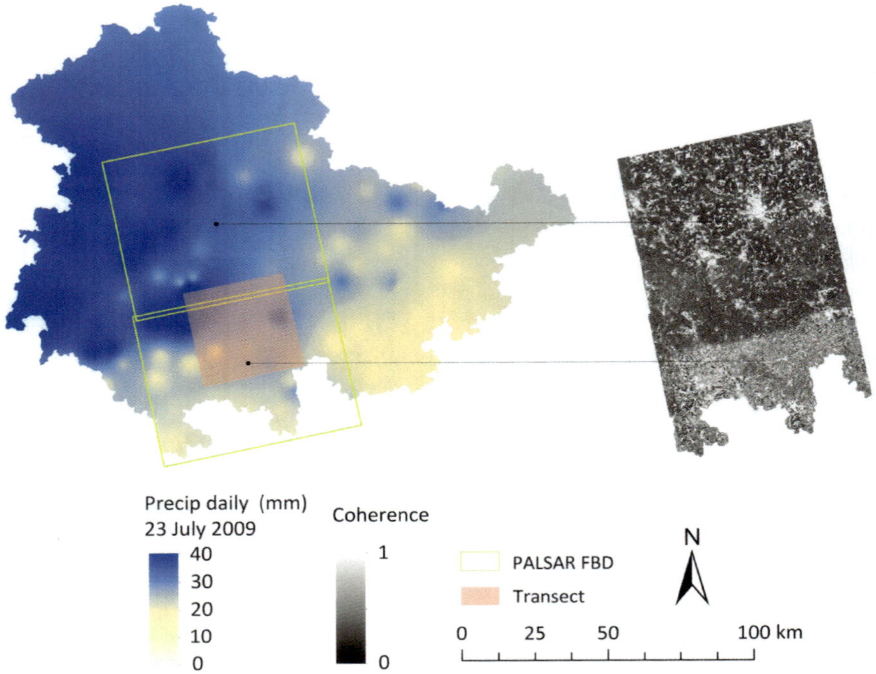

Fig. 5.31 Effects of precipitations on interferometric coherence (γ) measured by ALOS PALSAR FBD HH 39° Asc. $B_n = 620$ m InSAR data (frames FBD3,7). The PALSAR interferometric scenes were acquired on 23rd July/7th September 2009. The precipitation map represents the summation of 4 days' precipitations occurring before the radar acquisition on 23rd July 2009. The acquisition frames are depicted in *yellow* and the transect discussed in Fig. 5.32 in *red*

10 mm in the South (see also the climatic diagrams in Appendix A.1). In contrast to the first radar acquisition, only 2–3 mm precipitations were measured for the 4 days before the second radar acquisition. To gain a deeper insight into the given coherence and precipitation values, a 2D transect was drawn (see Fig. 5.31), and the underlying data values were plotted in the form of a histogram. Figure 5.32 depicts the histogram with azimuth values on the abscissa, interferometric coherence on the primary (left) ordinate and precipitations on the secondary (right) ordinate. The transect values represent the averaged values of the data in the range direction.

As shown in Fig. 5.32, the rain/hail precipitations gradually decreased from about 40 mm to approximately 26 mm, while the PALSAR interferometric coherence increased from 0.2 to 0.35. Some maximas and minimas can be distinguished in the coherence signal. These extremes are mostly explained by regions showing high coherence such as urban areas and low coherence such as forested areas or topographic areas facing away from the radar (see Sect. 5.1).

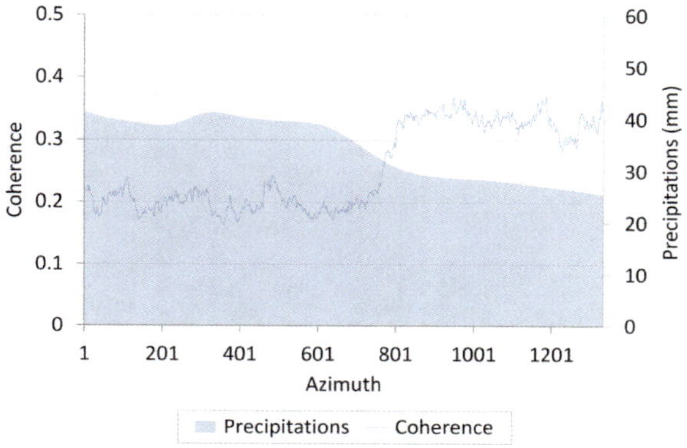

Fig. 5.32 Effects of precipitations on interferometric coherence (γ) measured by ALOS PALSAR FBD HH 39° Asc. $B_n =$ 620 m InSAR data (frames FBD3,7). The PALSAR interferometric scene was acquired on 23rd July and 7th September 2009. The digital numbers represent the values from the transect depicted in Fig. 5.32

5.3.2.2 Multipolarisation

The analyses of the different interferometric datasets presented above focused on HH polarisation, as most of the available radar data were acquired in this polarisation. However, TSX and PALSAR also provided other polarisations such as cross-polarisation HV for the PALSAR FBD mode.

HH and HV polarisations from PALSAR datasets were examined in order to describe the potentially different decorrelation mechanisms occurring in these two configurations. Owing to the significant temporal decorrelation shown by TSX, the coherence of this sensor for different polarisations was not investigated. Comparisons of PALSAR polarisations over the forested areas are presented in Fig. 5.33. The figure depicts several scatterplots with PALSAR HH coherence on the abscissa and PALSAR HV InSAR coherence on the ordinate. The main outcomes were shown to be similar for the three investigated species. Thus, only spruce species were selected for the illustration. The scatterplots processed for the other species are provided in Appendix B.3. Also, only interferometric acquisitions from frame FBD3 were selected because this frame (see Appendix A.1) covered most of the spruce forest stands. The plots for the remaining acquisition frames are also provided in Appendix B.3. Finally, areas with steep slopes (slope of >5°) were removed in this analysis, as they were shown to decorrelate the radar signal (see Sect. 5.1).

The plots shown in Fig. 5.33a compare the HH and HV components of PALSAR coherence. It is observed that these components are slightly different. In fact, referring to the 1:1 line, the clouds of points have the same orientation as that of the line with a little eccentricity towards the right, which indicates a slightly higher coherence for the HH component. The difference between HH and HV is about 0.1. This observation suggests that the microwaves see different targets, and the phase center is at a different position for the two polarisations. More specifically, the phase center for the HH and HV components should both be located in the canopy with a slight difference for HH, which should be prone to less extinction than HV polarisation. The plots in Fig. 5.33c, d referring to 6th May/21st June 2008 depict a few outlier points. These

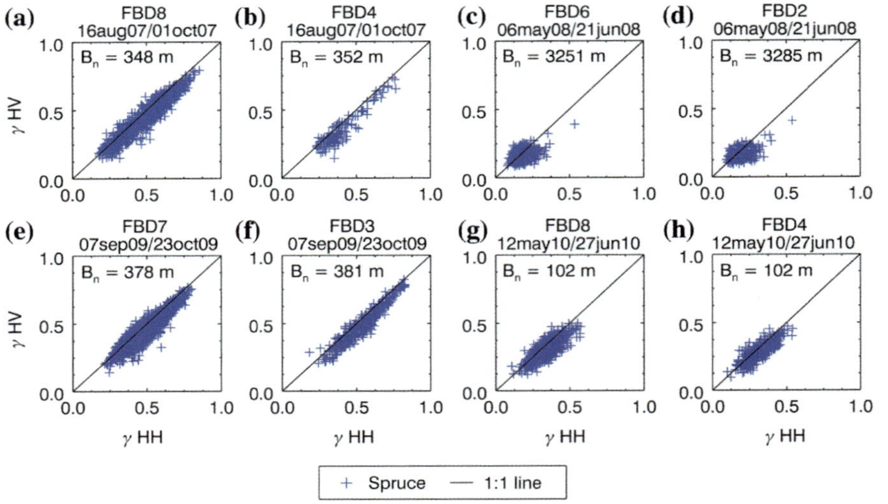

Fig. 5.33 Polarisation analysis with PALSAR FBD HH 39° Asc. versus PALSAR FBD HV 39° Asc. interferometric coherence (γ) (frames FBD2–4 and FBD6–8). *Blue points* represent Norway Spruce forest stands. Slopes of <5° were selected to limit topographic effects

outliers were found to be related to felled stands which were not updated in the inventory or very young regenerative stands.

In addition to HH and HV polarisations shown in Fig. 5.33, the VV components were investigated using the PALSAR PLR data. Figure 5.34 depicts the scatterplots comparing the HH and VV polarisations.

As shown in the subplots of Fig. 5.34, the InSAR coherence appeared to be well aligned on the 1:1 line. This observation suggested that the coherences are equivalent in the two like-polarisations. As with FBD coherence in Fig. 5.33, the outliers showing high HH and VV coherences were related to young regenerative trees. The few outliers showing relatively high HH ($\gamma > 0.4$) but low VV ($\gamma < 0.3$) coherence were shown to be mature stands. No particular reason could explain the deviation of the coherence of these stands from the overall trend.

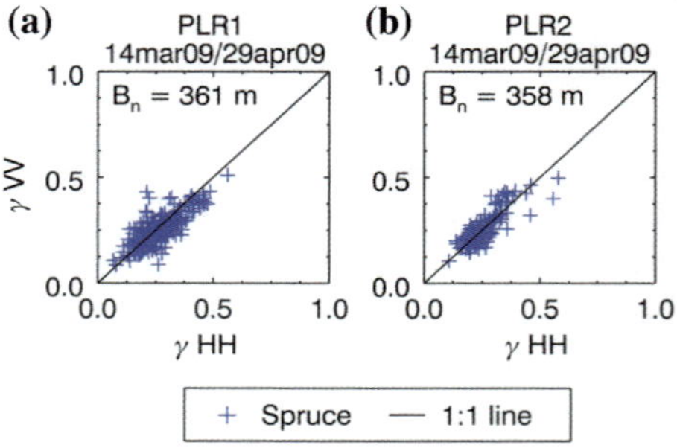

Fig. 5.34 Polarisation analysis with PALSAR PLR HH 22° Asc. versus PALSAR PLR VV 22° Asc. interferometric coherence (γ) (frames PLR1,2). *Blue points* represent Norway Spruce forest stands. Slopes of <5° were selected in order to limit topographic effects

5.3.3 *Sensitivity of InSAR Coherence to Forest GSV*

Several studies have shown that interferometric coherence is closely related to forest biomass (see Sect. 2.1). In this framework, after analyzing the parameters which potentially influence phase coherence, some investigations were performed in order to underline the potential of coherence for estimating forest biomass in the Thuringian Forest. Similar to the case of backscatter intensity presented in Sect. 5.2, *GSV* was the main parameter of interest for these investigations.

5.3.3.1 Perpendicular Baseline

The first analysis involved the examination of the relationship between phase coherence and *GSV* together with perpendicular baseline. In this regard, simple scatterplots, each of which represents one interferometric acquisition, were processed and sorted by the increasing normal baseline. Figures 5.35 and 5.36 depict the scatterplots processed for the CSK and PALSAR data, respectively, with *GSV* on the abscissa and interferometric coherence on the ordinate. For the sake of clarity, the plots show only Norway Spruce, as this species was well represented over all datasets. According to the results described in the last section and Sect. 5.1, precipitations as well as steep slopes may affect the interferometric coherence. To limit the potential influence of these effects as much as possible, only interferometric acquisitions with daily (p_{d}) and hourly (p_{h}) precipitations below 25 and 1 mm, respectively, were selected in addition to forested stands with slopes below 5°. It should also be noted that for a better comparison between the different spatial baseline configurations, in the case of PALSAR sensor (Fig. 5.36), only the FBD data were selected, and in the case of CSK sensor (Fig. 5.35), only acquisitions measured in the ascending mode at an incidence angle of 49.7° were retained. The reader is referred to Appendix B.3 for the plots processed for the overall acquisitions.

The CSK data in Fig. 5.35 indicate that interferometric coherence is slightly sensitive to *GSV* in some of the acquisition configurations. Indeed, some of the scatterplots depict a linear or exponential decrease in coherence with an increase in *GSV*. However, the relationship between the different acquisitions with their respective baseline configurations appeared to remain unclear, as some acquisitions with low baseline (i.e. $B_{\mathrm{n}} = 1$ m) depicted a better R^2 than the ones with a higher baseline (i.e. $B_{\mathrm{n}} = 25$ m, $B_{\mathrm{n}} = 196$ m). The weather may play a significant role in the mentioned scenario. For example, considering the acquisition pair 13th/14th July 2011 where the correlation is null, it is observed that very few hourly ($h_1 = 0.0$ mm, $h_2 = 0.3$ mm) but many daily precipitations ($p_{\mathrm{d}1} = 13.1$ mm, $p_{\mathrm{d}2} = 24.6$ mm) were registered. The daily precipitations affected the radar acquisitions, thus affecting the interferometric coherence. The highest coefficient of determination ($R^2 = 0.51$) is shown for the interferometric pair measured on 30th/31st October 2010, which was acquired with a normal baseline of 296 m. This baseline was the largest available baseline for the CSK interferometric dataset (see Appendix A.2). Further interpretations of the

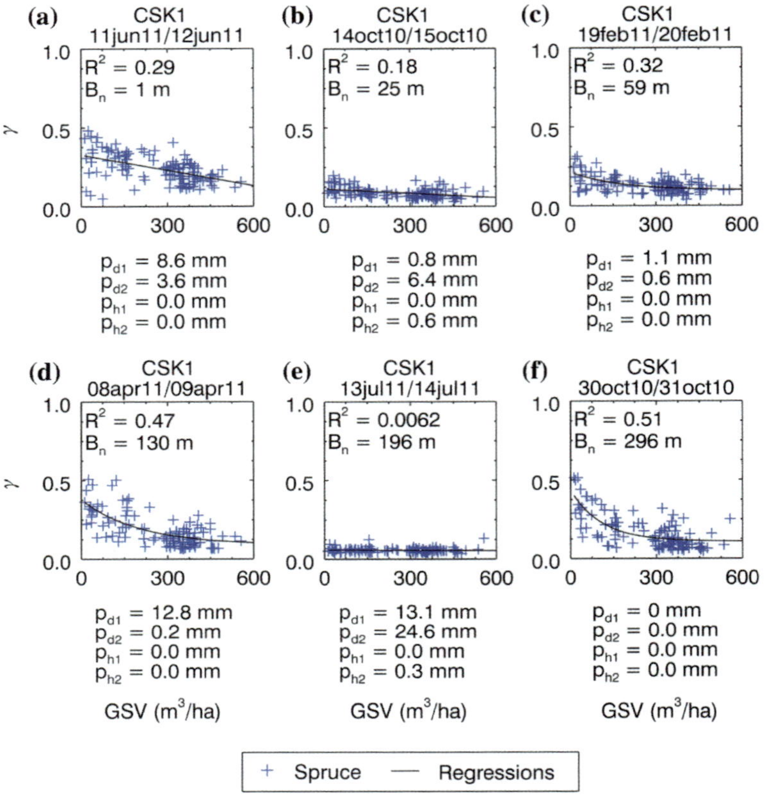

Fig. 5.35 *GSV* sensitivity analysis for different perpendicular baselines with CSK SM Himage HH 50° Asc. interferometric coherence (γ) (frame CSK1). *Blue points* represent Norway Spruce. Slopes of <5° were selected to limit topographic effects

relation coherence versus *GSV* are discussed below with the analysis of PALSAR coherence results.

The PALSAR FBD interferometric acquisitions in Fig. 5.36 are sorted by increasing normal baseline. Interferometric pairs showing twice the number of acquisition dates were measured on the same day over different frames. The plots indicate that similar to the case of CSK, PALSAR coherence may be related to *GSV* with a negative linear or exponential relationship. Additionally, the plots show that the sensitivity of coherence to *GSV* increases with increasing normal baselines until a spatial baseline distance is attained, after which the signal is totally decorrelated. These observations may require some physical explanations. In fact, the negative coherence trends are governed by two main decorrelation mechanisms, namely temporal and volume decorrelations. Temporal decorrelation may be characterised by the stability of the trees. In the case of young regenerative forests, owing to their rigid structure, in

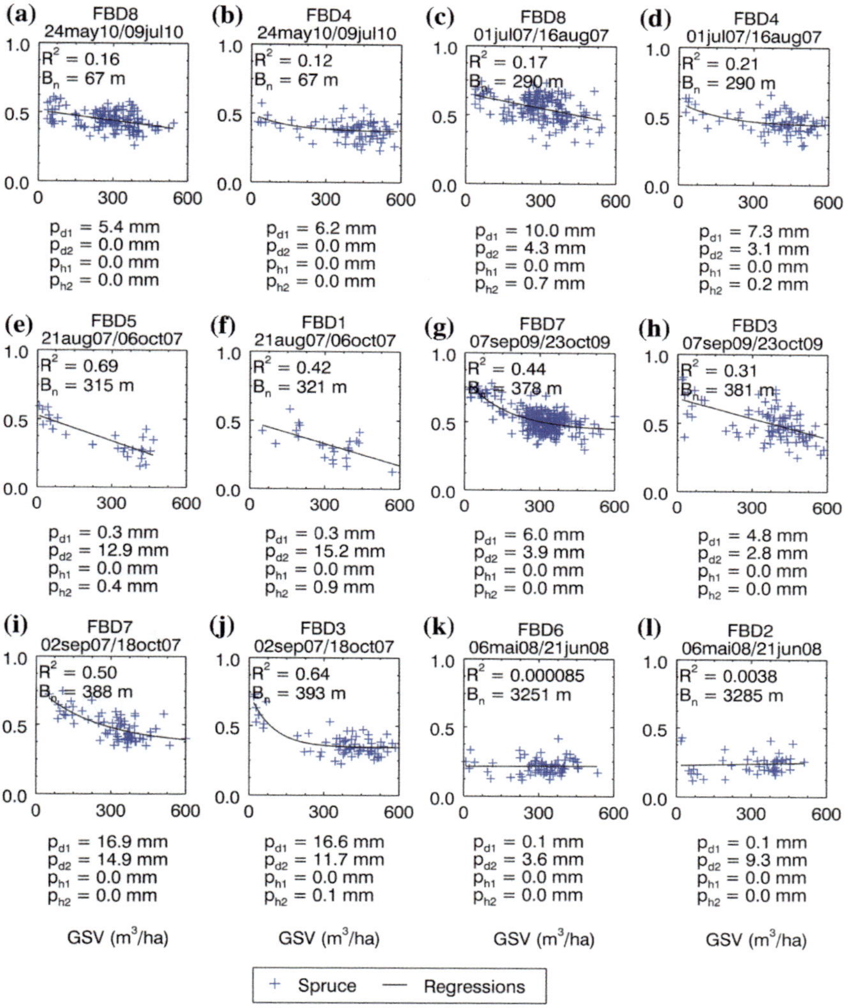

Fig. 5.36 *GSV* sensitivity analysis for different perpendicular baselines with ALOS PALSAR FBD HH 39° Asc. interferometric coherence (γ) (frames FBD1–8). *Blue points* represent Norway Spruce. Slopes of <5° were selected to limit topographic effects

windy conditions, the trees tend to show high stability and therefore relatively high coherence, while in the case of mature forests, under the same weather conditions, the trees tend to depict oscillations and therefore relatively low coherence owing to their high vertical extension. Following the above discussions on comparison of open areas with forests in Fig. 5.30, the higher PALSAR coherence for forest compared to open area may be even more pronounced when considering young spruce forests such as that in Fig. 5.36. This result may suggest that the scatterers for young spruces are consistently more stable than that for open area. The volume decorrelation may be related to the different volumes shown by young and mature forests. Larger and higher tree canopies exhibit greater volume causing greater volume decorrelation. Volume decorrelation may also be closely related to the perpendicular baseline of the interferometric system configuration. Briefly, the increasing baseline results in a greater difference in the incidence angle between the two radar acquisitions. This difference may also be directly related to the path of the microwaves taken into the forest canopy. The greater the angle, the greater is the probability that the microwaves follow a different path into the canopy, covering a larger part of the volume. Thus, the greater the normal baseline, the greater the sensitivity of the microwaves to the volume of the canopy. However, as observed above, the perpendicular baseline can be limited. According to the theoretical critical baseline of 7.8 km calculated for the PALSAR FBD data in Sect. 4.2 and the measured normal baseline of 3,251 m for the PALSAR acquisition 6th May/21st June 2008 for FBD6, about 40 % of the critical baseline is sufficient to significantly decorrelate the signal. Upon observation of the R^2 statistics, it is noted that the best correlations are shown for the large normal baselines (B_n = 300–3,251 m), while low correlations are depicted for the small baselines (B_n = 0–300 m). These results are explained by the higher sensitivity of large baseline interferometric systems to the volume of the canopies. The highest R^2 coefficients shown in Fig. 5.36 reach 0.64 for the acquisition 2nd September/18th October 2007 on frame FBD3 and 0.69 for 21st August/6th October 2007 on frame FBD5.

5.3.3.2 Species Comparisons

Till now, the InSAR coherence investigations have focused on Norway Spruce because this species was dominant in the different datasets. In addition to spruce, it may be interesting to investigate the other main species of the test site (beech and pine), as they show specific characteristics and may exhibit particular decorrelation mechanisms. In this context, spruce, beech and pine were compared and their relationship with *GSV* was analysed. Figure 5.37 presents three different scatterplots representing (a) CSK, (b) TDX and (c) PALSAR interferometric acquisitions, respectively. The plots of CSK and PALSAR were constructed on the basis of their perpendicular baselines, which were shown to be the most optimal ones from the available datasets in Figs. 5.35 and 5.36, respectively. *GSV* is shown on the scatterplots on the abscissa and the phase coherence on the ordinate. For each plot, comparisons are drawn for spruce, beech and pine. For each species in the different plots, a specific slope criterion was defined in order to prevent the effects of topography. In this respect, spruce, beech and pine forest stands with slopes below 5°, 10° and 5°, respectively, were selected. These thresholds consisted of a good compromise between the influence of topography and the remaining number of stands.

The species comparison for the CSK 1-day repeat-pass in Fig. 5.37a indicates that the coherence of spruce is higher than that of beech in the range 0–300 $m^3 ha^{-1}$. Above 300 $m^3 ha^{-1}$, the coherence appears to be equivalent for the different tree

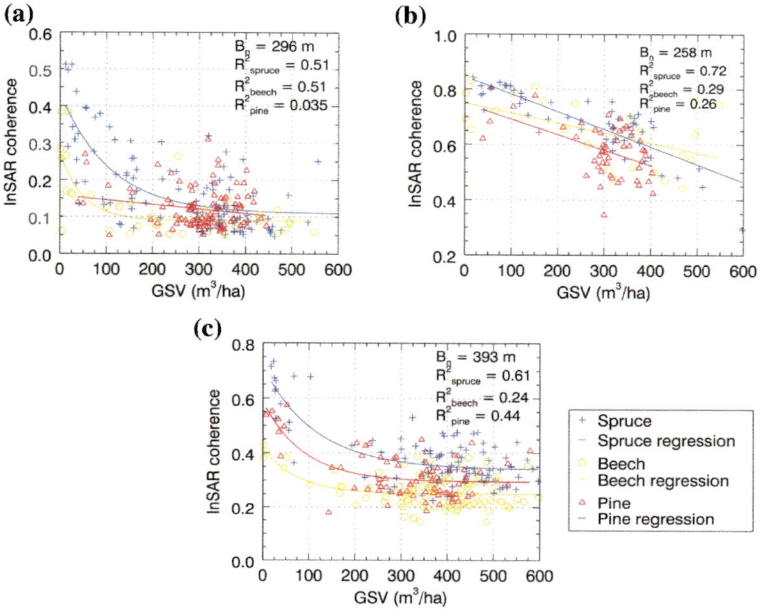

Fig. 5.37 *GSV* sensitivity analysis for tree species composition with **a** CSK SM Himage HH 50° Asc. $B_n = 296$ m, **b** TDX SM HH 38° Asc. $B_n = 258$ m and **c** ALOS PALSAR FBD HH 39° Asc. $B_n = 393$ m interferometric coherence (γ) (frames CSK1, TDX3 and FBD3). The CSK, TDX and PALSAR interferometric scenes were acquired on 30th/31st October 2010, 30th August 2011, 2nd September/18th October 2007, respectively. The tree species spruce, beech and pine are shown as *blue*, *yellow* and *red*, respectively. Slopes of <5° were selected in order to limit topographic effects

species compositions. The higher coherence shown by spruce can be explained by the different foliage compositions, namely needles for spruce and leaves for beech. Branches composed of leaves are generally well exposed to wind, while branches structured by needles are normally more stable in windy conditions. In other words, temporal decorrelation is more pronounced in broadleaves than in conifer forested areas. The plots depicting pine stands are difficult to interpret because a lack of stands appears in the first 200 m^3 ha^{-1}. The given relationship for spruce and beech appears to follow a nonlinear law expressed in this case by an exponential function. The coefficients of determination suggest that spruce and beech have the same correlation with $R^2 = 0.51$. However, the saturation appears to occur at different levels, namely around 100 and 250 m^3 ha^{-1} for beech and spruce, respectively. Considering the TDX single-pass acquisition in Fig. 5.37b, it can be observed for the three investigated species compositions that the level of coherence is much higher than that for the CSK interferometric configuration. The range of TDX coherence values does not differ from the one introduced for spruce in Fig. 5.29. For the three species, the interferometric coherence decreases linearly with an increase in *GSV*. It is also observed that the X-band single-pass coherence does not reach saturation.

The best correlation is shown by spruce, followed by beech and then pine, with R^2 values of 0.72, 0.29 and 0.26, respectively. To date, the high R^2 for *GSV* of spruce and TDX coherence is the best result obtained in the present study. The low R^2 for pine is explained by the large dispersion of the data, especially around 300 m^3 ha^{-1}. The last scatterplot in Fig. 5.37 illustrates the relationship between L-band PALSAR interferometric coherence and *GSV* for the three investigated species. The level of coherence appears to clearly differ for the three species with spruce showing the highest coherence, followed by pine and finally beech. As mentioned above, beech is more sensitive to temporal decorrelation owing to the structure of its branches and the composition of its foliage. In the case of spruce and pine, other explanations must be given for understanding the observed difference of $0.1 - 0.2$ in amplitude phase coherence. In fact, although they are both conifers, the branching arrangements of spruces are different from those of pines. Schematically, spruces can be considered as conical structures, while pines can be considered as trapezoidal structures (see Section test site 4.1 and Appendix A.1). The latter is clearly less consistent and is therefore less aerodynamic than the conical shape exhibited by spruces. In other words, the branching structure of spruce enables increased stability and particularly better resistance to windy conditions. The correlation results obtained for PALSAR coherence are $R^2 = 0.61$, 0.24 and 0.44 for spruce, beech and pine, respectively. The saturation point appeared at around 200 m^3 ha^{-1} for the three different species but varied significantly according to the perpendicular baseline. Thus, the observed values are only an indication and may considerably change from one interferometric acquisition to another.

5.3.4 Loci Coherence

Canopy penetration and ground visibility were discussed in Sects. 5.1 and 5.2. In particular, it is shown that the penetration of L-band microwaves in the Thuringian Forest canopies increases in specific conditions such as the presence of steep slopes facing the SAR platform or steep incidence angles of the SAR measurements. According to Sect. 3.2, another indicator which may provide additional information about the penetration of microwaves is the variation in the interferometric phase. Phase variations may be investigated using polarimetric interferometric SAR systems by combining phase difference resulting from various polarisations and two interferometric acquisitions. The greater the phase variation, the greater is the vertical distance among scattering mechanisms returning from the forest.

ALOS PALSAR performed two PolInSAR acquisitions over the Thuringian Forest. In this framework, the ground visibility through the forest canopies of the Thuringian test site could be further studied using PolInSAR techniques. To investigate the ground contribution from the SAR signal, loci coherence representations were considered. The loci coherence was depicted as a region on a unitary complex diagram. The region represents the analysis results for complex coherences computed for all polarizations on the basis of a single averaging window (see Eq. (3.67) in Sect. 3.2). In this regard, selection of one pixel corresponding to a specific area was undertaken for each loci diagram and a smoothing window of 7 × 7 pixels was used for the estimation of coherency matrices.

A key point in the investigation of the phase is the magnitude of coherence. According to Eq. (3.34) in Sect. 3.2, the decrease in coherence increases the standard deviation of the phase. Unfortunately, in addition to the volume decorrelation for mature forests discussed above, the investigated PALSAR PLR data appeared to be affected by significant temporal decorrelation, which may have been related to the particularly high precipitation that occurred before and during the two interferometric acquisitions ($d_1 = 18.9$ mm, $d_2 = 17.3$ mm). Thus, in most cases, the observed coherence ranged between 0.1 and 0.5, which is not high enough to allow reliable interpretations of the phase. Although the analysis of the loci diagrams was limited, the presence of two plots with a coherence higher than 0.5 could be considered for the analysis. These plots referred to open areas and young dense spruce forests. Illustrations of the performed loci plots for open areas and young dense spruce forest are given in Fig. 5.38a, c and b, d, respectively. The two plots per class represent

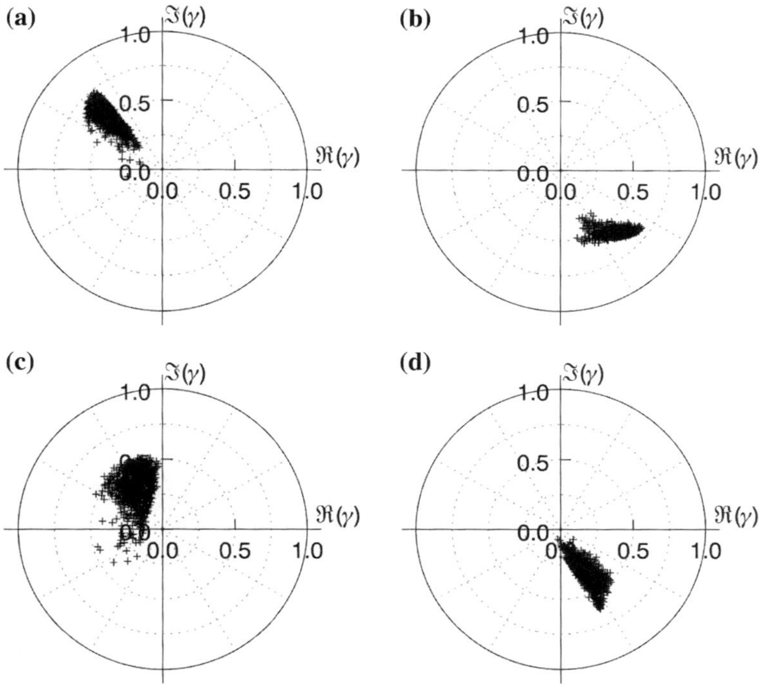

Fig. 5.38 Loci coherence analysis for **a**, **c** open area and **b**, **d** young dense spruce with ALOS PALSAR PLR 22° Asc. $B_n = 359$ m complex interferometric coherence ($\tilde{\gamma}$) (frame PLR2). The PALSAR polarimetric interferometric scene was acquired on 14th March/19th April 2009. Each point describes the complex coherence for one polarisation state. The height of ambiguity (h_a) is approximately 90 m

two different selected pixels. The pixels in flat areas were chosen in order to limit the effects of topography. The reader is referred to Appendix B.3 for information regarding the remaining processed loci coherence plots.

The four examples presented in Fig. 5.38 were chosen to illustrate the various results found from the loci coherence plots. In the case of open areas, it is expected that there are minor phase variations with high coherence. With a coherence higher than or equal to 0.5, the phase difference variation $\Delta\phi$ was found to be generally smaller than $\pi/6$. However, as shown with the phase angle of about $\pi/3$ in Fig. 5.38c, this was not always the case. As regards young spruce forests, assuming that the open area and the ground beneath the canopy depict similar scattering characteristics, it is expected that with high coherence, there would be a phase variation of at least $\pi/6$. The phase variation for a coherence value greater than 0.5 was found to be larger (as in (b)), but it was also smaller than $\pi/6$. According to these observations, although some data with moderate coherence could be depicted, it appeared that with the given PolInSAR dataset, it was difficult to draw any clear conclusion about the relative canopy penetration using loci coherence.

5.3.5 Summary

This section examines the decorrelation mechanisms which occur in the Thuringian Forest. For this, the TDX single-pass interferometric system as well as the CSK, TSX and ALOS PALSAR repeat-pass systems were considered with temporal baselines of 1, 11 and 46 days, respectively. Simple visualisation comparisons of interferometric coherence were first performed. They highlighted the various temporal and potential spatial decorrelations which occurred over the forested areas and for other landcovers in the different interferometric configurations. After visualizing the data, several quantitative analyses of the parameters potentially affecting phase coherence were undertaken. Comparisons of boxplots processed over open areas highlighted the large temporal decorrelation occurring in X-band 1-day, 11-day and L-band 46-day repeat-pass systems. Also, considering the boxplots performed over forested areas, a decrease in coherence for all the investigated interferometric systems could be observed. This variation was explained by temporal and volume decorrelation, which is intrinsically related to the forest canopies. Among the different weather parameters affecting the InSAR coherence, precipitation was shown to be an important variable. An obvious example could be presented with apparition on one of the ALOS PALSAR interferometric acquisitions of a locally marked drop in coherence, which was referred to a heavy storm event, including convective precipitations. The investigations focused on acquisitions measured with HH polarisation. However, some analyses were also undertaken for comparing HH with HV and VV polarisations. As expected, the amplitude of the phase coherence for HV was slightly lower than that for the copolarisations and was equal in the HH and VV polarisations. Hence, the HH component was considered for the remaining investigations. One of the objectives of this section was to evaluate the potential of interferometric coherence in the estimation of forest biomass. Scatterplots comparing interferometric coherence and *GSV* for the ALOS PALSAR system were developed for Norway Spruce forest stands. The sensitivity of coherence to the volume of the trees was first shown to increase with increasing perpendicular baseline until the signal is completely decorrelated for the acquisitions where the normal baseline exceeds the critical baseline by around 40 %. Then, comparing interferometric coherence to *GSV* for the different species and sensors for CSK and PALSAR systems, it was found that the coherence of spruces was the highest, followed by those of pines and beeches. This difference was mainly explained by the different branching structures and foliage compositions of the trees. The relationship between the InSAR coherence of the CSK and PALSAR sensors and *GSV* was shown to follow an exponential law, with saturation levels for the CSK sensor appearing at around 100 and 250 $m^3\,ha^{-1}$ for beech and spruce, respectively, and at around 200 $m^3\,ha^{-1}$ in the case of the PALSAR sensor for the three investigated species compositions. However, these saturation points varied widely, especially with the variation in perpendicular baseline. The sensitivity of the TDX coherence to *GSV* was shown to be surprisingly high with a negative linear relationship without any saturation till 600 $m^3\,ha^{-1}$. With a R^2 of 0.72 for spruce, TDX exhibited the best correlation compared to other sensors and species. The penetration of microwaves could not be completely clarified in Sect. 5.2 with the analysis of the backscatter intensity and polarimetric decompositions of the signal. Therefore, further investigations focusing on the phase center location were undertaken in this section. For this, loci coherence diagrams based on the PALSAR PLR PolInSAR data were considered. Unfortunately, owing to high temporal decorrelation and the resulting low coherence of the PALSAR data, it was not possible to draw any conclusions based on these diagrams.

5.4 Forest GSV Estimation

5.4.1 Parametric Modelling

The results of the empirical regression modelling are presented in this section. Three different topics are considered, namely training of the models, models' inversions and testing and error investigations. According to the results pertaining to backscatter intensity and interferometric coherence presented in Sects. 5.2 and 5.3, respectively, ALOS PALSAR and TDX interferometric coherence in HH polarisation constituted the most suitable datasets for retrieving *GSV* in the Thuringian Forest. In addition, only spruce appeared to show satisfactory correlations. In this framework, the focus for the parametric modelling of the Thuringian Forest *GSV* was performed on the basis of ALOS PALSAR and TDX InSAR HH coherence as well as on the composition of spruce. As introduced in Sect. 4.4 and described in Sect. 5.3, according to the SAR sensor and acquisition configuration, the relationship between PALSAR interferometric coherence and *GSV* appeared to be both linear and nonlinear. Assuming no a priori information on the suitable models, a methodology (Sect. 4.4) was adopted for modelling the available coherence data using linear and nonlinear models, and the model showing the lowest *RMSE* was selected. For this, the forest stands were subdivided into three datasets, namely a training set for retrieving *GSV*, a testing set for evaluating and choosing the suitable model and a testing set for estimating the final accuracy with the selected model. The number of stands from the inventory described in Sect. 4.3 and used in Sects. 5.1, 5.2 and 5.3 was too limited for obtaining a reliable number of stands for the three classes. Thus, for the purpose of regression modelling, a forest inventory based on the criteria defined in Sect. 4.3 was considered with a different definition of the tree species mixity (*BAAN*) and relative stocking (*VG*) parameters. Further information can be found in Sect. 4.4.

5.4.1.1 Model Training

The training of the models was undertaken using the methodology described in Sect. 4.4. This approach involved selection of a training dataset as well as an iteration process for removing potential outliers. For PALSAR HH coherence, Fig. 5.39 illustrates the model fitted with standard deviations (2σ). In the given example, the fit corresponded to the third iteration. The PALSAR scene was acquired on frame FBD1 on 21st August/6th October 2007.

As observed in Fig. 5.39, the nonlinear model appeared to fit the coherence trend well. However, the spread of the data suggests that for a given coherence value, the *GSV* values were rather widely dispersed, indicating that the estimations would not be accurate. This observation was particularly noted for spruces with a *GSV* above 200 $m^3\ ha^{-1}$, namely mature spruce stands, which did not show any significant decrease in coherence with increasing *GSV*. The difference in coherence dispersion between young and mature spruces may be explained by several reasons, such as the

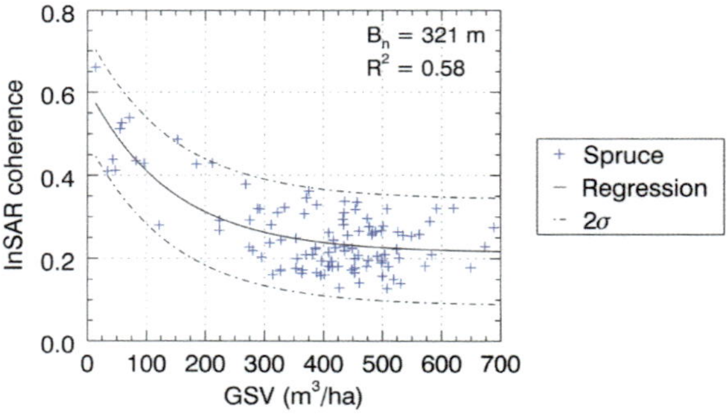

Fig. 5.39 Training process with a nonlinear model fitted to PALSAR FBD HH 39° Asc. (frame FBD1) interferometric coherence. The PALSAR interferometric data was acquired on 21st August/6th October 2007. Coherence was measured over spruce forest stands (*blue points*). Standard deviations (2σ) are depicted by *dashed lines*. The fit corresponds to the third iteration

Table 5.2 Estimated parameters for PALSAR FBD HH 39° Asc. (frames FBD1–4 and 6–8) and TDX SM HH 38° Asc. (frame TDX3) coherence data

Frame	Acquisition dates	Model	A	B	C
FBD1	21aug07–06oct07	Non-linear	0.52	−0.00943	0.20
FBD1	23may08–08jul08	Linear	−0.00001	0.16	–
FBD2	06may08–21jun08	Linear	0.00000	0.18	–
FBD2	12may10–27jun10	Linear	−0.00036	0.46	–
FBD3	02sep07–18oct07	Linear	−0.00050	0.57	–
FBD3	07sept09–23oct09	Linear	−0.00074	0.83	–
FBD3	10jun10–26jul10	Linear	−0.00055	0.50	–
FBD3	23jul09–07sept09	Non-linear	0.27	−0.00727	0.13
FBD4	01jul07–16aug07	Linear	−0.00022	0.55	–
FBD4	16aug07–01oct07	Linear	−0.00066	0.65	–
FBD4	18may08–03jul08	Linear	0.00002	0.12	–
FBD4	24may10–09jul10	Linear	−0.00022	0.46	–
FBD6	06may08–21jun08	Linear	−0.00011	0.38	–
FBD6	12may10–27jun10	Linear	−0.00003	0.17	–
FBD7	02sep07–18oct07	Linear	−0.00056	0.66	–
FBD7	07sept09–23oct09	Linear	−0.00052	0.71	–
FBD7	10jun10–26jul10	Linear	−0.00043	0.42	–
FBD7	23jul09–07sept09	Linear	−0.00077	0.57	–
FBD8	01jul07–16aug07	Linear	−0.00024	0.63	–
FBD8	16aug07–01oct07	Non-linear	0.51	−0.00275	0.26
FBD8	18may08–03jul08	Linear	0.00000	0.12	–
FBD8	24may10–09jul10	Linear	−0.00023	0.55	–
TDX3	30aug11	Linear	−0.00047	0.77	–

Three parameters were considered for the nonlinear models and two parameters for the linear models. The parameters for spruce forest stands were derived

saturation of the signal, the various structures of the forest discussed in Sect. 5.2 or the bias intrinsic to the coherence. In the latter case, referring to the description of Relation (3.33) in Sect. 3.2, the bias increases with decreasing coherence. In other words, for a lower coherence, the bias is higher, and therefore, according to the coherence trend with *GSV* in Fig. 5.39, the dispersion of the mature spruces' coherence is greater.

The above fitting operation allowed the estimation of the model variables. Table 5.2 presents an example of the estimated parameters A, B and C derived for spruce forest stands with the PALSAR FBD HH and TDX SM HH coherence acquisitions. The acquisitions with a value provided for parameter C referred to the data processed with a nonlinear model. The remaining acquisitions, where only parameters A and B were provided, referred to a linear model. As described in the methodology in Sect. 4.4, the choice of a linear or nonlinear function was based on the *RMSE* computed from the 30 % testing stands allocated for model selection. In the given example, for the sake of conciseness, the parameters directly referred to the selected models.

Table 5.2 indicates that most of the models used for the processing of the final accuracy were linear, indicating that the coherence trends were more linear than non-linear. As for the estimated parameters, A and B representing the dynamic range of nonlinear and linear models, respectively, showed that coherence reached approximately 0.5 and 0.77 in the best case. The decreasing rate expressed by the parameters A and B for the linear and nonlinear models, respectively, was negative and very small, as expected. Finally, the phase coherence offset values described by parameter C for the nonlinear model did not exceed 0.26.

5.4.1.2 Model Inversion and Testing

As described in Sect. 4.4, after estimating the regression parameters, inversion of the models was performed to retrieve *GSV* images from the coherence scenes. On the basis of the *GSV* images and forest stands belonging to the testing datasets, descriptive statistics were then computed. Table 5.3 describes the statistics derived for spruce forests stands over the PALSAR FBD HH and TDX SM HH coherence data. The given statistics were computed using the testing set of stands dedicated to the evaluation of the final accuracy.

From Table 5.3, it is first observed that the estimated *RMSE* for the different interferometric acquisitions ranges between 100 and 220 $m^3\,ha^{-1}$. This error is markedly high with respect to the overall range of *GSV* (0–700 $m^3\,ha^{-1}$), and thus, this accuracy is not sufficient for achieving the objectives of forestry and climatic modelling (see Sect. 2.1). The PALSAR data in Table 5.3 indicates that the highest accuracy was $RMSE = 112\,m^3\,ha^{-1}$, which was obtained with the coherence acquired on frame FBD3 on 7th September/23rd October 2009. The lowest accuracy was $RMSE = 218\,m^3\,ha^{-1}$, which was measured on frame FBD4 on 18th May/8th July 2008. The difference in the retrieval accuracies of the investigated acquisitions may be explained by several factors. Among these factors, the perpendicular baseline was shown to be significant. By comparing the statistics with the perpendicular baselines, it was observed that the normal baselines ranging between 300 and 700 m depicted an *RMSE* of less than 150 $m^3\,ha^{-1}$ and a bias of less than 20 $m^3\,ha^{-1}$, while the remaining acquisitions comprised *RMSE* and bias values above 150 and 20 $m^3\,ha^{-1}$, respectively. However, some exceptions could be observed where high hourly precipitations appeared to have affected the accuracy, such as 10th June/26th July 2010 and 23rd July/7th September 2009 acquisitions, which showed a bias of −28 and 32 $m^3\,ha^{-1}$, respectively. The descriptive statistics for the estimated *GSV* from the PALSAR FBS and PLR coherence data were also processed and are provided in Appendix B.4.

Concerning TDX, the *RMSE* was found to be high. Some discussions about this result are provided later in this section during the examination of topography error contribution.

Table 5.3 Statistics for modelling with parametric regressions of PALSAR FBD HH 39° Asc. (frames FBD1–4 and 6–8) and TDX SM HH 38° Asc. (frame TDX3) coherence in the Thuringian Forest

Frame	Acquisition dates	Baseline (m)	p_{h1} (mm)	p_{h2} (mm)	R^2_{training}	RMSE ($m^3\,ha^{-1}$)	Bias ($m^3\,ha^{-1}$)
FBD1	21aug07–06oct07	321	0.9	0.0	0.73	158	−9
FBD1	23may08–08jul08	−3,790	0.0	1.1	0.01	162	28
FBD2	06may08–21jun08	−3,285	0.0	0.0	0.01	217	31
FBD2	12may10–27jun10	102	0.0	0.0	0.52	170	−18
FBD3	02sep07–18oct07	393	0.0	0.1	0.50	142	2
FBD3	07sept09–23oct09	381	0.0	0.0	0.53	112	5
FBD3	10jun10–26jul10	363	0.0	5.0	0.61	119	20
FBD3	23jul09–07sept09	621	14.3	0.0	0.69	157	32
FBD4	01jul07–16aug07	290	0.0	0.2	0.17	201	8
FBD4	16aug07–01oct07	352	0.2	0.0	0.53	138	15
FBD4	18may08–03jul08	−3,874	0.0	5.0	0.01	218	79
FBD4	24may10–09jul10	67	0.0	0.0	0.22	206	26
FBD6	06may08–21jun08	3,251	0.0	0.0	0.04	189	37
FBD6	12may10–27jun10	102	0.0	0.0	0.13	201	19
FBD7	02sep07–18oct07	388	0.0	0.0	0.73	121	18
FBD7	07sept09–23oct09	378	0.0	0.0	0.69	117	7
FBD7	10jun10–26jul10	363	0.0	12.8	0.68	132	−28
FBD7	23jul09–07sept09	619	5.1	0.0	0.73	162	−22
FBD8	01jul07–16aug07	290	0.0	0.7	0.17	185	22
FBD8	16aug07–01oct07	−348	0.7	0.0	0.27	175	4
FBD8	18may08–03jul08	−3,852	0.0	9.5	0.01	192	43
FBD8	24may10–09jul10	67	0.0	0.0	0.21	177	−10
TDX3	30aug11	258	0.0	0.0	0.53	183	−17

Three statistics were retained, namely R^2 for training, *RMSE* and bias. Modelling was performed over spruce forest stands

In addition to the statistics presented above, it is useful for visually comparing the measured ground and SAR estimated *GSV* on a scatterplot. In this respect, a qualitative assessment of the estimation based on 2D scatterplots with measured *GSV* on the abscissa and estimated *GSV* on the ordinate was performed. Figure 5.40 depicts two illustrations of the processed scatterplots with PALSAR coherences acquired on (a) 7th September/23rd October 2009 and (b) 23rd July/7th September 2009. Both coherence acquisitions were measured over frame FBD7. To better interpret the plots, a 1:1 line and two dashed lines representing the inventory accuracy ($\sigma = \pm 20\,\%$) were overlaid.

Figure 5.40a shows that although there were some variations, the cloud of points extended in the same direction as the 1:1 line. This indicated that the selected

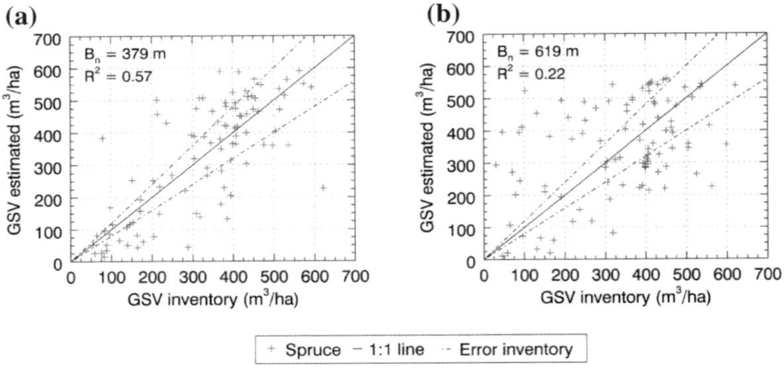

Fig. 5.40 Testing process visualisation for two PALSAR FBD HH 39° Asc. (frame FBD7) coherence data. Inventory *GSV* is given on abscissa and coherence estimated *GSV* on ordinate. The PALSAR interferometric data were acquired on **a** 7th September/23rd October 2009 and **b** 23rd July/7th September 2009. The coherences were measured over spruce forest stands. A 1:1 *line* and two *dashed lines* representing the inventory accuracy ($\sigma = \pm 20\,\%$) are overlaid for the interpretations

model was suitable for examining the relationship between *GSV* and interferometric coherence. Then, comparing the dispersion of the points with the different *GSV* reference values, it was observed that the dispersion for mature stands was higher than that for young regenerative stands. This is explained by the same reasons as those presented in relation to Fig. 5.39. Then, considering the plot in Fig. 5.40b, it was observed that the points were misaligned with the 1:1 line. In particular, the points referring to low *GSV* on the reference inventory data were retrieved as high estimated *GSV*. In fact, the acquisition shown in plot (b) corresponded to the coherence measured during the convective precipitation event, introduced in Sect. 5.3. In this case, the high coherence, which is usually observed for young spruces, decreased owing to precipitations and was interpreted by the model as high *GSV*. Consequently, the retrieved *RMSE* for the acquisition in Fig. 5.40b was larger than the *RMSE* for the acquisition depicted in plot (a).

5.4.1.3 Investigation of Sources of Errors

The *RMSE* for ALOS PALSAR and TDX coherences in Table 5.3 was found to be abnormally high when compared with other studies (see Sect. 2.1). This high error may be due to several factors, which were further investigated in this study. Equation (5.1) summarises the potential error contributions

$$\varepsilon = \varepsilon_{\text{topo}} + \varepsilon_{\text{inventory}} + \varepsilon_{\text{model}} + \varepsilon_{\text{radar}} + \varepsilon_{\text{weather}}\,. \tag{5.1}$$

Section 5.1 showed that the coherence signal varied with topography. Thus, the first factor to be mentioned is the sloping terrain of the Thuringian Forest test site. In addition to the topography, the forest inventory used here as a reference for the training and testing processes may also have included potential errors. In this respect, the comparison shown in Sect. 4.3 between the *GSV* measured during the project field campaigns and the *GSV* provided in the forest inventory showed large variations, which should not be neglected. These errors may have contributed to the fluctuation in the signal, resulting in inaccurate *GSV* retrieval. Also, the processing scheme and parametric regression method used for the modelling might not have been adapted or might have required some improvements. Finally, as mentioned above, the error might have increased because of the weather conditions or the radar configuration, such as an inappropriate perpendicular baseline.

The results of error sources investigations are presented below. First, distinctions between the estimations with and without slopes are described. Then, the comparison between the coherence-retrieved *GSV* and the *GSV* measurements obtained from the forest inventory and forest campaigns are discussed. Finally, elements related to the processing of spatial resolution of the data are given. For assessing the modelling method used in this section, the results of another approach based on non-parametric modelling are presented in Sect. 5.4.2 (*k*-NN).

Topography Error Contribution

The topography was considered by performing the modelling of PALSAR FBD HH and TDX SM HH coherence in forested areas with limited topography. Slopes below 10° and 5° were selected for PALSAR and TDX, respectively, in order to keep the number of stands consistent and preserve the quality of the stands. Figure 5.41 presents a histogram which compares the *RMSE* of the PALSAR and TDX data. The different acquisitions are depicted on the abscissa and the *RMSE* on the ordinate. The data processed without the slope criteria are shown in green, while the coherences modelled with the consideration of limited slopes are depicted in blue. For better readability, the acquisition dates given on the secondary abscissa refer only to the first acquisition of the coherence pairs. Also, some of the available acquisitions did not provide a sufficient number of stands for the selection of limited slopes. The latter data were not considered in Fig. 5.3.

In Fig. 5.41, the comparison of *RMSE* between stands with slopes below 10° and without slope selection shows clear differences. The acquisitions measured under frames FBD6–8 generally showed that the *RMSE* for modelling with limited slopes is lower than that for modelling without any slope criterion. However, the opposite was observed for frames FBD1–4. In the latter cases, the *RMSE* values for *GSV* estimated over forest stands with limited slopes were generally higher than that for stands without any limited gradient. These observations suggested that another factor potentially more influential than topography may have affected the modelling process with the selection of slopes. This factor was found to be related to the number of available stands and the quality of the stands. In the case of PALSAR acquisitions acquired

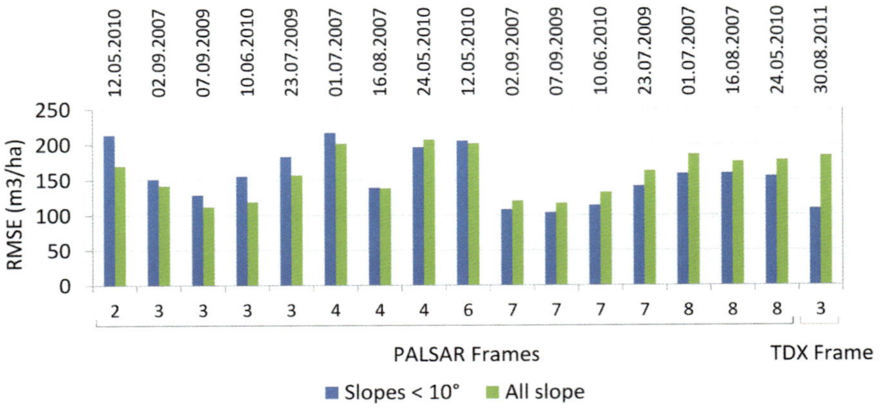

Fig. 5.41 Comparison of statistics between *GSV* estimated with topography and with limited topography for PALSAR FBD HH (frames FBD1–4 and FBD6–8) and TDX SM HH 38° Asc (frame TDX3). The modelling was performed over spruce forest stands. The comparison is based on *RMSE*

on frames FBD1–4, the number of available stands was generally limited compared to the number of stands for the acquisitions measured on frames FBD6–8. As mentioned above and in Sect. 4.4, the developed retrieval methodology consisted of fixing a minimum number of 300 stands per acquisition by adjusting the quality of the forest stands, with the dominant species percentage in a stand (*BAAN*) and relative stocking (*VG*) parameters. Consequently, on the basis of the available number of stands and the retrieval approach employed, the reliability of the stands for the scenes from frames FBD1–4 was drastically reduced compared to those from frames FBD6–8. Referring to these observations, it may be concluded that in the case of frames FBD6–8, the stands were reliable and the selection of slopes below 10° improved the estimation accuracy when compared to the no-slope selection. However, in the case of frames FBD1–4, the potential improvements with reduced topography were masked by the poor-quality stands. Therefore, the selection of slopes below 10° increased the *RMSE* for the data acquired over these frames instead of decreasing it.

With regard to TDX, Sect. 5.1 showed that the topography significantly affected the TDX coherence, particularly for high slopes facing the sensor. As shown in Fig. 5.41, the consideration of slopes below 5° considerably improved the modelling accuracy. Indeed, the *RMSE* decreased from 183 $m^3\,ha^{-1}$ in hilly terrains to 109 $m^3\,ha^{-1}$ without any topography. In the case of TDX, the frame of the data extended over the numerous spruce stands of the Thuringian Forest (see Appendix A.1). Thus, a large number of stands were available, and therefore, the quality of the stands was not affected.

In summary, the observations made in relation to Fig. 5.41 indicated a tradeoff between the quality of stands and topography. On the basis of the modelling results, the quality of stands appeared to be as significant as topography.

Forest Inventory Error Contribution

The *GSV* of the reference provided by the forest inventory was evaluated in Sect. 4.3 using the *GSV* values measured within the undertaken forest campaigns. At this stage, the *GSV* estimated from coherence may also be compared with the available measurement results of the forest campaigns. For this, histograms based on the *GSV* of the forest inventory, forest campaigns and coherence regression were established for the selected forest stands. The comparison of the estimated *GSV* with the campaign measurements did not present any particular trend. The histograms are provided in Appendix B.4.

In addition to the analysis of the histograms, spatial variations and potential systematic errors of the estimated coherence *GSV* were investigated by computing stand-wise difference maps which illustrate the differences between the forest inventory and coherence estimated *GSV*. The difference maps showed that *GSV* estimation was not always uniform over the coherence scenes but did not show any obvious systematic errors. The variations in *GSV* for the stand-wise difference maps may have been related to different factors such as topography or local winds, which affect the interferometric coherence. Other factors such as the time difference between the reference (inventory) and the SAR estimated *GSV* may have also contributed to the variations, but this was not believed to be significant.

Spatial Resolution Error Contribution

It is possible to address the question of an optimal processing spatial resolution because the choice of spatial resolution may affect not only the computation speed but also the *GSV* retrieval accuracy. Considering the effect on accuracy, despite the forest stands buffer of 25 m performed during pre-processing (Sect. 4.3), downsampling the coherence scenes to a lower spatial resolution till a certain point may result in a lower precision of the border of the forest stands within the images, thus resulting in more errors in estimation owing to the admittance of a larger number of non-forest or other forest scatterers in the stands.

In this context, the processed spatial resolution was investigated for the modelling approach defined in Sect. 4.4. In particular, the *GSV* estimation process was repeated for different spatial resolutions for the selected PALSAR coherence datasets. Each process involved a minimum stand size of 5 ha. At the end of the processing, the retrieved accuracy was compared to the spatial resolution. The results showed that starting from a spatial resolution of 50 m, the *RMSE* tends to increase with greater pixel sizes (see Appendix B.4). This increase may confirm the border effect of the stands, which was introduced above. However, it started only after 50 m and did not appear to significantly affect the *RMSE*. Thus, the spatial resolution may not be an important contributor of the errors presented in Eq. (5.1).

5.4.2 Non-parametric Modelling

Non-parametric modelling was performed using the k-NN approach. The k-NN technique was essentially tested in this study to compare the results of the regression method presented above with a different and potentially more accurate modelling technique. For this, three different topics were treated: selection of optimal k-NN modelling parameters, testing of the model with descriptive statistics and visualisation of the k-NN and regression-estimated *GSV* images. These topics are discussed in this section. The study conducted in the Thuringian Forest applying k-NN modelling consisted of a master thesis supervised within the framework of this doctoral thesis. The results presented in this section refer to some of the findings cited in the master thesis. For a comprehensive overview of the k-NN approach adopted for the Thuringian Forest, the reader is referred to the master thesis study Zhu (2012). The comparison of the models was limited to PALSAR acquisitions and spruce species.

5.4.2.1 Evaluation of k-NN Parameters

As described in the methodology section (Sect. 4.4), the k-NN approach includes several parameters which need to be defined. Among these parameters, the k numbers and weighting functions were investigated in order to determine the optimal configuration for PALSAR coherence. The different configurations were compared using the *RMSE* and bias metrics for k numbers ranging from 1 to 20. The *RMSE* and bias were calculated using the LOOCV method mentioned in Sect. 4.4. The first examinations were performed by comparing the various weighting functions using single PALSAR coherences. The Mahalanobis and Manhattan distances differ from the Euclidean distance only when two or more channels are used for k-NN processing. Because the principal objective of this section was to compare the retrieved *GSV* obtained with regressions and single coherence images (Sect. 5.4.1), the Mahalanobis and Manhattan distances were not considered in the following investigations. Also, to process the k-NN model, the forest inventory described in Sect. 4.3 was used. The number of stands was sufficient for achieving the purpose of this study. Figure 5.42 depicts an example of the obtained results with the *RMSE* and bias processed for a k number between 1 and 20. The k numbers are displayed on the abscissa, while the accuracy metrics are shown on the ordinate (RMSE on the primary ordinate and Bias on the secondary ordinate). The Euclidean distance was combined to three weighting functions: mean, inversed and inverse-square distance functions. The PALSAR coherence scene was acquired on frame FBD7 (Appendix A.1) on 2nd September/18th October 2007.

First, the behaviour of the curves in green indicates that the error rapidly decreases with increasing k number until it reaches a sufficient number of nearest stands for obtaining reliable accuracy. This tendency, which is typically observed with the prediction of k-NN errors (Franco-Lopez et al. 2001; Baffetta et al. 2009), is explained

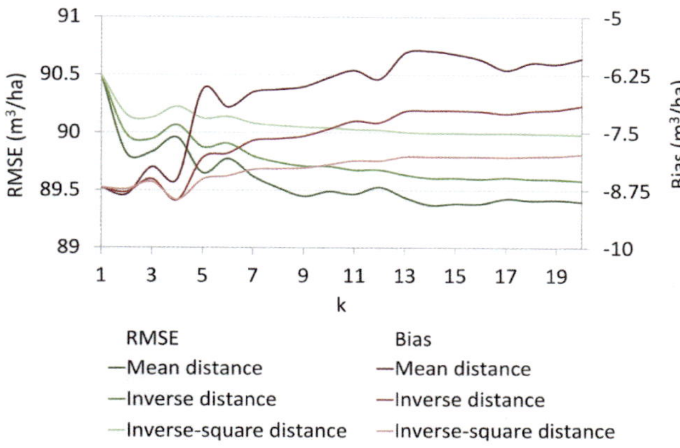

Fig. 5.42 Comparisons of *k*-NN modelling parameters with *RMSE* and bias statistics. One PALSAR FBD HH 39° Asc. (frame FBD7) coherence scene and *k* numbers between 1 and 20 were considered for the illustration. The PALSAR scene was obtained on 2nd September/18th October 2007. The statistics were derived for spruce forest stands. The different curves are distinguished in *green* for the *RMSE* and in *red* for the Bias. The curves represent with different color tones the Euclidean distance combined to mean, inversed and inverse-square distance weighting functions

by the increased number of forest stands. However, as shown by the curves in red, the increase in the number of nearest neighbours also increases the bias. This result can be explained by the greater number of data involved in the *k*-NN processing; this typically increases the standard deviation, resulting in increased bias, as shown in several studies (Franco-Lopez et al. 2001; Chung et al. 2009). Then, a comparison of the curves of the weighting functions in different color tones indicate that the mean weighting function depicted the best accuracy in terms of *RMSE*. However, considering the *RMSE* range, it should be noted that the differences between the three weighting functions remained insignificant. Finally, on a more close observation of the *RMSE* and *k* numbers for the different curves, the variations in accuracy appeared to stabilise after reaching about 15 nearest neighbours. This observation agreed with other findings, which reported a stability point between $k = 10$ and 15 (Kajisa et al. 2008; Franco-Lopez et al. 2001). On the basis of this result and the abovementioned outcomes, the overall available PALSAR coherence scenes of the Thuringian test site was processed with a Euclidean distance, a mean weighting function and a *k* number corresponding to 15 nearest neighbours ($k = 15$).

5.4.2.2 Model Testing

The available interferometric PALSAR datasets were processed with the above-mentioned *k*-NN configuration and compared to the regression modelling accuracy in Table 5.3. Table 5.4 depicts the *k*-NN statistics derived for the same PALSAR

Table 5.4 Statistics for modelling with k-NN algorithm of PALSAR FBD HH 39° Asc. coherence in the Thuringian Forest performed over spruce forest stands

Frame	Acquisition dates	Baseline (m)	p_{h1} (mm)	p_{h2} (mm)	RMSE ($m^3\ ha^{-1}$)	Bias ($m^3\ ha^{-1}$)
FBD1	21aug07–06oct07	321	0.9	0.0	100	−6
FBD1	23may08–08jul08	−3790	0.0	1.1	144	−5
FBD2	06may08–21jun08	−3285	0.0	0.0	135	−12
FBD2	12may10–27jun10	102	0.0	0.0	119	−2
FBD3	02sep07–18oct07	393	0.0	0.1	92	−1
FBD3	07sept09–23oct09	381	0.0	0.0	88	−9
FBD3	10jun10–26jul10	363	0.0	5.0	96	−5
FBD3	23jul09–07sept09	621	14.3	0.0	94	−2
FBD4	01jul07–16aug07	290	0.0	0.2	120	1
FBD4	16aug07–01oct07	352	0.2	0.0	84	−1
FBD4	18may08–03jul08	−3874	0.0	5.0	135	−4
FBD4	24may10–09jul10	67	0.0	0.0	130	0
FBD6	06may08–21jun08	3251	0.0	0.0	119	2
FBD6	12may10–27jun10	102	0.0	0.0	113	−3
FBD7	02sep07–18oct07	388	0.0	0.0	89	−6
FBD7	07sept09–23oct09	378	0.0	0.0	88	−9
FBD7	10jun10–26jul10	363	0.0	12.8	96	−5
FBD7	23jul09–07sept09	619	5.1	0.0	94	−2
FBD8	01jul07–16aug07	290	0.0	0.7	113	−3
FBD8	16aug07–01oct07	−348	0.7	0.0	107	−6
FBD8	18may08–03jul08	−3852	0.0	9.5	120	2
FBD8	24may10–09jul10	67	0.0	0.0	115	−6

Two statistics were retained: *RMSE* and bias

acquisitions shown in Table 5.3. For a better comparison of the k-NN and regression statistics, the acquisitions were sorted with the same order.

First, the k-NN accuracy results in Table 5.4 indicate that the *RMSE* generally ranged between 80 and 150 $m^3\ ha^{-1}$. Thus, compared to the results of the regression method described in Sect. 5.4.1, the k-NN approach was approximately 50 $m^3\ ha^{-1}$ more accurate than the regression approach. Fehrmann and Kleinn (2007) and Fehrmann et al. (2008) also compared the two approaches and reported the same general observations. He mainly explained the higher k-NN accuracy compared to the regression technique on the basis of the ability of the k-NN algorithm to be more flexible when representing local trends. Although it achieves better accuracy than the regression approach, the *RMSE* obtained with the k-NN algorithm still remained unsatisfactory for the applications discussed in the introduction of this thesis. Then, comparing the perpendicular baseline to the *RMSE*, it could be observed that the best accuracy was achieved with normal baselines ranging between 300 and 700 m.

The same result was shown in Table 5.3 for the regression method. Finally, a comparison of the precipitations to the *RMSE* indicated that the interferometric acquisitions which presented high precipitations or dry conditions as well as equivalent normal baselines (i.e. 10th June/26th July 2010 and 2nd September/18th October 2007, frame FBD7) depicted close *RMSE* values. This result suggested that the *k*-NN method might have been affected by precipitations to a lesser extent than the regression technique. As mentioned above, this observation may be explained by the capability of the *k*-NN algorithm to exploit the local variations.

In processing, the use of an extremely small number of stands may affect the estimation accuracy. Investigations in Zhu (2012) showed that using 100 stands would lead to insignificant changes in accuracy, and 500 forest stands would be optimal. In the given case, the number of stands for the forest inventory described in Sect. 4.3 ranged from 105 (frame FBD1) to 2012 (frame FBD7). These numbers were adequate for this study.

5.4.2.3 Visual Comparisons of *k*-NN and Regression-Estimated *GSV* Images

In addition to the comparison of the statistics in Table 5.4, it is useful to visualise the estimated *GSV* retrieved from the *k*-NN and regression methods. In this framework, the *GSV* images retrieved with the *k*-NN algorithm were computed and examined with the *GSV* images based on the regression approach. Figure 5.43 depicts the resulting images for comparisons with (a) the *k*-NN *GSV* image, (b) the regression-based *GSV* image, (c) the forest inventory and (d) a digital orthophoto. The *GSV* images were retrieved from a PALSAR FBD HH coherence acquired on 7th September/23rd October 2009. The covered area is the Goldisthal Region, which is denoted by its upper reservoir facility. It should be noted that the *GSV* images in (a), (b) and (c) were masked using a forest mask and a spruce mask, which were both established within the framework of Sect. 5.5.

By comparing the (a) *k*-NN and (b) regression *GSV* images in Fig. 5.43, it is observed that the image derived using the *k*-NN algorithm is brighter than the image processed with the regression technique. This difference is explained by a property of the *k*-NN method. With the weighting of the *k* nearest forest stands, the algorithm tends to be more accurate towards the mean of the total *GSV*, while it tends to overestimate the low *GSV* and underestimate the high *GSV*. Thus, in most cases, the *GSV* values shown in Fig. 5.43a were around 350 $m^3\,ha^{-1}$, which represented the mean of the *GSV* values. This particularity of *k*-NN was also reported by several authors (Kajisa et al. 2008; Holmgren et al. 2000). Then, comparing the *GSV* images with the forest inventory in Fig. 5.43c, it was observed that the general *GSV* patterns were not very well reproduced, except where low *GSV* values were estimated (brown colours). Given the topography of the area shown in Fig. 5.43d, and the related effects on coherence described in Sect. 5.1, these results appeared to be realistic.

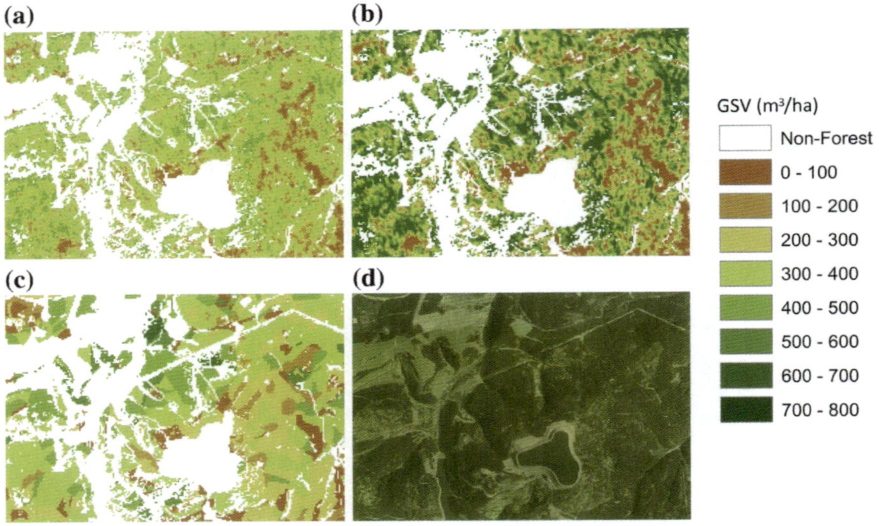

Fig. 5.43 Visual comparison of *GSV* images established from **a** *k*-NN and **b** regression modelling techniques, **c** *GSV* image derived from the forest inventory and **d** digital orthophoto provided for comparison purposes. The forested areas are located around the Goldisthal upper reservoir facility

5.4.3 Summary

This chapter focused on the modelling of SAR interferometric coherence. Two different modelling approaches were discussed: parametric and non-parametric. Parametric modelling consisted of empirical regressions performed for spruce using ALOS PALSAR and TDX coherences. Simple linear and nonlinear fits were computed for each interferometric scene. The fits were based on training and testing sets of spruce stands. The model showing the lowest *RMSE* was considered for the estimation of the final accuracy, which was calculated on the basis of a separate testing set of stands. The results showed that for almost all acquisitions, the linear model presented a lower *RMSE* than the nonlinear model. Therefore, the final *RMSE* estimations were based on linear models in most cases. The final *RMSE* values ranged from 100 to 220 $m^3\,ha^{-1}$. This achieved accuracy was clearly not sufficient for applications such as forestry management or climate modelling. To understand the potential sources of errors, the topography, forest inventory and processing approach were investigated. These examinations indicated that modelling accuracy is affected by topography, especially for TDX coherence. The investigations also highlighted the importance of selecting reliable forest stands from the perspective of homogeneity. The non-parametric technique was based on the *k*-NN method and was focused on the ALOS PALSAR data and spruce species. It was tested for comparison with the regression approach. The *k*-NN algorithm was processed using different *k* numbers and weighting functions in order to select the optimal parameters. On the basis of the analysis of the parameters, the Euclidean distance metric with a mean weighting function and a k number of 15 were shown to be suitable for this study and were considered for the modelling of the available coherence datasets. The results obtained from the k-NN algorithm showed a 50 $m^3 ha^{-1}$ improvement in the *RMSE* compared to the regression approach. However, this improvement was still not sufficient for envisaging any application in the field of forestry or climate-change modelling.

5.5 Fusion of SAR Information

5.5.1 Extraction of Forested Areas and Species Separation

As mentioned in Sect. 4.4, to separate forested areas from other landcover classes and differentiate between the different tree species in the Thuringian Forest, the derivation of thresholds based on histogram analysis was proposed. In addition to establishing thresholds between the classes, a cross-correlation analysis for quantifying the separation quality of the classes was suggested. Illustrations of the histogram analysis are provided below. Figure 5.44 depicts one plot showing normalised frequency histograms (a) and two associated plots representing the normalised difference (b) and cross-correlation between forest and other classes (c). The graphs in plots (b) and (c) refer to Eqs. (4.14) and (4.16), respectively.

As shown in Fig. 5.44a, crop, open area and water classes can be easily distinguished from the forest class using the PALSAR HV backscatter intensity. However, the urban class cannot be separated from the forest class using this sensor configuration. The normalised difference histogram in Fig. 5.44b derived from the normalised frequency histogram in (a) depicts the same information in a different way so that thresholds may be derived. The thresholds separating other landcovers from forested areas can be found at locations where their corresponding functions are equal to zero, i.e. at approximately -10, -11 and -13 dB for crop, open are and water, respectively. The corresponding reliability of separation between other landcovers from forest is shown in Fig. 5.44c. At a lag $h = 0$ dB, the correlation is less than 0.2 for open area, open area and water, while it reaches approximatively 1.0 for urban. The high correlation between urban and forest suggests that these classes are not separable, while the low correlation between forest and other classes indicates that they can be easily separated from forest. For computational optimisation, cross-correlation was performed using Fourier transformation. Therefore, the functions of the cross-correlation analysis were periodic and indicated a return to their initial state at $R(0)$.

On the basis of the histogram analysis, the thresholds of the classes were derived and the separability between the classes was quantified for the available radar datasets by using correlations. This process considered information on not only backscatter intensity but also interferometric coherence. Of the different processed histograms, two particular configurations required attention. The first one concerned

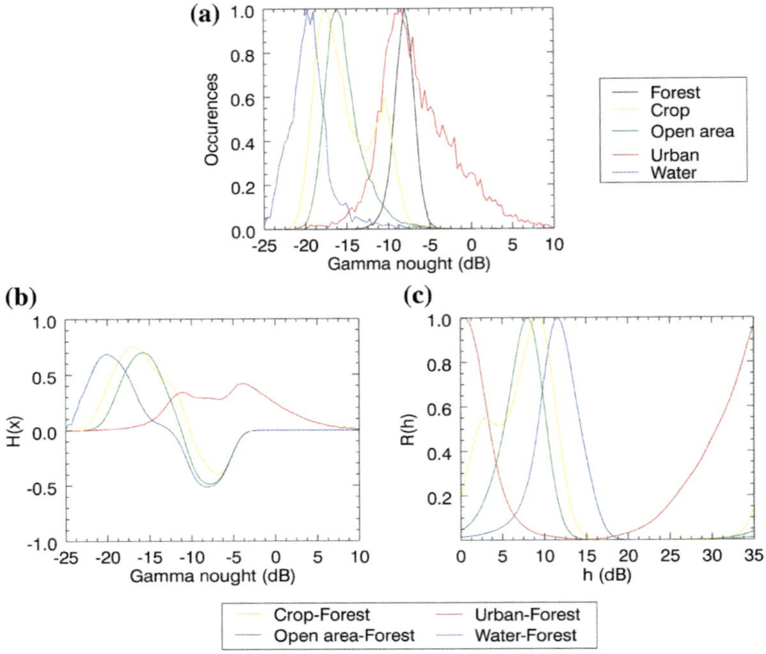

Fig. 5.44 Histogram analysis for separation of forest from other landcover classes; **a** shows the normalised frequency histogram, while **b** and **c** depict the associated normalised difference and cross-correlation functions, respectively. Normalised difference is used to determine the thresholds between forest and other landcover classes. Cross-correlation provides a separability quantification for assessing the quality of separation between the classes. This histogram analysis was performed with ALOS PALSAR FBD HH 39° Asc. backscatter intensity (frame FBD7). The data were acquired on 2nd September 2007

the distinction of the forest class from other landcover classes using the ratio of L-band HV intensity to X-band HH 1 day coherence. The second one, which was introduced in Sect. 5.2 within the frequency analysis, involved the separation of tree species using the ratio of L-band HH intensity to X-band HH intensity. These two configurations are illustrated in Fig. 5.45, which depicts (a)–(d) the normalised occurrence histograms processed independently for each configuration and (e)–(f) the fusion of the data using the abovementioned ratios.

The plots depicting landcover classes in Fig. 5.45a, c suggest that crop, open area, urban and water had relatively similar configurations, while forest showed contrasting trends between PALSAR intensity and CSK coherence. Therefore, the ratio is high for forest and low for the rest of the classes, as depicted in Fig. 5.45e. Similarly, the discrimination between conifers, i.e. spruces and pines, and broadleaves, i.e. beeches, can be improved using the ratio of PALSAR HH intensity (Fig. 5.45b) to CSK HH intensity (Fig. 5.45d), as the classes depict two diametrically opposite situations within these two datasets. The obtained ratio is depicted in Fig. 5.45f.

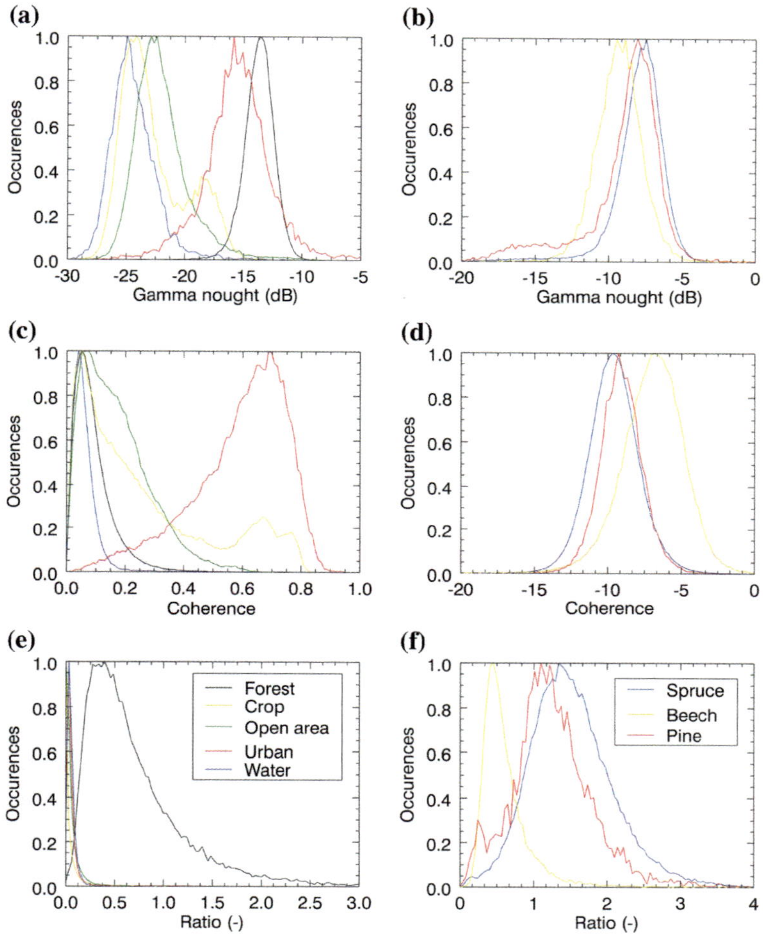

Fig. 5.45 Histogram analysis for **a**, **c** and **e** depicting separation of forest from other landcover classes and **b**, **d** and **f** depicting discrimination among tree species compositions. **a** was computed with ALOS PALSAR FBD HV 39° Asc. backscatter intensity (frame FBD7), **c** with CSK SM Himage HH 37° Asc. $B_n = 83\,\text{m}$ interferometric coherence (frame CSK1) and **e** with the ratio between the latter two datasets. **b** was processed with ALOS PALSAR FBD HH 39° Asc. backscatter intensity (frame FBD7), **d** with CSK SM Himage HH 37° Asc. backscatter intensity (frame CSK1) and **e** with the ratio between the latter two datasets. The PALSAR and CSK intensity data were acquired on 2nd September 2007 and 2nd June 2010, respectively, and CSK coherence data on 2nd/3rd June 2010

5.5.2 Suggested Data Configurations

On the basis of the fusion methodology described in Sect. 4.4, three different products were processed within the developed approach: forest/non-forest map, species map

Table 5.5 Data configuration suggestions for processing a forest/non-forest map

Configuration	Forest discrimination to	
	Crops	Open areas
Recommended	γ^0 L HV	γ^0 L HV
($S > 80$)	γ^0 L HH/γ X HH 1d	γ^0 L HH
	γ X HH single-pass	γ X HH single-pass
	γ^0 L HH	
Minimal	γ^0 L HH leaf-off/γ X HH 1d	γ^0 L HH frozen/γ X HH 1d
($50 < S < 80$)	γ^0 L HH frozen	γ^0 L HH frozen
	γ^0 L HH leaf-off	γ^0 L HH leaf-off
	γ^0 L HH/γ^0 X HH VV	
Not recommended	γ X HH 1d	γ X HH 1d
($S < 50$)	γ^0 X HH VV HV VH	γ^0 X HH VV HV VH

The considered landcover classes are crop, open area and forest. The suggestions were derived from the investigations performed in the Thuringian Forest. The data configurations are sorted in the decreasing order of importance

and *GSV* map. As mentioned in Sect. 4.4, the quality of these products intrinsically depended on the provided dataset. Accordingly, with respect to the investigations performed for the backscatter intensity (Sect. 5.2), interferometric coherence (Sect. 5.3), *GSV* modelling (Sect. 5.4.1) and the above histogram analysis, the most suitable data configuration can be identified. This point is treated in this section.

To summarise the data configurations qualitatively, Tables 5.5, 5.7 and 5.8 illustrate the general results of this thesis in three categories: recommended configuration, minimal configuration and non-recommended configuration. The recommended configuration represents situations where the data yield a score (see Eq. 4.18) higher than 80 for landcover and species maps and *GSV* is retrieved with an *RMSE* of less than $20\,m^3\,ha^{-1}$ for *GSV* maps. In this case, the reliability of the forest/non-forest map or species map is relatively high, and the *GSV* map is relatively accurate with reference to the foresters' requirements (see Sect. 2.1). The minimal configuration would designate a score of 50–80 and an *RMSE* of $20–100\,m^3\,ha^{-1}$. In this case, the reliability of the masking products and the accuracy of the *GSV* image are acceptable. Finally, the non-recommended configuration refers to the investigated datasets which show scores lower than 50 for the masks and an *RMSE* of greater than $100\,m^3\,ha^{-1}$ for the *GSV* estimates. In this case, the products are neither reliable nor accurate. The datasets were assigned to the three established configurations according to the findings obtained from the Thuringian test site. It should be noted that the corresponding results obtained from another test site may be different, particularly for another forest type. For each suggested category, the data configurations were sorted in the decreasing order of importance.

First, the data configurations established from the histogram analysis for the forest/non-forest map are presented. Tables 5.5 and 5.6 describe the three suggested configurations.

Table 5.6 Data configuration suggestions for processing a forest/non-forest map

Configuration	Forest discrimination to	
	Urban	Water
Recommended ($S > 80$)	γ X HH 1d	γ^0 X HH VV HV VH
	γ^0 L HV/γ^0 X HH VV	γ^0 L HH HV
		γ X HH single-pass
		γ^0 L HH leaf-off
		γ^0 L HH frozen
Minimal ($50 < S < 80$)	γ L HH 46d $B_n < 300$ m	–
	γ L HH 46d $B_n > 300$ m	
	γ^0 L HH/γ^0 X HH VV	
	γ L HV 46d $B_n < 300$ m	
	γ L HH 46d leaf-off	
	γ L HV 46d $B_n > 300$ m	
	γ X HH single-pass	
Not recommended ($S < 50$)	γ^0 L HH leaf-off/γ X HH or VV 1d	–
	γ^0 L HH leaf-off/γ^0 X HH or VV	
	γ^0 X HH	
	γ^0 L HV	
	γ^0 L HH frozen	
	γ^0 X VV	
	γ^0 L HH	
	γ^0 X HV VH	

The considered landcover classes are urban, water and forest. The suggestions were derived from the investigations performed in the Thuringian Forest. The data configurations are sorted in the decreasing order of importance

Tables 5.5 and 5.6 indicate a general trend. Forest is differentiated from other landcovers on the basis of backscatter intensity. Some exceptions are observed for the urban class which may be distinguished from forest on the basis of interferometric coherence. However, the coherence depicted less stable and therefore less consistent results than the backscatter intensity. A more detailed observation in Table 5.5 shows that L-band information appears to be the most suitable for separating crop and open area from forest. The integration of X-band 1 day (d) coherence information (and to a lesser extent X-band HH VV backscatter intensity) in a ratio with L-band measures may also be envisaged in order to improve the classifications. The most recommended configurations for separating crop and open area from forest would be to use the L-band HV backscatter intensity. This result makes sense, as the scattering mechanisms for crops and open areas with surface returns are different from forest scatterings, which are mostly multiple volume reflections. Considering the urban class in Table 5.6, the summarised results show that the X-band HH 1d coherence is expected to provide the best discrimination of this class from forest. The ratio of L-band HV backscatter intensity to X-band HH or VV intensity may also provide satisfactory classification results. These two configurations having scores higher than 80 are recommended for the separation of forest from urban. These results may be conceived because urban areas generally depict higher coherence and greater double bounce scatterings than those shown by forests. Finally, water may be easily distinguished from forest on the basis of L-band or X-band backscatter intensity (see Table 5.6). The water class is represented in the Thuringian Forest mainly by calm artificial lakes. Hence, from the visualisations in Sect. 5.2.1, the microwave reflections are mostly specular, leading to poor-amplitude signals returning to the sensor compared to those from forests, thus allowing forests to be easily distinguished from water.

The results shown in Tables 5.5 and 5.6 should be analysed with a few additional remarks. First, the given data recommendations were based on the investigations of a large number of data. The results which did not encounter a sufficient number of ROIs or remote sensing data in an equivalent configuration were not considered. Data configurations which also showed high instability, such as coherence, were also rejected from the proposed configurations. Then, some data configurations were deliberately not provided, such as the γ L-band HH 46d in the crop class or the γ^0 L-band HV/γ X-band HH 1d ratio. For the sake of clarity, data configurations which did not improve the discrimination of the other classes from forest were excluded. In the case of the γ L-band HH 46d for the crop class, the γ^0 L-band HH showed better classification results than those obtained from the corresponding coherence. Thus, the processing of the coherence would not be necessary here. In the case of γ^0 L-band HV/γ X-band HH 1d, the use of γ X-band HH 1d would proportionally include more misclassifications as compared to the scenario considering only the L-band HV intensity. Thus, the processing of such a ratio would also not be necessary. In other words, provision of all potential data combinations is logically unnecessary, and only the ones which are capable of improving the classification were considered. Finally, it is observed that in some cases, details on the normal baseline or incidence angle are given for the configurations shown in Tables 5.5 and 5.6. In the absence of these details, no significant changes were observed in these parameters. Also, the γ^0 L-band HH, L-band HH leaf-off and γ^0 L-band HH frozen configurations were differentiated. These data configurations represented the acquisitions made by PALSAR in the FBD and FBS modes. The γ^0 L-band HH, L-band HH leaf-off and γ^0 L-band HH frozen refer to the vegetative period, winter acquisitions with leaf-off and winter acquisitions with leaf-off and frozen conditions, respectively.

Then, after describing the configurations for the derivation of a forest/non-forest map product, Table 5.7 presents the configurations related to the computation of a tree species map product.

As shown in Table 5.7, according to the defined data configurations, it can be generally established using the investigated radar information that the separation between the tree species is limited. In the case of discrimination of spruce from beech, the γ^0 L-band HH/γ^0 X-band HH or VV leaf-off ratio appeared to be the most conclusive; however, the score in this configuration did not exceed 80. Although similar to spruces, pines bear conifers, the separation between beech and pine is very limited. The best score was computed with the γ^0 L-band HH/γ^0 X-band HH or VV leaf-off ratio; however, it did not exceed 50. Finally, the differentiation of spruce

Table 5.7 Data configuration suggestions for processing a tree species composition map

Configuration	Tree species composition	
	Spruce versus Beech	Pine versus Beech
Recommended (S > 80)	–	–
Minimal (50 < S < 80)	γ^0 L HH/γ^0 X HH VV leaf-off	–
	γ^0 L HV/γ^0 X HH VV	
	γ^0 L HV/γ^0 X HH VV	
Not recommended (S < 50)	γ^0 L HH leaf-off/γ^0 X HH VV	γ^0 L HH/γ^0 X HH VV leaf-off
	γ^0 X HH VV leaf-off	γ^0 L HV/γ^0 X HH VV
	γ^0 X HH VV	γ^0 L HH leaf-off/γ^0 X HH VV
	γ^0 L HH	γ^0 X HH VV
	γ^0 L HH leaf-off	γ^0 L HH
	γ^0 L HV	γ^0 L HV
	γ^0 L HH frozen	γ^0 L HH frozen
	γ^0 X HV VH	γ^0 L HH leaf-off
		γ^0 X HV VH

The considered tree species are spruce, beech and pine. The suggestions were derived from the investigations performed in the Thuringian Forest. The data configurations are sorted in the decreasing order of importance

from pine was found to be impossible with the available radar datasets. The best score in this regard was shown with coherence information, suggesting that although both spruce and pine are conifers, they present distinct canopy structures. This result was introduced and discussed in Sect. 5.3. The specific remarks made above for Tables 5.5 and 5.6 are also applicable here to the separation between the three tree species.

Finally, the *GSV* map products are discussed by referring to the investigations described in Sect. 5.4. Table 5.8 lists the data according for the three established configurations.

As shown in Table 5.8, no data fits the recommended or minimal configuration. In fact, the accuracy obtained from the parametric modelling process in Sect. 5.4 did not fall below 100 m^3 ha^{-1}; this means that only non-recommended configurations can be processed for the Thuringian Forest using regressions. In the case of the k-NN modelling approach, *RMSE* between 50 and 100 m^3 ha^{-1} was achieved with L-band 46 days HH coherence and normal baselines greater than 300 m. This result indicated a minimal configuration.

5.5.3 Product Examples and Validation

This section presents product examples which were derived from the fusion approach described in Sect. 5.4. Noting that the recommended data configuration introduced

Table 5.8 Data configuration suggestions for processing a *GSV* map

Configuration	*GSV* retrieval regressions	*k*-NN
Recommended ($RMSE < 50\,\mathrm{m^3\,ha^{-1}}$)	–	–
Minimal ($50 < RMSE < 100\,\mathrm{m^3\,ha^{-1}}$)	–	γ L HH 46d $B_n > 300$ m
Not recommended ($RMSE > 100\,\mathrm{m^3\,ha^{-1}}$)	γ L HH 46d $B_n > 300$ m $p_h < 5$ mm	γ L HH 46d $B_n < 300$ m
	γ L HH 46d $B_n > 300$ m $p_h > 5$ mm	
	γ L HH 46d $B_n < 300$ m $p_h < 5$ mm	
	γ L HH 46d $B_n < 300$ m $p_h > 5$ mm	
	γ X HH Single-pass	

Two modelling techniques were considered, namely a regression method and the *k*-nearest neighbour (*k*-NN) algorithm. The suggestions resulted from the investigations performed in the Thuringian Forest and with spruce species. The data configurations are sorted in the decreasing order of importance

above may be available only partially, it would be interesting to examine the appearance of the products in the absence of the suitable datasets. The first part of this section focuses on product examples which were processed with the recommended as well as minimal or non-recommended configurations. The second part of this section focuses on product examples derived mainly from all available datasets, thus consisting of the potentially best products obtained from the Thuringian Forest. It may be recalled that the terms synergy and fusion, which were defined in Sect. 3.3, designate two different ways of processing data. With respect to the methodology described in Sect. 4.4, these two terms refer to the combination of data.

To illustrate the outcomes of the fusion algorithm when the input data included minimal or non-recommended configurations, a forest/non-forest map was processed with three partially overlapping scenes. The three scenes included a single acquisition of PALSAR HV intensity, PALSAR HH coherence and CSK HH coherence. Figure 5.46a depicts the forest/non-forest map. Additionally, the 'score' maps providing the accuracy of the established forest/non-forest map are presented in Fig. 5.46b–e for crop, open area, urban and water, respectively.

Before interpreting the different products, it needs to be noted that some pink areas appear on the validation products. According to the defined legend, these regions do not provide any information. In fact, as described in Sect. 4.4, the minimum score to be achieved can be defined. In the given example, a score of 30 was defined for the fusion process. Thus, all areas with a score lower than this fixed value are depicted in pink in order to show that the separation of forest from the given landcover class over this region was too inaccurate for it to be considered in the final product.

Figure 5.46a shows that the derived forest/non-forest map varies significantly according to the provided input data. In Fig. 5.46c, the open areas are barely differentiated from forest by PALSAR HH coherence. This information is shown by the pink area. In Fig. 5.46b, the PALSAR HH coherence also shows a lower score (bright grey) compared to the PALSAR HV intensity (dark grey area) for the *crop*

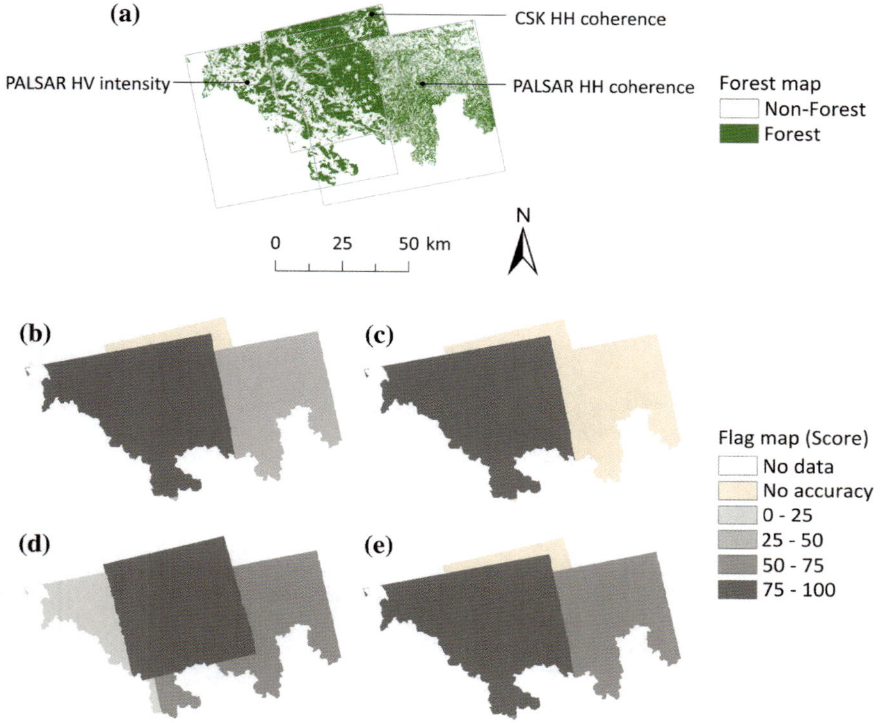

Fig. 5.46 Illustration of **a** a forest/non-forest map processed with a single acquisition of PALSAR FBD HV 39° Asc. intensity, PALSAR FBD HH 39° Asc. $B_n = 363$ m coherence and CSK SM Himage HH 20° Asc. $B_n = 61$ m coherence data (frames FBD6,7 and CSK1). The scenes were acquired on 21st June 2008, 10th June/26th July 2010 and 31st August/1st September 2010 for PALSAR intensity, PALSAR coherence and CSK coherence, respectively. The additional maps **b–e**, respectively represent the scores for crop, open area, urban and water, which are provided to the user to spatially quantify the accuracy of the forest/non-forest map

class, thus indicating that the PALSAR HH coherence is less reliable than the PALSAR HV intensity for the separation of *crop* from *forest*. In Fig. 5.46e, *water* was relatively well separated (dark grey) from *forest* using the different datasets, except for CSK coherence, which is depicted in pink (no information). In summary, the scores shown by the products depicted in Fig. 5.46 indicate the reliable areas of the forest/non-forest map. According to the products, the most reliable area is located at the centre of the forest/non-forest map, where the available datasets intersect. To an extent, this result reflects the established data configurations presented in Tables 5.6 and 5.7.

Similar to the forest/non-forest map presented above in Fig. 5.46, an example of a species map was processed using the same data configuration, with the exception of the PALSAR and CSK intensity scenes, which were provided in HH polarisation

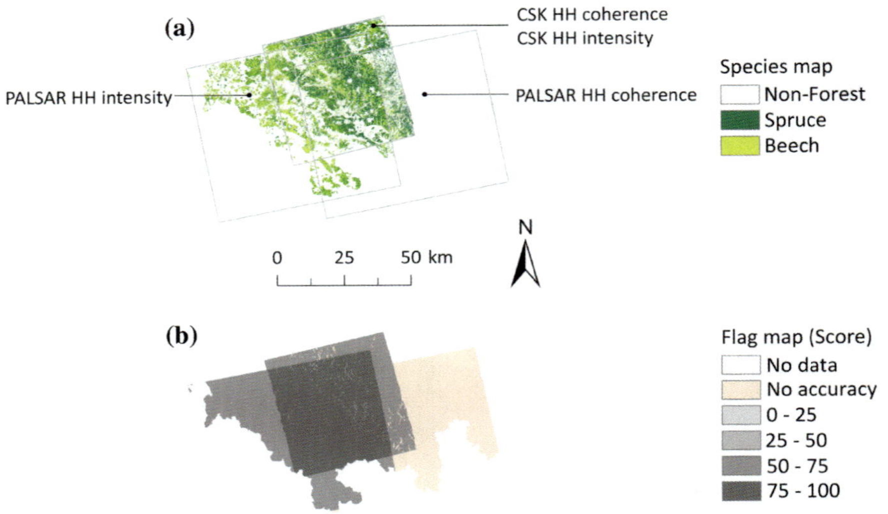

Fig. 5.47 Illustration of **a** a species map processed with a single acquisition of PALSAR FBD HH 39° Asc. intensity, PALSAR FBD HH 39° Asc. $B_n = 363$ m coherence, CSK SM Himage HH 20° Asc. intensity and CSK SM Himage HH 20° Asc. $B_n = 61$ m coherence data (frames FBD6,7 and CSK1). The scenes were acquired on 21st June 2008, 10th June/26th July 2010, 31st August/1st September 2010 and 31st August 2010 for PALSAR intensity, PALSAR coherence, CSK intensity and CSK coherence, respectively. The additional map **b** represents a score provided to the user for spatially quantifying the accuracy (score) of the species map

for this example. Figure 5.47a depicts the derived species map, and Fig. 5.47b shows its accuracy in terms of the score.

As shown in Fig. 5.47a, spruce and beech are distinguished. In fact, the established algorithm does not include pines because their separation from the other species was extremely limited (see Sect. 5.2 and Table 5.7). The accuracy map in Fig. 5.47b indicates that the score varies geographically. When the PALSAR HH and CSK HH intensities are available, the score is highest, while when only PALSAR HH intensity is available, the score is relatively low. Similar to the forest map described in Fig. 5.46, the minimum accuracy here was fixed at 30. According to this configuration, PALSAR and CSK coherences having a score lower than 30 were not considered in the computation of the species map and were processed as 'no information' (pink) on the validation map. Upon a closer look to the tree species map in Fig. 5.47a, the product appears to be noisy, especially where the ratio L-band HH/X-band HH intensity is not available. This observation shows that the use of mono-temporal datasets is not reliable.

The product examples provided above illustrated the fusion approach for the case where the available radar data were limited. The forest/non-forest, species and *GSV* products were also processed with the numerous available datasets of the Thuringian Forest. The species and *GSV* maps derived using the large number of datasets given in the Thuringian test site are presented below.

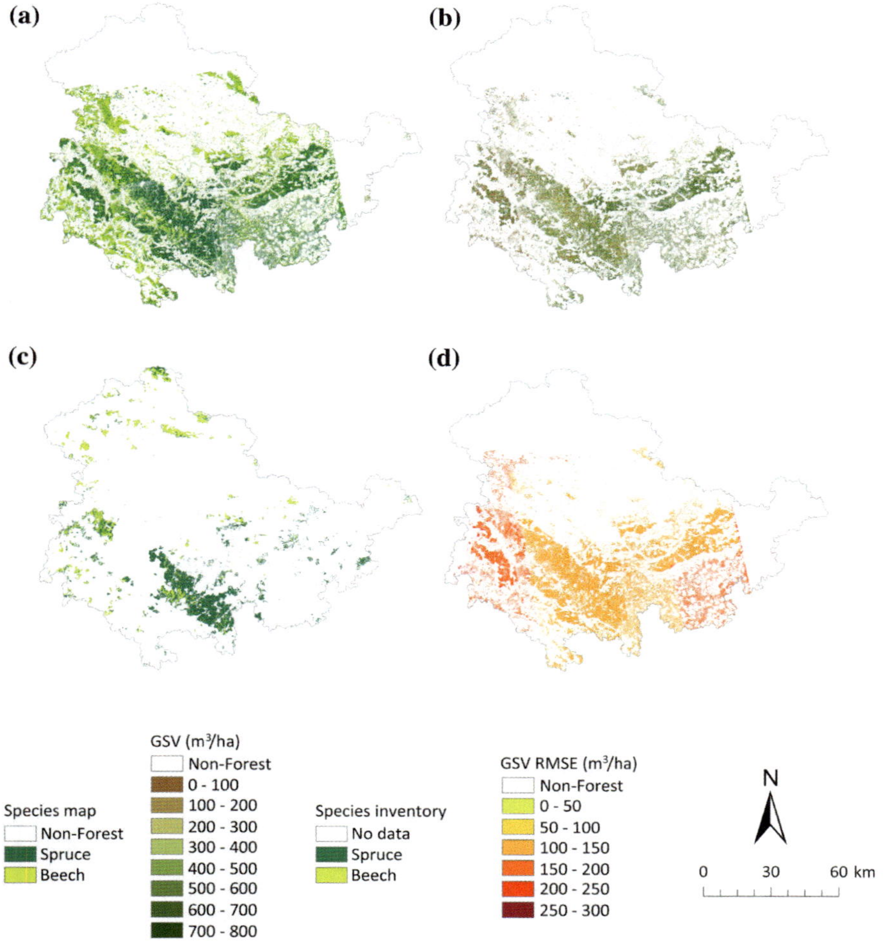

Fig. 5.48 Illustration of **a** processed SAR species map, **c** inventory species map, **b** processed SAR *GSV* map and **d** its corresponding *RMSE* accuracy map. The processed SAR species map was based on the following datasets: 220 PALSAR FBD intensity; 47 PALSAR FBS intensity; 39 PALSAR PLR intensity; 19 PALSAR coherence; 24 CSK intensity; 12 CSK coherence; 8 TDX intensity; 4 TDX coherence; and 122 TSX intensity. In addition to these datasets, the processed *GSV* map included 35 *GSV* images derived from the regression modelling shown in Sect. 5.4. The species and *GSV* maps were processed at a spatial resolution of 25 m. The *GSV* map product represented only spruce, as the modelling was based on spruce. Beech was masked using the species map depicted in (**a**). Pine was not masked as it was not possible to separate it from spruce

Figure 5.48 presents the (a) species and (b) *GSV* maps processed for Thuringia. In addition, the species provided by the forest inventory and the *RMSE* map derived to evaluate the *GSV* map are depicted in (c) and (d), respectively. A number of 460 intensity and 35 coherence scenes acquired under various configurations were used for the processing of the species map. In addition to these datasets, 35 *GSV* images derived from the regression modelling shown in Sect. 5.3 were considered for the processing of the *GSV* map. The SAR-derived maps were processed at a spatial resolution of 25 m. In other words, the X-band datasets and the PALSAR FBS data pre-processed at 10 and 12.5 m, respectively, were downsampled to 25 m. Downsampling was performed using the pixel aggregation method. Also, the *GSV* map product presented in Fig. 5.48b only represented spruce. Spruce depicted the highest sensitivity to coherence and was therefore the only species chosen for the modelling described in Sect. 5.4. Because spruce was the only species considered for the *GSV* map, beech was masked using the species map presented in Fig. 5.47a. However, pine was not masked because its separation from spruce was not possible. The investigations is this thesis showed that the leaf-off period is more suitable than the leaf-on period for separating broadleaves and conifers (see Table 5.7). Due to time constraint, this characteristic was not taken into account for the products depicted in Fig. 5.48. Finally, to process the forest/non-forest information, the distinction between images acquired before and after the Kyrill storm was not considered. In fact, it was chosen to keep a maximum datasets available for the processing of the end products.

A comparison between the processed species map in Fig. 5.48a and the one resulting from the forest inventory (Figure c) shows that the forest coverage differs. As mentioned in Sect. 4.2, information concerning private forested areas could not be delivered for political reasons. However, the missing information on the forest inventory may still be displayed using remote sensing data. Comparing the species' extension between Fig. 5.48a, c, it is observed that the dominant spruces and beeches in the South and North regions, respectively, can be easily distinguished on the processed species map (Fig. 5.48a). In Fig. 5.48b, similar to the species map shown in Fig. 5.48a, the spruce forests of the Thuringia are easily recognizable with their Northwest to Southeast extension. The accuracy (*RMSE*) map given in Fig. 5.48d indicates that the *RMSE* generally ranges between 100 and 200 m^3 ha^{-1}, as expected from the modelling results from Sect. 5.4 summarised in Table 5.8. The lower accuracy of the *GSV* map in the West and Southeast than that in other regions is indicated in dark orange. In terms of the spatial baseline and weather conditions, the acquired configurations of the PALSAR coherence scenes were not optimal for these parts of the site, which explains the decreased accuracy in these regions. According to the fusion algorithm described in Sect. 4.4, the *GSV* map in Fig. 5.48b represents, the *GSV* estimates having the lowest *RMSE* among the overlaying *GSV* images. In the case of multitemporal *GSV* images, instead of considering the *GSV* values with the lowest *RMSE*, the available multitemporal *GSV* images could have been merged. This process was not considered in the proposed fusion approach because such an operation would have improved the quality of the *GSV* estimates only if a large number of multitemporal datasets were available (see Santoro et al. 2011) and if the configuration of most of these datasets was optimal, i.e. with a large baseline and no particular precipitations (see Sect. 5.3). In the case of the Thuringian Forest, these conditions were not met; hence, the averaged *GSV* images would have decreased the quality of the estimates instead of improving it.

The above visualisations present a general overview of the products but do not enable the local assessment of their quality. To provide a critical perspective of the generated products, specific areas were selected and compared with some reference data. Figure 5.9 illustrates some of the selected areas for the species and *GSV* products. The different columns of Fig. 5.9 show these areas in a digital orthophoto, together with the SAR species product, inventory species information, the SAR *GSV* product and inventory *GSV* information.

The first area was chosen in the Schönbrunn Lake region, as presented in Sects. 5.1 and 5.2, as this area presented numerous spruce and beech stands as well as a sloping terrain. A comparison of Table 5.9a–c with the SAR species map, the orthophoto and the inventory information indicates that spruces and beeches are well distinguished. Hence, the combined ratio of the L-band and X-band information appeared to work efficiently, which is consistent with the SAR data available for this region (see Appendix A.1) and the data configurations presented in Table 5.7. The aerial views shown on orthophoto were acquired during the leaf-off (April) and leaf-on (May) seasons, which explains why in Table 5.9a the image present a separation and beeches are mostly depicted in brown soil and branches. The *GSV* map in Table 5.9d indicates that the general patterns between the SAR *GSV* spruce map and the forest inventory appeared to match only for young spruces shown in brown. These observations are similar to the ones depicted in Sect. 5.4.2. Finally, comparing the forest/non-forest regions in the species map (Table 5.9b) with the orthophoto and inventory map, it is observed that the forested areas agreed well with the reference data.

The second area was selected to illustrate some misclassifications in the case of the species product or inaccuracies in the *GSV* product observed at forests edges adjacent to an open area facing the radar system.

With regards first to the species map in Table 5.9g, these specific conditions enhanced the SAR backscatter intensity with double reflections and contributed to the assignment of beech instead of spruce. Further investigations on forest edges showed that the dihedrals were mostly noticed at X-band frequencies and at this frequency, their intensity extended in the same range as that of beech (−8 to 0 dB). Consequently, as the X-band data covered the given area and were used for the separation of spruce from beech in the present case (according to Table 5.7), the spruce forest edges were classified as beech. Also, X-band interferometric data were available for the area shown in Table 5.9e. However, the scatterers of forest edges facing the radar presented low coherence, and therefore, the misclassified beeches could not be masked by coherence. SAR acquisitions are performed in ascending and descending mode. A potential solution for solving the edges miss-classifications would be to distinguish in the developed fusion approach the different acquisition passes. With reference to urban landcover, it should be noted that the high intensity owing to forest-edge dihedrals is also close to the intensity shown by urban areas. This landcover could be optimally masked by X-band or L-band coherence. In addition to double bounces at forest edges facing the radar, the edges facing away from the radar produced a shadow. However, the shadow was not problematic in the present case because its backscatter intensity was close to the intensity of the *water* landcover class, which could be easily masked using the available satellite information (see Table 5.8, recommended data).

Table 5.9 Comparisons of the SAR species and *GSV* end products with digital orthophotos and forest inventory information

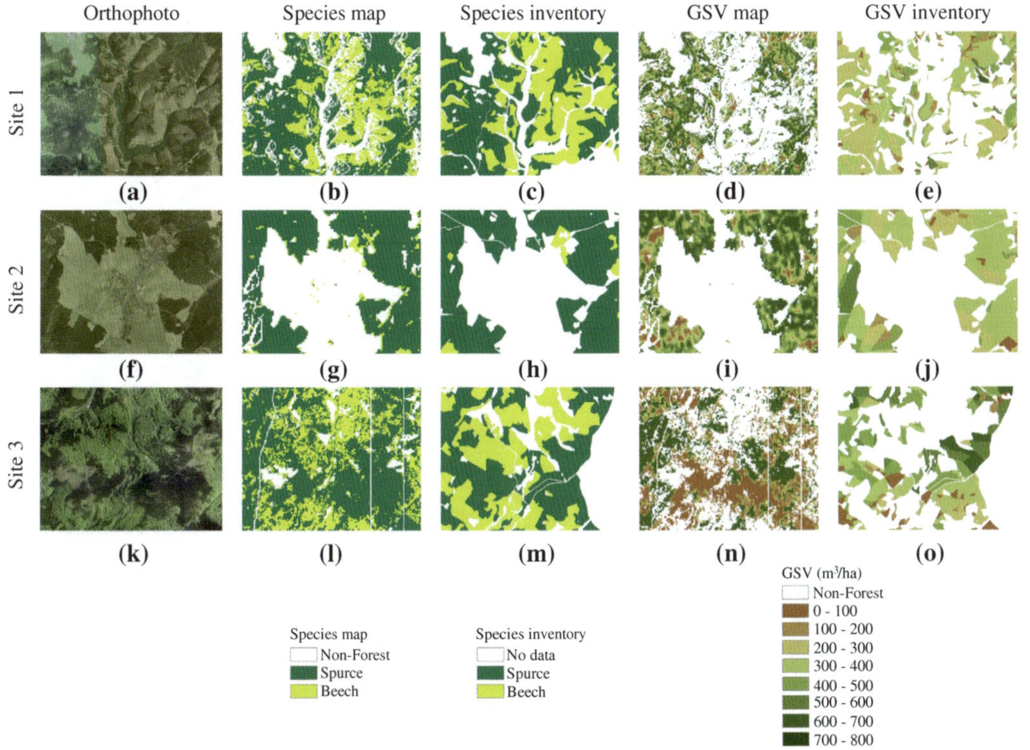

The comparisons are based on the products presented in Fig. 5.48. The images aligned in lines ((a)–(e), (f)–(j), (k)–(o)) show different areas. The first area confirms the good classification of spruce and beech in a complex terrain with topography. The second area demonstrates some effects specific to forest-stand edges. The last area represents particular processing artefacts

Observing then the *GSV* map in Table 5.9i, it can be seen that the forest edges classified as beech were masked, as the *GSV* map was modelled only for spruces. In fact, if beech *GSV* values were modelled with coherence, the decorrelated edges would have been depicted in dark green, suggesting high *GSV* values. This is explained by the parametric models used in Sect. 5.4 which would have interpreted the low coherence of the edges as areas presenting significant vegetation. Disregarding the aforementioned artefacts, it is observed that as for the first selected area (Table 5.9a–e), the species classification apparently agree well with the reference information and *GSV* values seemed to match to the inventory data only for young spruces.

The third example of the selected area focused on a specific artefact which was observed for the different products. As depicted in Table 5.9l, n, some lines representing non-forested areas are visible. Comparing the area with the orthophoto or inventory data, these lines clearly appear as a non-physical effect. Some examinations showed that this effect was represented only by the resampled data, such as X-band datasets and PALSAR FBS data. Indeed, the resampling of these datasets to 25 m, suggested merging of the data contained in the border pixels of the images with nodata values, leading to misclassifications of the forests. This problem might be solved in future with an improved resampling method. In contrast to the first two examples, in the third example, the general patterns of the SAR species and *GSV* maps did not agree with the reference data. This was explained by the limited SAR data available in this given area.

Quantitative results based on error matrices may be presented after visual assessments of the species and *GSV* SAR products. Two confusion matrices were created: one referring to the forest/non-forest classification and the other referring to the separation between spruce and beech. Table 5.10 describes the two confusion matrices which refer to the data presented above.

Before interpreting the matrices, several points should be made. First, the confusion matrices were computed on a pixel basis using the inventory information as reference. Then, a specific area was selected in the center of the site. Indeed, as shown in Fig. 5.47, the inventory data did not cover private forests; this would have led to an increase in omission and commission errors if considered in the error matrix, therefore leading to grossly inaccurate results. Finally, spruce and beech were two important species of the test site. However, other species such as pine, fir and oak were also present. To eliminate this potential bias from the accuracy results, the remaining species were masked on both SAR and inventory species maps.

The error matrix for the forest classification shown in Table 5.10a indicates that more than three-quarters of the forests were well classified. With regard to the available data covering the test site and the recommended datasets presented in Tables 5.5 and 5.6, this accuracy appears to be reasonable. As mentioned above, radar acquisitions measured before and after the Kyrill storm were considered for the computation of the forest/non-forest map. By taking this event into account in the selection of the radar data, the accuracy should be improved. With regard to the species classification, an accuracy of 78 % was achieved. This accuracy is fairly acceptable. However, as mentioned above, leaf-off and leaf-on periods for X-band data were not distinguished in this example. In this framework, some improvements may be performed by taking into account this vegetation phenology.

Table 5.10 Accuracy matrices for the processed (a) forest/non-forest and (b) species map products of Thuringia

(a)

Classified data \ Reference data	Non-forest	Forest
Non-forest	78446	64584
Forest	62983	473961

Overall classification Accuracy: 81%

(b)

Classified data \ Reference data	Non-forest	Spruce	Beech
Non-forest	93945	38912	16354
Spruce	89340	360217	26196
Beech	11820	14177	29013

Spruce vs. Beech classification Accuracy: 78%
Overall classification Accuracy: 71%

5.5.4 Summary

This section presents the results of the fusion approach developed in this study. After briefly presenting the performed histogram analysis, which enabled the extraction of forested areas and separation between tree species, the suggested data configurations for fusion processing were discussed and examples based on different data configurations were proposed. The data configurations were split into three categories: recommended, minimal and non-recommended configurations. On the basis of the histogram analysis, the recommended configurations for identifying forested areas referred to the L-band backscatter intensity because at this frequency, the differences between the scattering mechanisms of forested areas and other landcovers are significant. In addition to amplitude information, the interferometric coherence was shown to properly separate forested areas from *urban areas*, especially at X-band frequencies and with a 1-day repeat-pass. The discrimination between the tree species was found to be limited. According to the suggested data configuration, only a minimal configuration could be proposed by considering the ratio of L-band intensity to X-band HH intensity. This ratio only showed satisfactory results for the separation of conifers from broadleaves. The differentiation between the two conifers spruce and pine was not possible. On the basis of the different data configurations, product examples processed with an established fusion methodology were proposed. The first example highlighted the case in which the recommended data configurations were not available. In this case, it was shown that when dealing with various datasets having different extensions, the key is to keep track of the map quality. This issue was resolved by proposing quality-flag maps, which enabled the effective evaluation of the established products. The second example illustrated the case in which numerous and various datasets could be used. The *GSV*, forest and species maps for this example consisted of the end products processed for the Thuringian test site. To validate these end products, qualitative and quantitative examinations were performed. The qualitative assessment relied on the comparison between the species and *GSV* maps with reference data from three selected areas. Generally, the comparisons showed that the patterns depicted by the species map were similar to those shown by the reference data. In the case of the *GSV* image, only young spruces were shown to match the inventory data. Some artefacts could be also highlighted. One of these artefacts was found to be related to forest edges which enhanced the backscatter intensity with dihedrals and decreased the coherence due to the particular forest scatterers. Another artefact could be associated with processing issues. Quantitative product evaluation was performed using error matrices in order to assess the quality of the forest/non-forest and species classifications. As for forest identification, approximatively 81 % of the pixels were successfully classified, while as for the species map, about 71 % of the pixels were successfully classified. These results appeared to agree with the previous visual observations. The quantitative evaluation of the GSV product refers to the processed RMSE flag. In the given example, due to the limited number of multitemporal PALSAR coherence data acquired under optimal conditions (i.e. large baseline, dry conditions), it was chosen to not perform any merging fusion of the estimated GSV images. Therefore, the final accuracy of the GSV map performed for the Thuringian Forest corresponds to the statistics presented in Sect. 5.4. In future studies, the presented fusion algorithm may be tested with a large number of temporal GSV datasets. Furthermore, the merging fusion and synergy process proposed in this thesis may be improved by considering weighting functions such as the one presented in Santoro et al. (2011). The fusion approach demonstrated in this chapter shows that the combination of different datasets is promising for improving the mapping of forest biomass. This approach allows extending the spatial extension of the forest products, their updated frequency and potentially their accuracy. The fusion algorithm presented for the Thuringian Forest may be applicable at a national and global scale.

References

Allain C, Cloitre M (1991) Characterizing the lacunarity of random and deterministic fractal sets. Phys Rev A 44:3552–3558

Baffetta F, Fattorini L, Franceschi S, Corona P (2009) Design-based approach to k-nearest neighbours technique for coupling field and remotely sensed data in forest surveys. Remote Sens Environ 113(3):463–475

Brolly M, Woodhouse IH (2012) A 'Matchstick Model' of microwave backscatter from a forest. Ecol Model 237–238:74–87

Champion I, Dubois-Fernandez P Dupuis X (2011) Retrieving forest biomass from the texture of SAR images. In: EARSeL eProceedings, Czech Republic, Prague, pp 102–109

Chung SY, Yim JS, Cho HK, Shin MY (2009) Comparison of forest biomass estimation methods by combining satellite data and field data. In: Proceedings of IUFRO Division 4: extending forest inventory and monitoring 2009, Quebec, Canada

Cloude S (2009) Polarisation: applications in remote sensing. Oxford University Press, Oxford

Cloude SR, Papathanassiou KP (2003) Three-stage inversion process for polarimetric SAR interferometry. IEE Proc Radar Sonar Navig 150(3):125

Fehrmann L, Kleinn C (2007) A k-Nearest Neighbour (k-NN) Approach for Estimation of Single Tree Biomass. In: McRoberts RE, Reams GA, Van Deusen PC, McWilliams WH (eds) Proceedings of the seventh annual forest inventory and analysis symposium. U.S. Department of Agriculture Forest Service, Portland, pp 251–259

Fehrmann L, Lehtonen A, Kleinn C, Tomppo E (2008) Comparison of linear and mixed-effect regression models and a k-nearest neighbour approach for estimation of single-tree biomass. Can J For Res 38(1):1–9

Ferrazzoli P, Guerriero L (1995) Radar sensitivity to tree geometry and woody volume: a model analysis. IEEE Trans Geosci Remote Sens 33(2):360–371

Franco-Lopez H, Ek AR, Bauer ME (2001) Estimation and mapping of forest stand density, volume, and cover type using the k-nearest neighbors method. Remote Sens Environ 77(3):251–274

Haralick RM, Shanmugam K, Dinstein I (1973) Textural features for image classification. IEEE Trans Syst Man Cybern 8:610–621

Holmgren J, Joyce S, Nilsson M, Olsson H (2000) Estimating stem volume and Basal area in forest compartments by combining satellite image data with field data. Scan J For Res 15(1):103–111

Kajisa T, Murakami T, Mizoue N, Kitahara F, Yoshida S (2008) Estimation of stand volumes using the k-nearest neighbors method in Kyushu. Jpn J For Res 13(4):249–254

Moghaddam M, Saatchi S (1993) Analysis of scattering mechanisms in SAR imagery over boreal forest: results from BOREAS '93. IEEE Trans Geosci Remote Sens 33(5):1290–1296

Neumann M, Saatchi SS, Ulander LMH, Fransson JES (2012) Assessing performance of L- and P-band polarimetric interferometric SAR data in estimating boreal forest above-ground biomass. IEEE Trans Geosci Remote Sens 50(3):714–726

Ortiz SM, Breidenbach J, Knuth R, Kändler G (2012) The influence of DEM quality on mapping accuracy of coniferous- and deciduous-dominated forest using TerraSAR-X images. Remote Sens 4(12):661–681

Rauste Y (1990) Incidence-angle dependence in forested and non-forested areas in Seasat SAR data. Int J Remote Sens 11(7):1267–1276

Rauste Y, Häme T, Pulliainen JT, Heiska K, Hallikainen MT (1994) Radar-based forest biomass estimation. Int J Remote Sens 15(14):2797–2808

Santoro M, Beer C, Cartus O, Schmullius C, Shvidenko A, McCallum I, Wegmüller U, Wiesmann A (2011) Retrieval of growing stock volume in boreal forest using hyper-temporal series of Envisat ASAR ScanSAR backscatter measurements. Remote Sens Environ 115(2):490–507

Sun G, Ranson KJ (1998) Radar modelling of forest spatial patterns. Int J Remote Sens 19(9):1769–1791

Sun G, Simonett DS, Strahler AH (1991) A radar backscatter model for discontinuous coniferous forests. IEEE Trans Geosci Remote Sens 29(4):639–650

Villard L, Borderies P, Le Toan T, Koleck T, Albinet C (2010) Topography effects on forest radar scattering, consequences on biomass retrieval. In: 2010 IEEE international geoscience and remote sensing symposium, IEEE, pp 60–63

Westman WE, Paris JF (1987) Detecting forest structure and biomass with C-band multipolarization radar: physical model and field tests. Remote Sens Environ 22(2):249–269

Woodhouse IH (2006) Predicting backscatter-biomass and height-biomass trends using a macroecology model. IEEE Trans Geosci Remote Sens 44(4):871–877

Zhu J (2012) Multi-sensor estimation of forest stem volume using non-parametric method KNN over temperate forest. M.Sc. thesis, Friedrich-Schiller University of Jena

Chapter 6
Conclusions

The present thesis involved estimating the *GSV* throughout a temperate forest using the fusion of SAR information. After briefly introducing the thesis and reviewing the literature and theory related to SAR and forests' *GSV* retrieval, this document presented details on the test site, available data, and results of the investigations. To conclude this thesis, a summary of the results was prepared and perspectives for future studies were defined. In this chapter, the first section summarises the main outcomes of the thesis and the second section provides the outlook for future investigations.

6.1 Principal Outcomes of the Thesis

The research undertaken in this thesis focused on the Thuringian Forest, in the centre of Germany. The main characteristics of the forest are a marked topography, large range of *GSV* and pre-dominance of three different tree species, namely Norway spruce, European Beech and Scots Pines. The SAR instruments considered for the investigations consisted of four different sensors, namely ALOS PALSAR, TSX, CSK and TDX. These spaceborne platforms allowed the acquisition of an extensive set of SAR data measured in various sensor configurations (frequencies, polarisations, incidence angles). In addition to the satellite data, ancillary datasets such as a forest inventory, a LiDAR DEM, digital orthophotos and weather information were collected to support the processing and validation works. To enable the comparison of these different pieces of information, the SAR and ancillary datasets were pre-processed to high-level products. To this end, recent methods for calibration, geocoding, co-registration and topographic normalisation were used. The pre-processing of SAR data included the backscatter intensity as well as the interferometric coherence, polarimetric decompositions and loci coherence.

With respect to the investigations of the data, attention was first given to the topography of the test site. The effects of slopes on the backscatter intensity and the coherence were examined. It may be shown that topography leads to significant

N. Ackermann, *Growing Stock Volume Estimation in Temperate Forested Areas Using a Fusion Approach with SAR Satellites Imagery*,
Springer Theses, DOI 10.1007/978-3-319-13138-2_6

variations of the signal, even when modern topographic corrections were applied to the data. The topographic variations were found to be related to forests' specific scattering and decorrelation mechanisms which occur over sloped terrain. However, the examinations did not allow for a complete understanding of the ongoing phenomena.

After presenting the effects of topography, an in-depth analysis of the scattering mechanisms was performed. In this regard, time-series, frequency, incidence angle, polarisation and pass direction analyses were carried out. The results of these investigations mainly demonstrated that radar data have the potential to separate conifers from broadleaves. In particular, leaf-off period, a co-pol backscatter intensity ratio X-band/L-band, positive temperature and large incidence angle appeared to be optimal. However, the separation of two conifer species such as spruce and pines was shown to be limited. In addition to the comparison of the different radar parameters, the sensitivity of the radar backscatter to *GSV* was examined. The correlation between the SAR amplitude signal and forest *GSV* was found to be poor for the three investigated tree species ($R^2_{\text{Spruce}} = 0.26$, $R^2_{\text{Pine}} = 0.16$ and $R^2_{\text{Beech}} = 0.11$), even by using L-band frequency data. Further analyses based on air temperature, forest structure and scattering intensity decomposed into surface, volume and double bounce contributions were undertaken in order to understand this poor correlation. It was found that the high range of *GSV* and the horizontal structure of the forest played a determining role on the amplitude signal. In particular, because the Thuringian Forest is managed for commercial purposes, young forest stands appeared to be predominantly very dense, increasing the backscatter intensity and leading to an early saturation of the signal ($GSV_{\text{Saturation}} < 100\,\text{m}^3\,\text{ha}^{-1}$).

After examining the scattering mechanisms, the decorrelation mechanisms which are due to interferometric coherence were investigated. Comparisons of Whisker boxplots with TSX, CSK, TDX and ALOS PALSAR interferometric systems between open and forested areas showed a significant contribution of temporal decorrelation in X-band frequency with 1 day and 11 days repeat-pass. Using these sensors, the volume decorrelation could also be demonstrated, especially at L-band frequency with ALOS PALSAR. Among the different abiotic factors controlling the decorrelation mechanisms, the weather conditions with precipitations appeared to have the most significant effect. As for the backscatter intensity, the interferometric coherence sensitivity to *GSV* was measured. A decreasing trend with increasing coherence was found for the three investigated tree species. Comparing different normal baselines with PALSAR FBD data, it was shown that the increase in the spatial baseline improved the sensitivity of interferometric coherence to *GSV*. However, total decorrelation of the signal and consequently poor sensitivity were found with the normal baseline exceeding $B_n = 3.3\,\text{km}$, which corresponded to approximately 40 % of the PALSAR FBD critical baseline. Surprisingly, a comparison of the available SAR sensors showed that the best correlation was obtained for TDX, followed by PALSAR and CSK with $R^2_{\text{TDX}} = 0.72$, $R^2_{\text{PALSAR}} = 0.61$ and $R^2_{\text{CSK}} = 0.51$, respectively. TDX coherence exhibited a linear negative relationship without the presence of any particular saturation point up to $600\,\text{m}^3/\text{ha}$. In contrast to TDX, CSK and PALSAR presented various nonlinear and linear negative relations and a defined saturation point. The saturation in the optimal configuration (i.e. large normal baselines, little precipitations) occurred for CSK around 100–200 m^3 ha^{-1} for beech and spruce, respectively, and for PALSAR around 200 m^3/ha for the three species composition. The level of saturation was found to be highly dependent on the perpendicular baseline, and therefore varied significantly within the different interferometric acquisitions. Comparing the three investigated species, the highest coherence and best correlations were generally observed for Norway Spruce, followed by pine and beech. This difference between species was explained mainly by the dissimilar canopy structure and foliage composition of the trees.

Based on the investigation results of the scattering and decorrelation mechanisms, a modelling process for the retrieval of *GSV* was established using PALSAR and TDX interferometric coherence as the main source of information. The modelling was based on two different techniques: parametric and non-parametric. The parametric modelling technique involved simple regressions and was split into different steps, including model selection, training, inversion and testing. The achieved *RMSE* for ALOS PALSAR ranged from 100 to 200 m^3 ha^{-1} according to the interferometric acquisition configurations. This accuracy was clearly not sufficient for applications such as forestry management or climate modelling. To understand the large error, further investigations based on the topography of the site, forest inventory and processing approach were performed. It was found that the topographic variations significantly increased the error, in particular for TDX. It was also shown that the selection of reliable forest stands in terms of homogeneity was relevant. The non-parametric technique was based on the k-nearest neighbour (k-NN) and was conducted for comparisons with the regression approach. Optimal parameters in terms of k number, distance metrics and weights were investigated for PALSAR coherence. Based on a leave-one-out cross-validation technique, the *RMSE* was computed at stand level for the available data. The achieved accuracy with the k-NN approach was improved by about 50 m^3 ha^{-1} compared with the parametric technique. However, the estimates were still not sufficiently accurate for envisaging applications in the field of forestry or climate change research.

The final focus of this work was to develop a fusion methodology allowing the derivation of *GSV* from different SAR information. With regards to the results of the data investigations and the available data, the combination approach was focused on the established products, such as a *GSV* or a forest/non-forest map. By combining the different datasets, this study aimed at obtain the final *GSV* map from different sources of data in order to achieve an efficient combination of the available datasets. The objectives of this approach were to increase the accuracy of the *GSV* map, its spatial extension and the updated frequency. Such an approach was envisaged because of the numerous datasets and sensors which have become available in recent years. The developed fusion approach included the derivation of forest/non-forest and species masks. The latter were based on preliminary histogram analyses which allowed the automatic derivation of classes' thresholds and separability values. The histogram investigations led to the suggestion of different data configurations, namely a not recommended, a minimal and a recommended configuration. Among the recommended configurations for separating forest from other landcovers, the consideration of L-band HV intensity and X-band 1-day coherence was shown to be the optimal solution. The L-band frequency was useful primarily to separate forests from *crops* and *open areas*, whereas the X-band coherence was suitable mainly for discriminating between forested and *urban areas*. Concerning the differentiation of tree species, the ratio of the X-band/L-band HH intensity was generally shown to be the most efficient dataset, in particular for the discrimination of broad-leaves from conifers. However, according to the defined configurations, spruce and beech separation was limited to minimal configurations, while the distinctions between pine and beech and that between spruce and pine were limited to non-recommended configurations. The processing of forest *GSV* maps within the developed fusion algorithm was performed with two additional products, namely a forest/non-forest map and a species map. Each of these products was delivered with an accuracy flag map so that the quality of the products could be tracked. Accounting that the optimal data configuration may be only partially available or completely unavailable, examples of products computed with limited amount of data were first presented. These examples highlighted the usefulness of the quality flags accompanying the products. To illustrate the cases where numerous and various datasets could be used, the large quantities of data for the Thuringian Forest were computed using the developed fusion algorithm to generate a *GSV*, species and forest/non-forest map. The latter consisted of the end products for the Thuringian Forest. The assessment of these three end products was performed qualitatively using visual comparisons with reference data. In addition, quantitative evaluations based on error matrices were performed for the species and forest/non-forest maps. Quantitative assessments for *GSV* estimates were not considered for the potential improvement of the *GSV* accuracy, as consistent multitemporal data were required, and in this study, such datasets were not available. In this case, the quantitative assessment of the *GSV* product with the *RMSE* flag directly indicated the accuracy obtained with the regression modelling approach. The qualitative evaluation showed that the SAR derived species map and the reference data depicted similar patterns. In the case of the *GSV* estimates, young forests appeared to better match the inventory than did mature forests. The visual observations also showed that the products contained some artefacts. One could be related to the forest edges, while another could be explained by processing issues. The quantitative validation work established approximatively 81 and 71 % of well-classified pixels for the forest/non-forest and species maps, respectively. These accuracies appeared to match with the previous visualisations.

6.2 Perspectives for Future Studies

Based on the findings of this work, some perspectives regarding future studies were drawn. These perspectives were addressed by a central question which aims to determine how the *GSV* estimation could be improved in test sites such as the Thuringian Forest.

According to the results of this thesis, one major issue affecting the Thuringian Forest is the topography of the site. For forested areas, the analyses of the topography with the radar backscatter intensity and interferometric coherence in Sect. 5.1 clearly showed systematic variations of the SAR signal with the slopes' orientations. These variations could not be fully explained and will require further investigations. In this respect, it is suggested to differentiate between the influence of slopes on the physiology of trees from the influence of slopes on SAR microwaves. In the case of the tree physiology, observations made during the field campaigns clearly showed, for example, various trunk shapes and heights which were related to wind and topography. With respect to the SAR microwave interactions in topographic forested areas, physical modelling should be considered to understand the different scattering and decorrelation mechanisms. Some works have been presented in the domain of the scattering contributions such as in Amar et al. (1993) or more recently, in Villard et al. (2010) with the MIPERS model. In the case of the interferometric coherence, some studies introduced spectral-shifts compensations (Santoro et al. 2007; Lee and Liu 2001), but it appears that no study was performed to correct the volume decorrelation induced by slopes. The SAR signal returning from forests in topographic areas is relatively complex and will require further studies to enable the accurate estimation of *GSV* in topographic regions. The topographic correction will have to be specific to the SAR system configurations (frequency, polarisation, incidence angle) as well as to the forest characteristics (biome, species, structure). In addition to providing an understanding of the physics of the forest and the SAR microwaves, the processing of topographic normalisation may be improved. Recent papers proposed new correction methods for improving the radar signal in hilly areas. For example, Small (2011) and Frey et al. (2013) worked on the geometric distortions of SAR images, and suggested to improve the calculation of the pixel area based on the integration of DEM pixels using small triangular surface patches. This correction was not applied in this thesis, and may be considered for future studies.

In addition to the topography, another challenging issue of the Thuringian test site was the large range of *GSV* (up to $750\,\mathrm{m}^3\,\mathrm{ha}^{-1}$) and the horizontal structure of the forest, which contribute to the early saturation of the SAR amplitude signal (see Sect. 5.2). To address this problem and obtain accurate estimations of *GSV*, different solutions may be envisaged. A first solution would involve retrieving information about the forest distribution and incorporate it in a semi-empirical model. The forest distribution may be better understood by performing theoretical modelling, such as in Woodhouse (2006) or Brolly and Woodhouse (2012), or by deriving texture parameters. Computation of Haralick parameters and lacunarity were respectively shown, for instance in Kuplich et al. (2005) and Su and Krummel (1996), to be potential methods for the estimation of forests density. A second solution would rely on allometric relations and the interferometric height. TDX interferometric acquisitions were mainly used in this thesis to estimate the interferometric phase coherence. However, some preliminary studies (which are not discussed in this document) showed that it may be possible to retrieve forest stand heights using TDX data (see Appendix D). Forest height is one of the most relevant parameters for estimating *GSV* and would therefore warrant further investigations. A study undertaken by Rauste et al. (1994) showed that for a forest in Freiburg, Germany, P-band SAR backscatter data did not saturate, even for *GSV* above $700\,\mathrm{m}^3\,\mathrm{ha}^{-1}$ (Rauste et al. 1994). The Freiburg test site depicted almost equivalent characteristics to those of the Thuringian Forest, which suggest that similar results should be found with P-band frequency in the test site of this thesis. The consideration of P-band data for the estimation of *GSV* in the Thuringian Forest would be an asset, especially because of the high *GSV* range of the forest. At the time of writing this thesis, P-band frequency was not available on spaceborne platforms and no aerial data was available for the Thuringian Forest. However, the BIOMASS P-band mission was selected by the European Space Agency (ESA) 7th Earth Explorer Opportunity Mission and should be launched in 2019 for the retrieval of biomass across the whole globe (Le Toan et al. 2012).

The investigation of the scattering mechanisms in Sect. 5.2 denoted the difficulty involved in understanding the microwave interactions which occur in the Thuringian Forest. In particular, it was shown that some of the actual models discussed in the literature such as the Yamaguchi decomposition or the WCM were not suitable for this forest. In this context, it may be necessary to rethink the modelling approaches for the conditions given in the Thuringian Forest. Typical forest scattering models discretise the forest into one or two layers (see Sect. 2.1), the first layer representing the ground and the second layer representing a random homogenous volume. Certain species such as Norway Spruce may not have a random volume, but an oriented volume, as the branches are oriented concentrically along the stems and in equivalent directions. To gain a better understanding of the scattering mechanisms in the forest and the estimation of *GSV*, further analyses such as the use of PolSAR or PolInSAR data with a suitable temporal baseline may be envisaged. For example, the assumption of a random volume instead of oriented volume may be investigated using the RVoG model.

The outcomes of this thesis summarised in Tables 5.5, 5.7 and 5.8 in Sect. 5.5 showed that the use of a single sensor and acquisition mode was generally limited to the production of accurate and reliable forest *GSV* maps. In this regard, a final option for improving the *GSV* in the Thuringian Forest would be to combine the available information. The combination of SAR information was actually the main topic of Sect. 5.5. A fusion algorithm was conceived with certain flexibility, in the sense that it consisted of a core process which could be easily updated with new functionalities and datasets or which could also be replaced with improved methodologies. The main elements of the algorithm were implemented in IDL routines, which together formed a whole processing chain. Regarding the established fusion algorithm, some suggestions can be made for future works. First, new sensors, in particular optical sensors, may be integrated into the existing datasets. For example, the introduction within the framework of the ENVILAND2 project (see Sect. 1.2) of RE information in the fusion algorithm showed significant improvements regarding the separation of the tree species. The consideration of three or more species will require a few adaptations of the established fusion algorithm. In particular, the process used to retrieve the classes' thresholds will need to be revised and the *GSV* will have to be retrieved for each specific species, implying the need to fit *GSV* models to each of the investigated tree species. Second, as described in Sect. 5.5, the multitemporal merging of *GSV* data was not envisaged in the algorithm because of the limited amount of available multitemporal coherence data used for deriving the *GSV* images and the limited number of coherence images which were optimally acquired, namely with a large baseline and no precipitations. Hence, for the case under study, the fusion process did not improve the accuracy of the *GSV* estimates. In this regard, to increase the accuracy of *GSV* retrieval, a large number of multitemporal coherence images acquired under optimal conditions may be investigated. Also, considerations of the *GSV* according to criteria such as weather information, forest stands relative stocking or percentage of dominant tree species may be conceived. A multitemporal approach, such as the one presented in Santoro et al. (2011), may also be implemented within the proposed fusion algorithm. Third, disregarding the *GSV* accuracy, a major drawback of the proposed fusion approach is the supervised fashion of the algorithm. Indeed, the masking process part of the fusion approach required knowledge about landcover classes in addition to the modelling approach which required stands of reference for estimating *GSV*. To improve the transferability of the fusion approach, the extraction of a fully automatic training site would be required. It should be noted that this topic was studied in the framework of the ENVILAND2 project.

In conclusion, to derive a *GSV* map, this thesis has performed an in-depth analysis of several topics. The investigations confirmed some of the results established in the literature and provided new knowledge for future scientific studies focusing on the estimation of forest *GSV*. These new outcomes will serve as a foundation for the development of more accurate and above all more robust retrieval algorithms. Finally, while there remains much work for the realisation of accurate, global and fully automatic estimates of *GSV*, this thesis showed that the remote sensing techniques are promising for the operational mapping of forest *GSV*.

References

Amar F, Fung AK, De Grandi G, Lavalle C, Sieber A (1993) Backscattering from forest canopies over slanted terrain. In: Proceedings of IGARSS'93—IEEE international geoscience and remote sensing symposium, vol 2, pp 576–579

Brolly M, Woodhouse IH (2012) A 'Matchstick Model' of microwave backscatter from a forest. Ecol Model 237–238:74–87

Frey O, Santoro M, Werner CL, Wegmuller U (2013) DEM-based SAR pixel-area estimation for enhanced geocoding refinement and radiometric normalization. IEEE Geosci Remote Sens Lett 10(1):48–52

Kuplich TM, Curran PJ, Atkinson PM (2005) Relating SAR image texture to the biomass of regenerating tropical forests. Int J Remote Sens 26(21):4829–4854

Le Toan T, Ulander L, Papathanassiou K, Rocca F, Quegan S, Davidson M, Scipal K (2012) The BIOMASS mission retrieval algorithms: results from recent campaigns. In: 2012 IEEE international geoscience and remote sensing symposium, pp 5546–5549

Lee H, Liu JG (2001) Analysis of topographic decorrelation in SAR interferometry using ratio coherence imagery. IEEE Trans Geosci Remote Sens 39(2):223–232

Rauste Y, Häme T, Pulliainen JT, Heiska K, Hallikainen MT (1994) Radar-based forest biomass estimation. Int J Remote Sens 15(14):2797–2808

Santoro M, Askne J, Werner CL, Wegmüller U (2007) Observations, modeling, and applications of ERS-ENVISAT coherence over land surfaces. IEEE Trans Geosci Remote Sens 45(8):2600–2611

Santoro M, Beer C, Cartus O, Schmullius C, Shvidenko A, McCallum I, Wegmüller U, Wiesmann A (2011) Retrieval of growing stock volume in boreal forest using hyper-temporal series of Envisat ASAR ScanSAR backscatter measurements. Remote Sens Environ 115(2):490–507

Small D (2011) Flattening gamma: radiometric terrain correction for SAR imagery. IEEE Trans Geosci Remote Sens 49(8):3081–3093

Su H, Krummel J (1996) Lacunarity as a texture measure for a tropical forest landscape. In: The 1996 ERDAS users' group meeting. Atlanta, Georgiea

Villard L, Borderies P, Le Toan T, Koleck T, Albinet C (2010) Topography effects on forest radar scattering, consequences on biomass retrieval. In: 2010 IEEE international geoscience and remote sensing symposium, pp 60–63. IEEE, July 2010

Woodhouse IH (2006) Predicting backscatter-biomass and height-biomass trends using a macroecology model. IEEE Trans Geosci Remote Sens 44(4):871–877

Appendix A
Test Site and Data Specificities

A.1 Test Site Specificities

See Figs. A.1, A.2, A.3, A.4, A.5, A.6, A.7 and A.8.

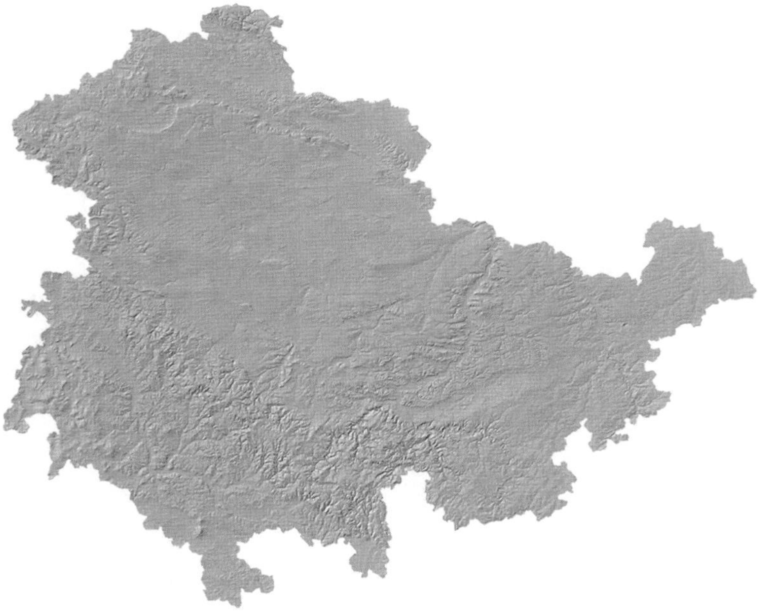

Fig. A.1 Overview of the Thuringian topography

N. Ackermann, *Growing Stock Volume Estimation in Temperate Forested Areas Using a Fusion Approach with SAR Satellites Imagery*,
Springer Theses, DOI 10.1007/978-3-319-13138-2

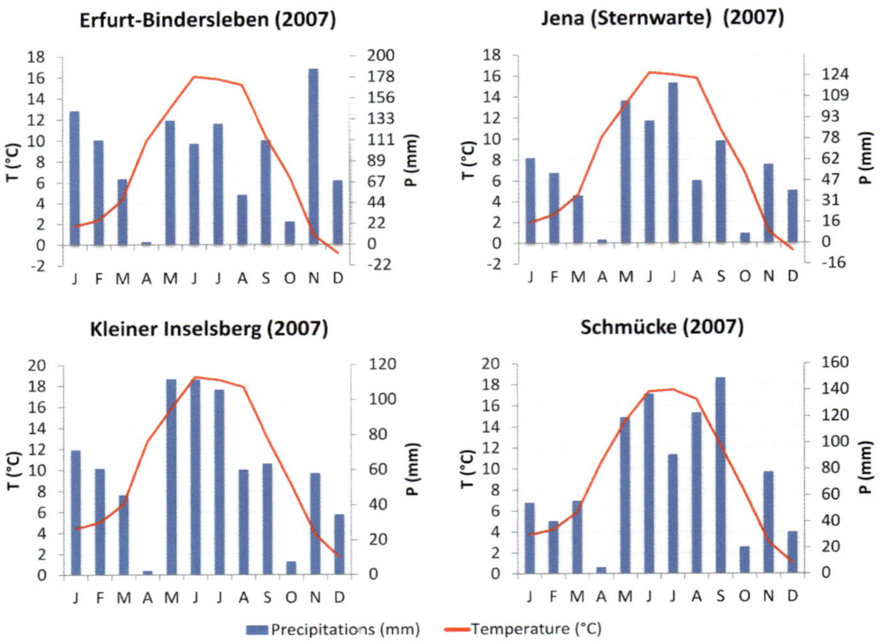

Fig. A.2 Climatic diagrams for the year 2007

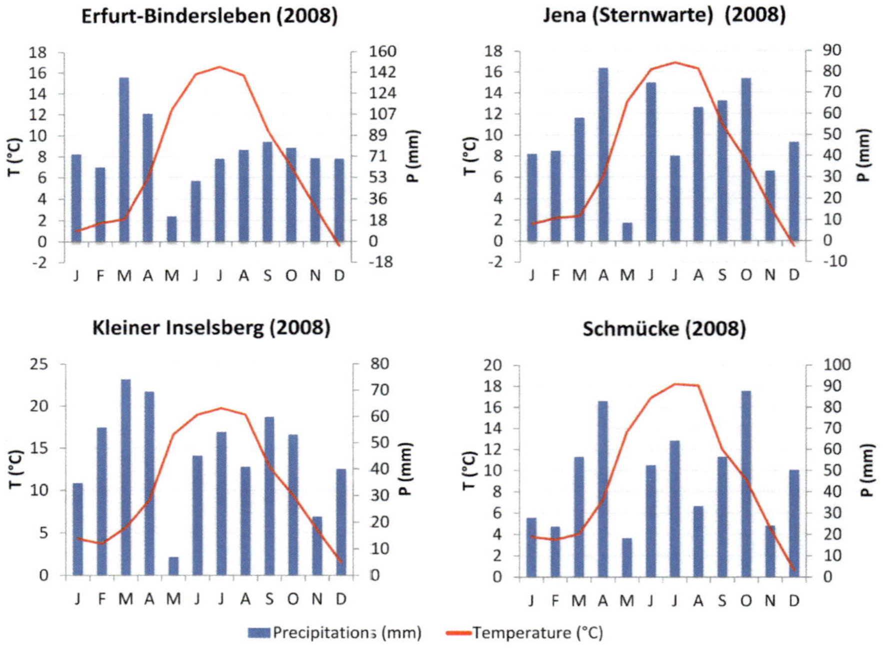

Fig. A.3 Climatic diagrams for the year 2008

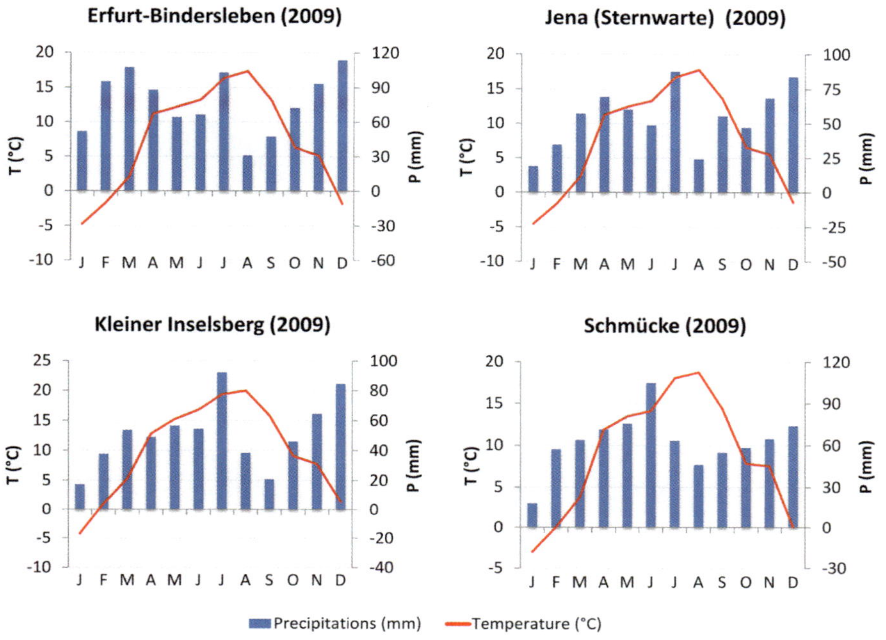

Fig. A.4 Climatic diagrams for the year 2009

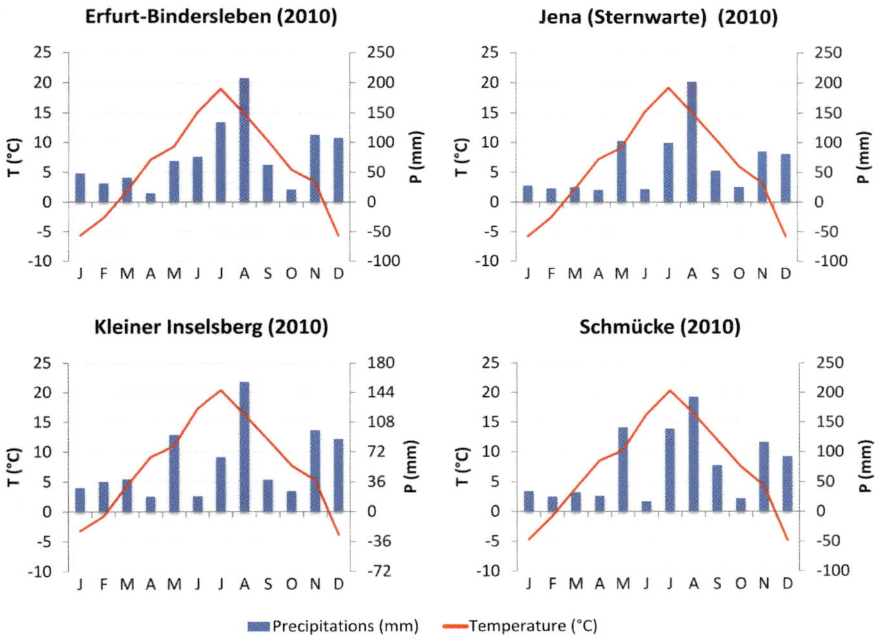

Fig. A.5 Climatic diagrams for the year 2010

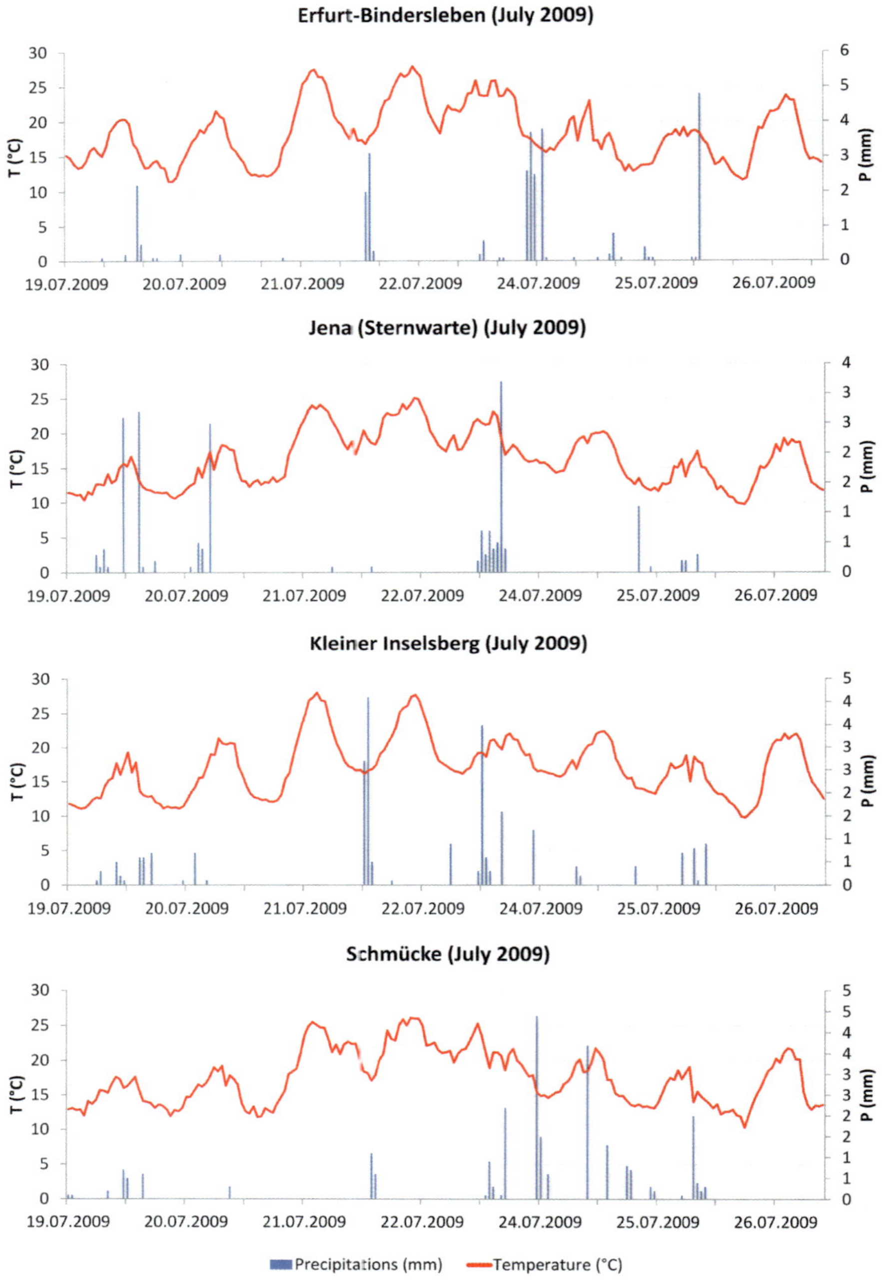

Fig. A.6 Climatic diagrams for the precipitation event in July 2009

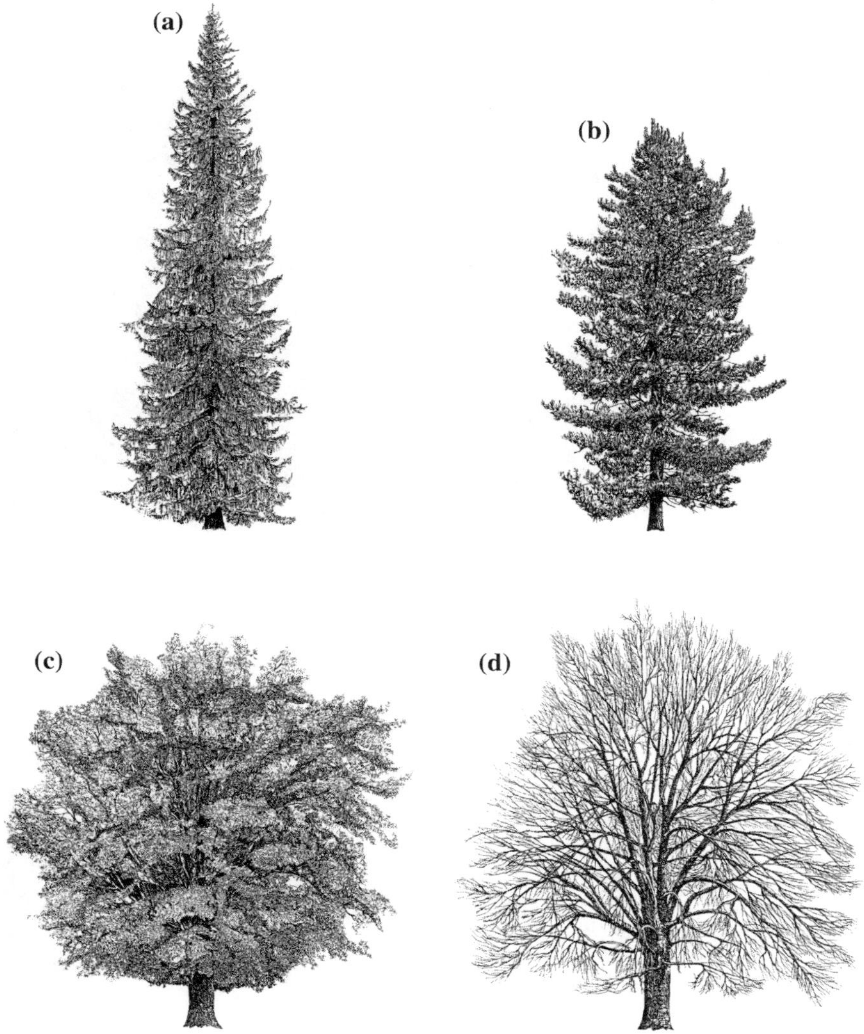

Fig. A.7 Illustration of **a** Norway Spruce, **b** Scots Pine, **c** European Beech in summer (leaf-on) and **d** European Beech in winter (leaf-off)

Fig. A.8 The Schönbrunn Lake region during Autumn season

A.2 SAR Data Overview

See Figs. A.9, A.10, A.11, A.12, A.13, A.14, A.15 and A.16.

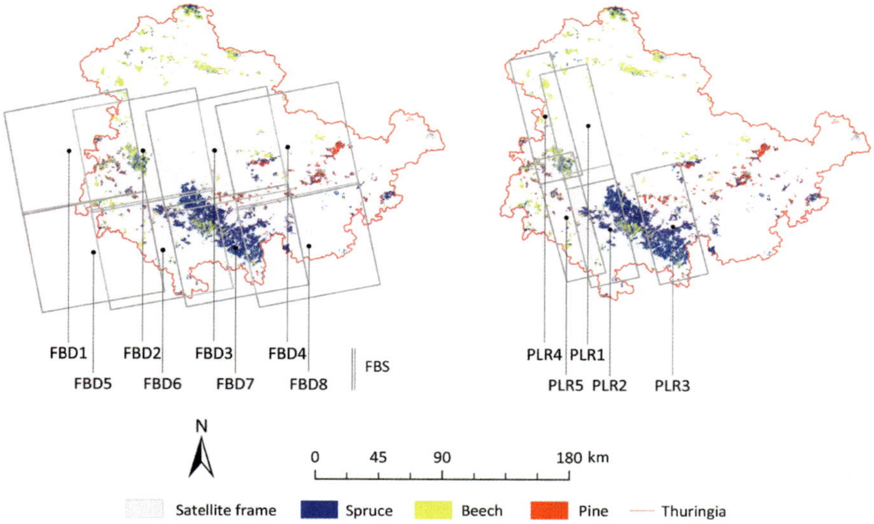

Fig. A.9 Satellite frames locations of the available PALSAR data

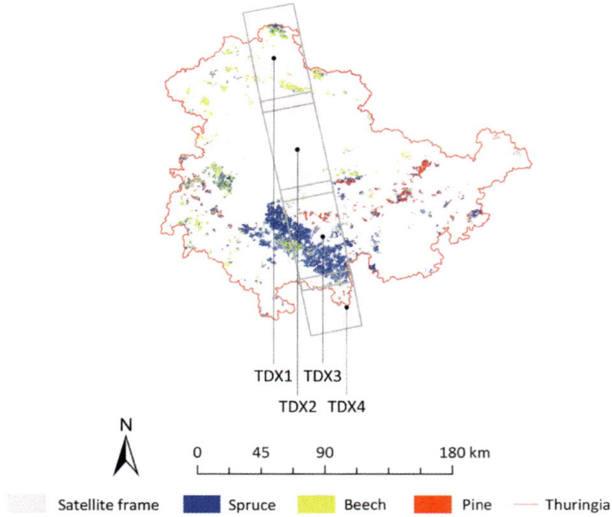

Fig. A.10 Satellite frames locations of the available TDX data

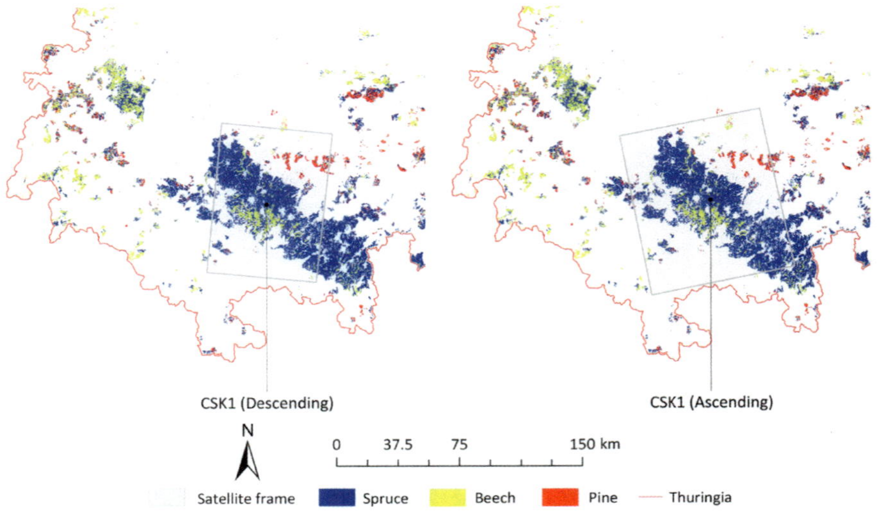

Fig. A.11 Satellite frames locations of the available CSK data

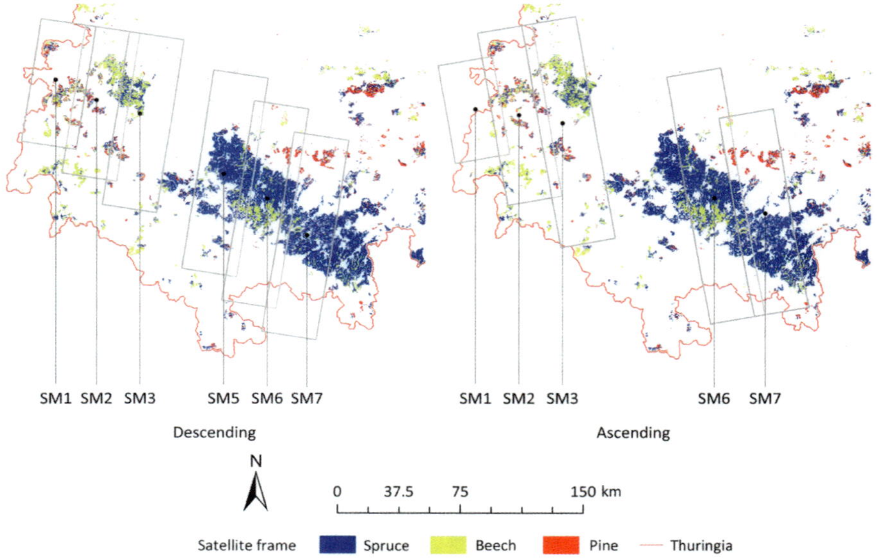

Fig. A.12 Satellite frames locations of the available TSX SM data

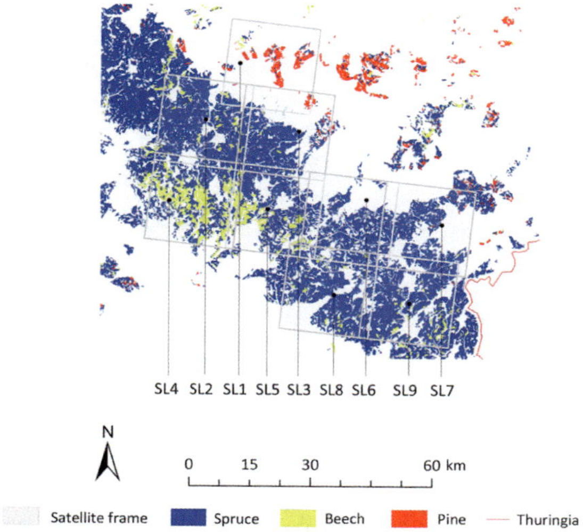

Fig. A.13 Satellite frames locations of the available TSX SL data

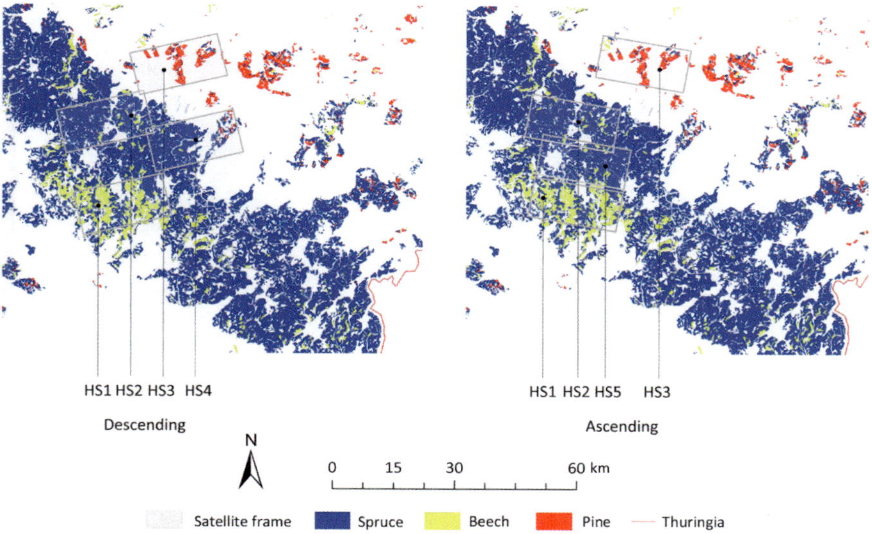

Fig. A.14 Satellite frames locations of the available TSX HS data

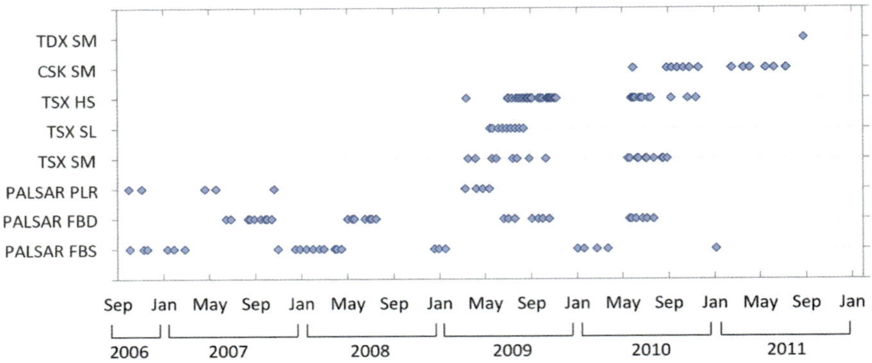

Fig. A.15 Temporal representation of the available SAR data

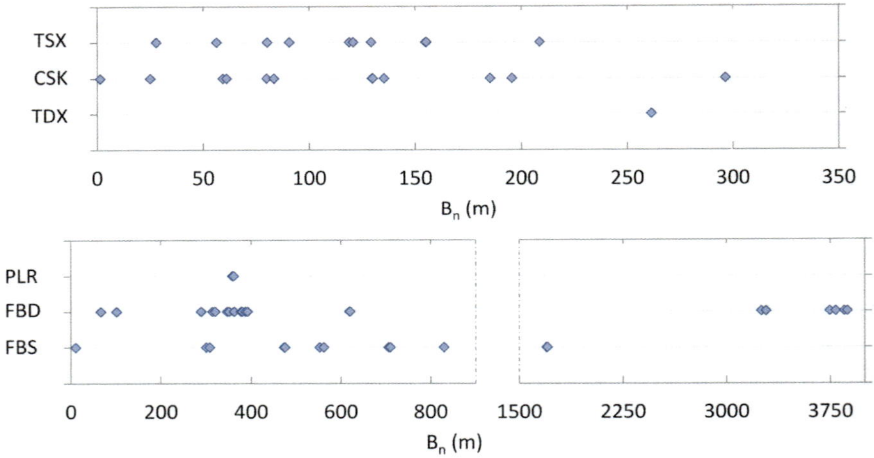

Fig. A.16 Representation of the SAR data perpendicular baselines

A.3 Forest Inventory and Forest Campaign Data

See Tables A.1 and A.2.

Table A.1 Simplified sample of the Thuringian Forest inventory

FOA	ETF	REV	ABT	UABT	TFL	FLAECHE	HBA_S	BAANT	ALT	HOEHE	BON	BHD	GF	VG	HAVOR_GES	STICHTAG
13	S	5	47	a	3	4.82	511	0.5	56	23	0.41	24	26	1.07	275	01.01.2010
58	S	1	1568	m	1	10.28	412	0.6	100	17	4.01	25	9.3	0.5	66	01.01.2008
57	S	8	504	a	5	4.34	311	0.5	76	32.6	0.91	37	34.5	0.79	441	01.10.2000
13	S	5	45	a	6	6.6	511	0.5	59	25	0.08	23	26	0.98	313	01.01.2010
13	S	5	49	a	4	4.31	511	1	118	39	0.31	45	31.4	0.95	623	01.01.2010
13	S	5	46	a	7	3.82	511	0.55	56	24	0.06	24	24	0.97	270	01.01.2010
13	S	5	50	a	4	13.14	511	0.75	93	31	1.12	32	30	1.02	450	01.01.2010
13	S	5	37	a	4	7.98	511	0.6	48	23	−0.71	19	25	1.03	276	01.01.2010
13	S	5	44	a	2	6.84	511	0.8	100	30	1.75	32	24	0.82	347	01.01.2010
55	S	3	345	a	3	5.68	621	0.55	30	12	1.83	12	13.6	0.9	56	01.01.2010
55	S	3	346	a	1	6.83	621	0.47	31	14	1.49	14	15.5	0.9	84	01.01.2010
55	S	3	345	a	2	6.31	511	1	171	27	3.86	46	11	0.36	145	01.01.2010
55	S	3	347	b	1	6.41	511	1	162	29	3.43	43	26	0.84	371	01.01.2010
55	S	3	343	a	2	18.85	511	1	148	32	2.66	44	19	0.6	302	01.01.2010
55	S	3	342	a	1	20.04	511	0.9	94	29	1.7	38	26	0.87	365	01.01.2010
55	S	3	330	a	3	10.82	511	1	134	31	2.66	44	22	0.7	338	01.01.2010
55	S	3	330	a	1	9.77	511	0.4	56	24	0.06	25	21	0.76	228	01.01.2010
13	S	9	87	a	2	13.07	511	0.48	65	27	0.08	27	27	0.98	342	01.01.2010
13	S	5	52	a	1	9.93	511	0.67	159	31	2.98	51	27.1	0.89	394	01.01.2010
51	S	8	812	a	1	6.62	511	0.77	66	25.4	0.68	30	28.2	1.14	325	01.10.2004
13	S	9	85	a	8	6.39	511	0.35	110	24	3.67	41	22.3	0.85	231	01.01.2010
13	S	9	85	a	1	10.03	511	0.72	74	28	0.62	29	27	0.99	353	01.01.2010

Table A.2 Forest campaign measured stands

#	Stands ID	Species	DSW 1 (Inventory)		DSW 2 (Inventory)		Campaign	
			Date (year)	GSV ($m^3\,ha^{-1}$)	Date (year)	GSV ($m^3\,ha^{-1}$)	Date (year)	GSV ($m^3\,ha^{-1}$)
1	34,S,2,1224,a,7	Spruce	2009	386	2010	408	2009	485
2	34,S,2,1234,a,1	Beech	2009	414	2010	540	2009	653
3	34,S,11,474,a,11	Pine	2009	177	2010	283	2009	388
4	21,S,7,714,a,2	Spruce	2006	117	2010	122	2010	169
5	21,S,12,1230,a,5	Spruce	2006	51	2010	37	2010	233
6	21,S,12,1205,a,1	Beech	2006	218	2010	242	2010	298
7	34,S,4,1008,a,1	Beech	2009	360	2010	431	2010	557
8	34,S,4,1016,a,5	Beech	2009	240	2010	75	2010	72
9	34,S,4,1016,a,6	Beech	–	–	2010	343	2010	320
10	34,S,4,1009,a,2	Beech	2009	438	2010	370	2010	597
11	34,S,4,1015,a,2	Beech	2009	521	2010	299	2010	536
12	16,S,6,65,a,12	Spruce	2008	114	2010	135	2011	263
13	16,S,6,73,a,1	Spruce	2008	121	2010	138	2011	276
14	16,S,7,105,b,2	Spruce	2008	108	2010	111	2011	250
15	16,S,7,127,a,10	Spruce	2008	88	2010	87	2011	249
16	34,S,8,777,a,7	Spruce	2009	542	2010	535	2011	504
17	34,S,5,1061,a,2	Spruce	2009	132	2010	103	2011	273
18	34,S,8,774,a,5	Spruce	2009	431	2010	462	2011	567
19	21,S,7,718,a,1	Beech	2006	378	2010	393	2011	454
20	21,S,12,1231,a,2	Beech	2006	310	2010	326	2011	372
21	34,S,8,747,a,3	Beech	2009	457	2010	457	2011	534
22	28,S,4,607,a,7	Beech	–	–	2010	1	2011	112

A.4 ROIs Overview

See Fig. A.17.

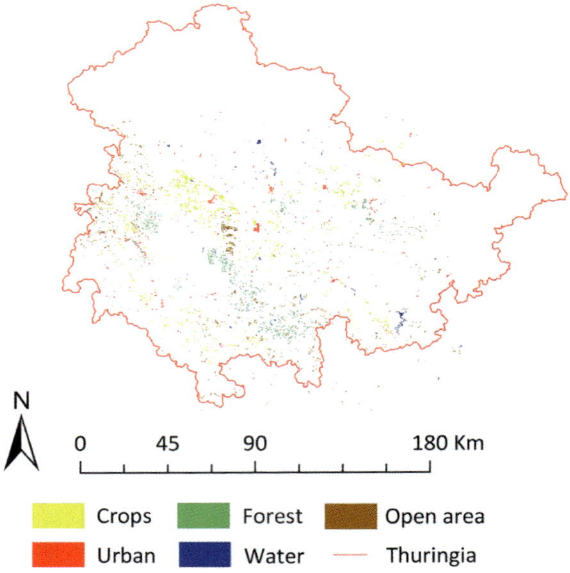

Fig. A.17 Overview of the selected ROIs

A.5 Weather Data Sample

See Tables A.3 and A.4.

Table A.3 Simplified sample of the weather data measured 4 days before the radar acquisitions

Sensor	Beam mode	Acquisition date	Track/ beam	Frame	$T°$ (°C)	Precip. (mm)	Humid. (%)	Wind speed (m s^{-1})
PALSAR	FBD	01.10.2007 21:31	638	1000	9.1	68.0	92.9	4.9
PALSAR	FBD	01.10.2007 21:31	638	1010	10.7	81.5	88.4	5.2
PALSAR	FBD	24.06.2009 21:37	640	1000	11.6	5.6	88.1	3.6
PALSAR	FBD	24.06.2009 21:37	640	1010	13.3	4.7	81.4	2.8
PALSAR	FBD	06.07.2009 21:33	638	1000	16.9	18.7	79.2	3.0
PALSAR	FBD	06.07.2009 21:33	638	1010	18.2	12.7	75.0	3.0

(continued)

Table A.3 (continued)

Sensor	Beam mode	Acquisition date	Track/ beam	Frame	$T°$ (°C)	Precip. (mm)	Humid. (%)	Wind speed (m s^{-1})
PALSAR	FBD	23.07.2009 21:35	639	1000	17.1	32.7	78.4	4.6
PALSAR	FBD	23.07.2009 21:35	639	1010	17.8	43.5	77.3	4.7
PALSAR	FBD	24.09.2009 21:37	640	1000	12.5	1.8	86.4	5.0
PALSAR	FBD	24.09.2009 21:37	640	1010	12.7	2.0	87.5	4.9
PALSAR	FBS	01.01.2008 21:30	638	1000	−1.5	5.7	93.5	3.4
PALSAR	FBS	02.04.2008 21:29	638	1000	5.7	20.3	81.7	3.9
PALSAR	FBS	03.01.2009 21:31	638	1000	−5.8	10.0	90.8	3.0
PALSAR	FBS	06.01.2010 21:33	638	1010	−9.0	0.4	89.6	2.2
PALSAR	FBS	08.04.2010 21:31	638	1000	17.7	0.1	61.7	3.4

Table A.4 Simplified sample of the weather data measured 4 hours before the radar acquisitions

Sensor	Beam mode	Acquisition date	Track/ beam	Frame	$T°$ (°C)	Precip. (mm)	Humid. (%)	Wind speed (m s^{-1})
PALSAR	FBD	01.10.2007 21:31	638	1000	11.8	0.0	85.8	2.4
PALSAR	FBD	01.10.2007 21:31	638	1010	12.6	0.0	85.6	2.4
PALSAR	FBD	24.06.2009 21:37	640	1000	14.8	0.0	85.5	3.8
PALSAR	FBD	24.06.2009 21:37	640	1010	15.8	0.0	77.2	3.2
PALSAR	FBD	06.07.2009 21:33	638	1000	15.2	6.4	89.8	2.5
PALSAR	FBD	06.07.2009 21:33	638	1010	16.6	5.4	85.5	2.4
PALSAR	FBD	23.07.2009 21:35	639	1000	16.1	4.8	92.1	5.5
PALSAR	FBD	23.07.2009 21:35	639	1010	16.4	9.1	90.2	5.4
PALSAR	FBD	24.09.2009 21:37	640	1000	9.7	0.0	95.8	3.5
PALSAR	FBD	24.09.2009 21:37	640	1010	9.5	0.1	97.5	2.8
PALSAR	FBS	01.01.2008 21:30	638	1000	−2.8	0.1	94.9	1.7
PALSAR	FBS	02.04.2008 21:29	638	1000	2.9	2.8	93.9	4.7
PALSAR	FBS	03.01.2009 21:31	638	1000	−6.4	0.0	84.3	3.7
PALSAR	FBS	06.01.2010 21:33	638	1010	−11.0	0.0	88.1	2.7
PALSAR	FBS	08.04.2010 21:31	638	1000	8.6	0.0	78.0	4.1

A.6 Topographic Effects on GSV

See Fig. A.18.

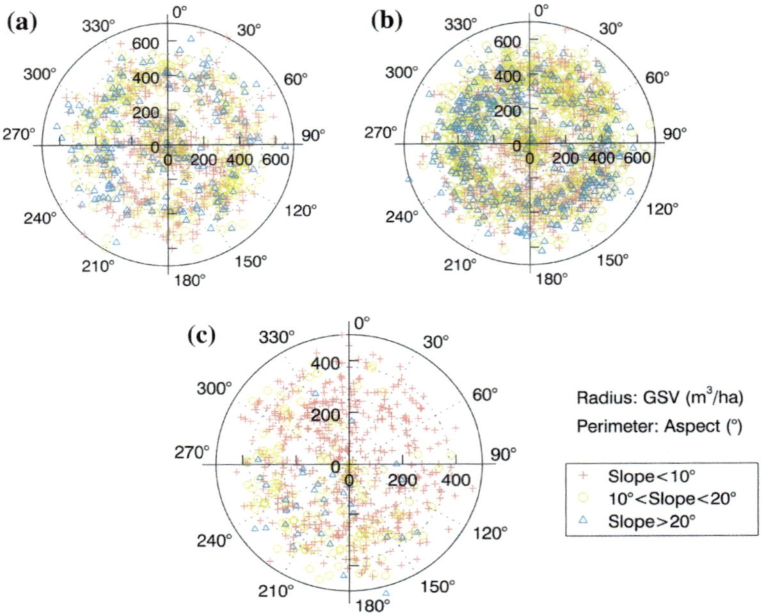

Fig. A.18 Topographic effects on *GSV* for **a** spruce, **b** beech and **c** pine tree species composition. The graphs consist of polarplots with the aspect angle (°) on the circumference and *GSV* ($m^3 ha^{-1}$) on the radius of the plots. The classes represent the inclinations of three slopes, namely below 10° (*blue*), between 10° and 20° (*yellow*) and above 20° (*red*)

A.7 SAR Pre-processing Specificities

See Figs. A.19 and A.20 and Table A.5.

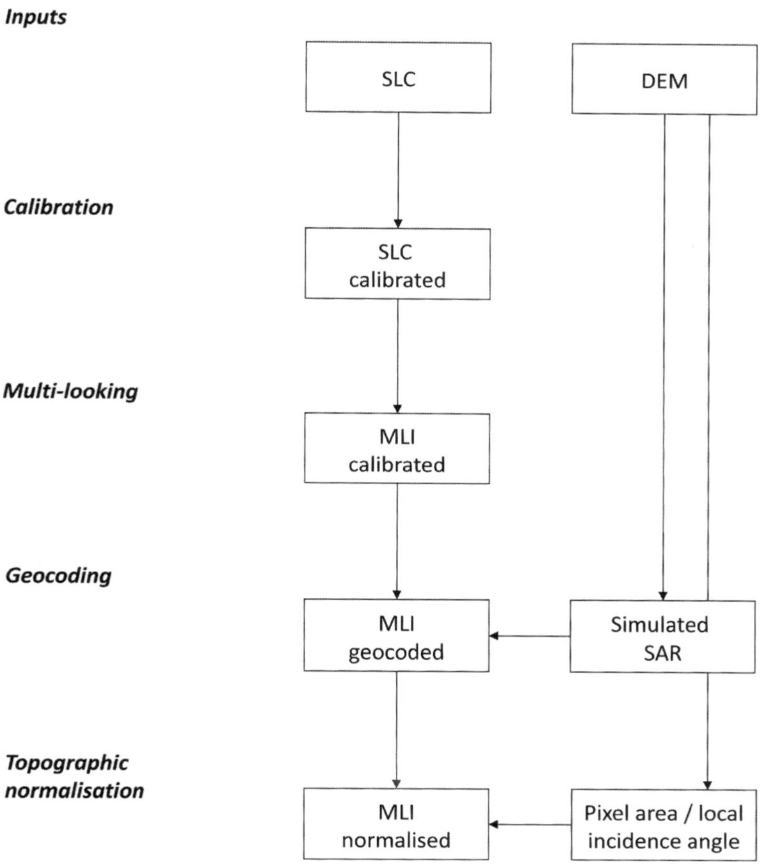

Fig. A.19 Simplified flowchart for the SAR backscatter pre-processing

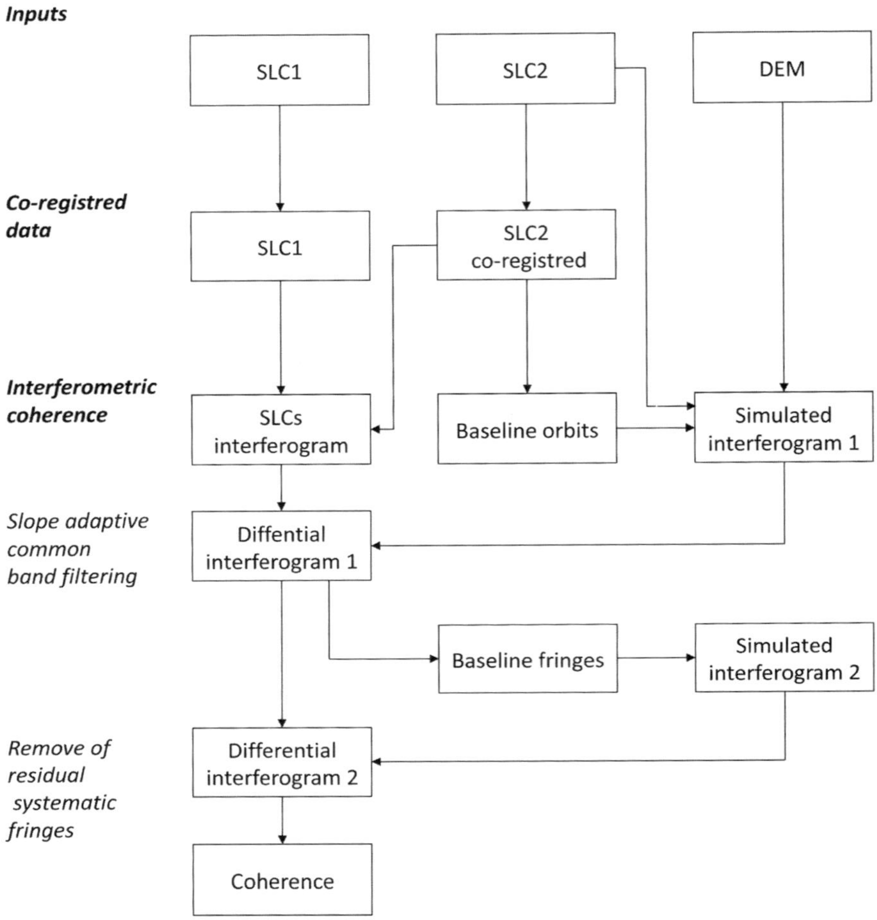

Fig. A.20 Simplified flowchart for the InSAR coherence pre-processing

Table A.5 Multi-looking factors used for pre-processing and spatial resolutions defined for data analysis

Sensor	Beam mode	L_a	L_r	δ_a (m)	δ_r (m)
PALSAR	FBS	4 (3)	2 (1)	12.5	12.5
PALSAR	FBD	8	2 (1)	25	25
PALSAR	PLR	1	7	25	25
TSX	SM	5–7	3–4	10	10
TSX	SL	7–8	3–4	10	10
TSX	HS (Single-pol)	4–5	6–8	10	10
TSX	HS (Dual-pol)	10–11	4–8	10	10
CSK	SM	5	5	10	10
TDX	SM	7	6	10	10

The different multi-looking factors shown by TSX are explained by the varying incidence angle of the acquisitions. In the case of PALSAR, a few datasets were processed with the multi-looking factors shown in brackets. The difference between these factors with the original ones was unsignificant. Therefore, the datasets concerned by the multi-looking factors shown in brackets were not re-processed

Appendix B
Results

B.1 Topography Analysis

See Figs. B.1, B.2 and B.3.

N. Ackermann, *Growing Stock Volume Estimation in Temperate Forested Areas Using a Fusion Approach with SAR Satellites Imagery*,
Springer Theses, DOI 10.1007/978-3-319-13138-2

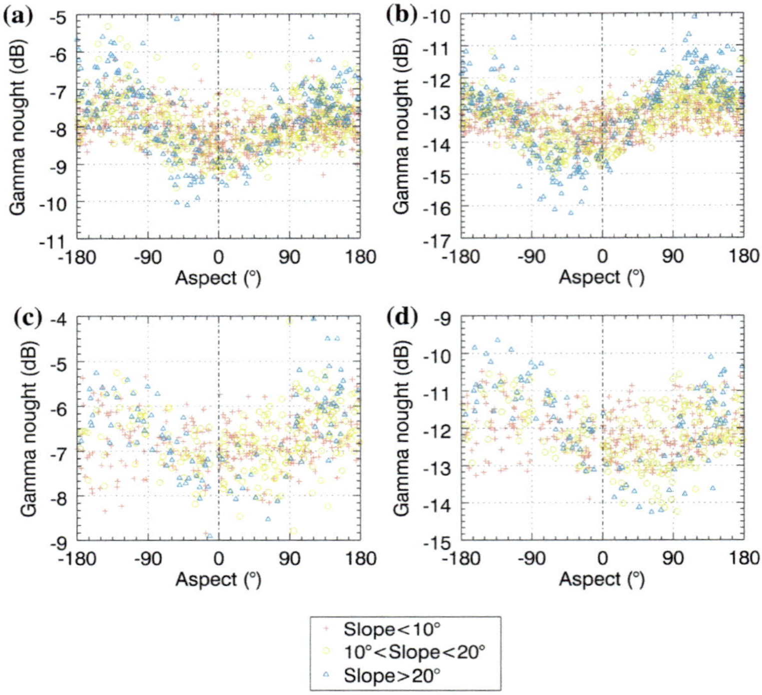

Fig. B.1 Topographic effects on the normalised backscatter intensity. The scatterplots depict the aspect angle on the abscissa between −180° and 180°, with azimuth slopes centred at 0° and backscatter intensity (γ^0) on the ordinate. Negative slopes are described between −180° and 0° and positive slopes between 0° and 180°. The classes represent the inclination of three slopes, namely below 10° (*pink*), between 10° and 20° (*yellow*) and above 20° (*blue*). The SAR data are composed of TSX SM 38° Asc. **a** HH and **b** HV (frame SM6) as well as TSX SM 36° Desc. **c** VV and **d** VH (frame SM5) backscatter intensity. The TSX SM scenes were acquired on 1st September 2010 and 29th July 2009, respectively. The forested areas represent Norway Spruce

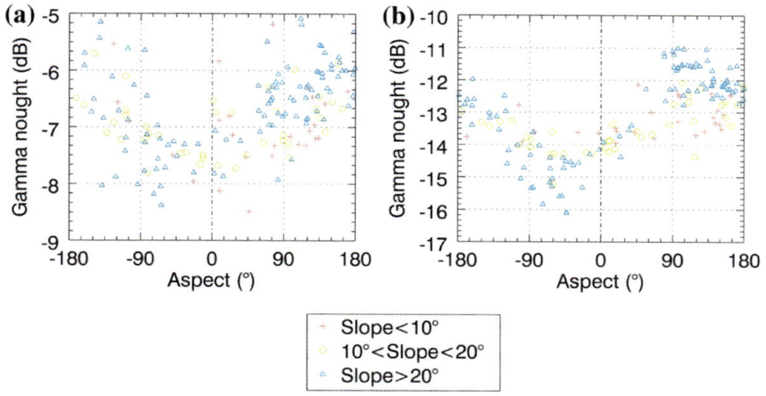

Fig. B.2 Topographic effects on the normalised backscatter intensity. The scatterplots depict the aspect angle on the abscissa between −180° and 180°, with azimuth slopes centred at 0° and backscatter intensity (γ^0) on the ordinate. Negative slopes are described between −180° and 0° and positive slopes between 0° and 180°. The classes represent the inclination of three slopes, namely below 10° (*pink*), between 10° and 20° (*yellow*) and above 20° (*blue*). The SAR data are composed of TSX SM 38° Asc. **a** HH and **b** HV (frame SM6) backscatter intensity. The TSX SM scenes were acquired on 1st September 2010. The forested areas represent European Beech

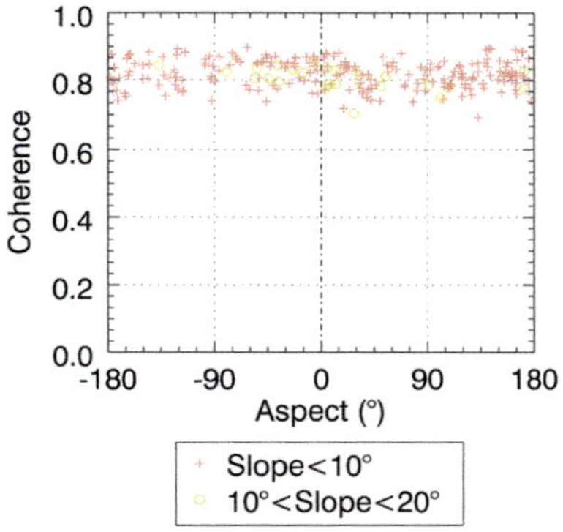

Fig. B.3 Topographic effects on interferometric coherence corrected for spectral shift (adaptive slope common band filtering). The scatterplots depict the aspect angle on the abscissa between −180° and 180° with the azimuth slopes centred at 0° and the interferometric coherence (γ) on the ordinate. Negative slopes are described for the range of −180° to 0° and positive slopes between 0° and 180°. The classes represent the inclinations of the two slopes, namely below 10° (*pink*) and between 10° and 20° (*yellow*). The InSAR data is TDX SM HH 38° Asc. $B_n = 258$ m (frame TDX3) coherence acquired on 30th August 2011. The points represent open areas

B.2 Scattering Mechanisms Analysis

B.2.1 Incidence Angle

See Figs. B.4, B.5, B.6, B.7, B.8, B.9, B.10, B.11 and B.12.

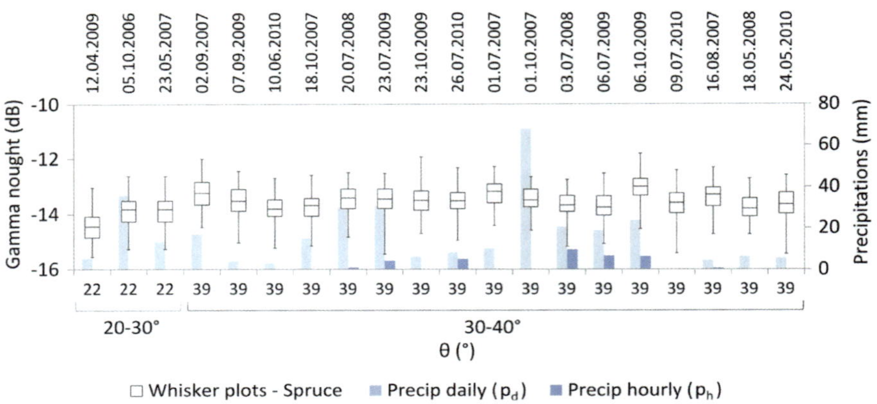

Fig. B.4 Incidence angle analysis with ALOS PALSAR FBD HV 39° Asc. and PLR HV 22° Asc. backscatter intensity (γ^0) (frames FBD7-8, PLR3). The boxplots represent spruce species with the sample minimum, *lower quartile* (Q1), median (Q2), *upper quartile* (Q3), and sample maximum. Weather information is depicted in *bright blue* and *dark blue* for daily (p_d) and hourly (p_h) precipitations, respectively. A slope of <5° was selected to limit topographic effects

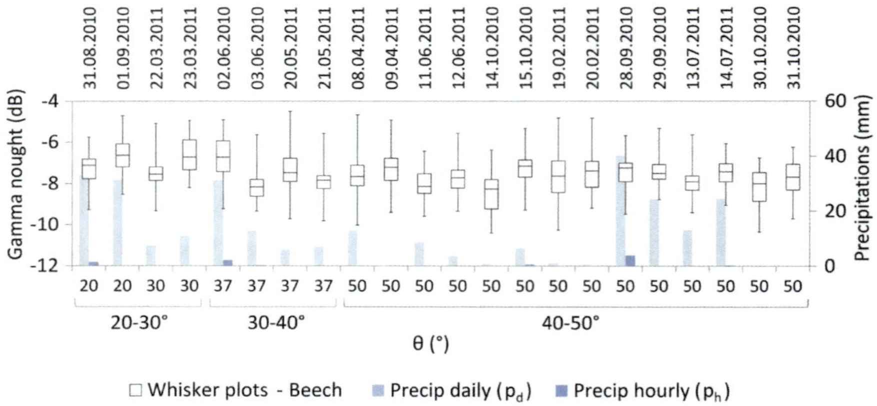

Fig. B.5 Incidence angle analysis with CSK SM Himage HH 20°–50° Asc. backscatter intensity (γ^0) (frame CSK1). The boxplots represent beech species with the sample minimum, *lower quartile* (Q1), median (Q2), *upper quartile* (Q3) and sample maximum. Weather information is depicted in *bright blue* and *dark blue* for daily (p_d) and hourly (p_h) precipitations, respectively. A slope of <5° was selected to limit topographic effects

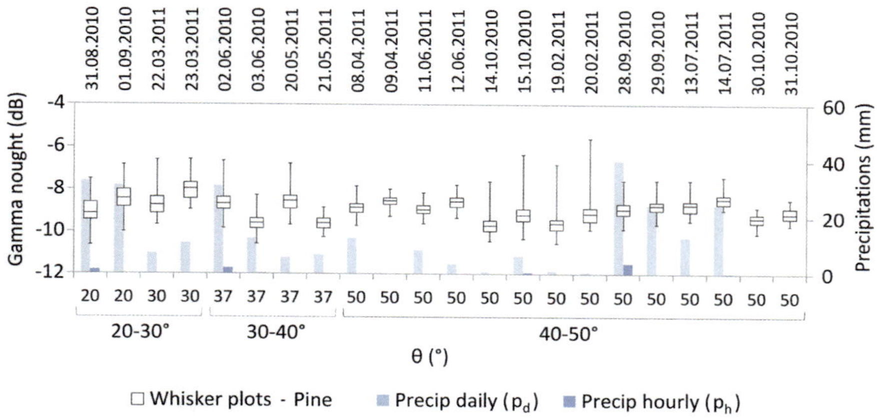

Fig. B.6 Incidence angle analysis with CSK SM Himage HH 20°–50° Asc. backscatter intensity (γ^0) (frame CSK1). The boxplots represent pine species with the sample minimum, lower quartile (Q1), median (Q2), *upper quartile* (Q3) and sample maximum. Weather information is depicted in *bright blue* and *dark blue* for daily (p_d) and hourly (p_h) precipitations, respectively. A slope of <5° was selected to limit topographic effects

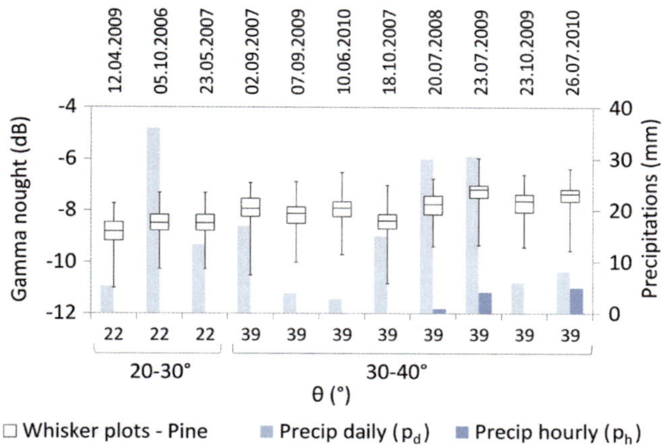

Fig. B.7 Incidence angle analysis with ALOS PALSAR FBD HV 39° Asc. and PLR HV 22° Asc. backscatter intensity (γ^0) (frames FBD7, PLR3). The boxplots represent pine species with the sample minimum, *lower quartile* (Q1), median (Q2), *upper quartile* (Q3), and sample maximum. Weather information is depicted in *bright blue* and *dark blue* for daily (p_d) and hourly (p_h) precipitations, respectively. A slope of <5° was selected to limit topographic effects

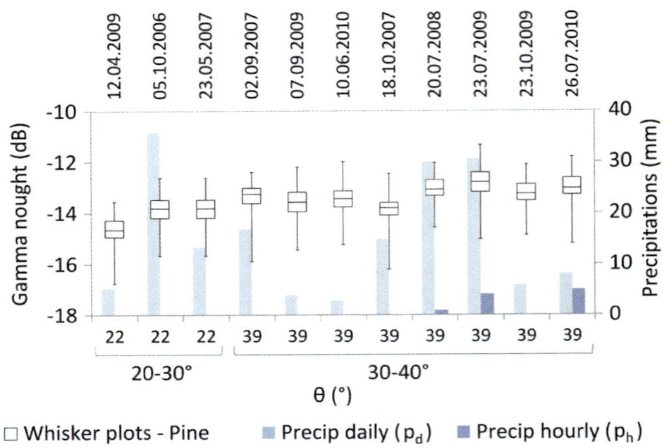

Fig. B.8 Incidence angle analysis with ALOS PALSAR FBD HV 39° Asc. and PLR HV 22° Asc. backscatter intensity (γ^0) (frames FBD7, PLR3). The boxplots represent pine species with the sample minimum, *lower quartile* (Q1), median (Q2), *upper quartile* (Q3), and sample maximum. Weather information is depicted in *bright blue* and *dark blue* for daily (p_d) and hourly (p_h) precipitations, respectively. A slope of <5° was selected to limit topographic effects

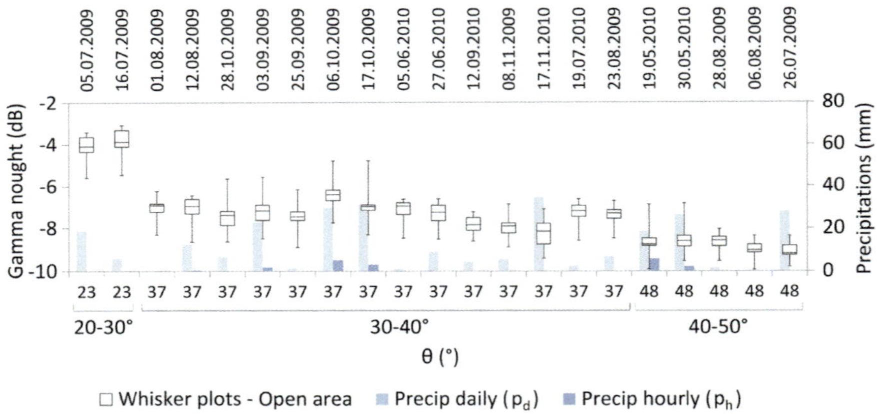

Fig. B.9 Incidence angle analysis with TSX HS HH 23°–48° Asc. backscatter intensity (γ^0) (frame HS1). The boxplots represent open areas with the sample minimum, *lower quartile* (Q1), median (Q2), *upper quartile* (Q3) and sample maximum. Weather information is depicted in *bright blue* and *dark blue* for daily precipitation (p_d) and hourly precipitation (p_h), respectively

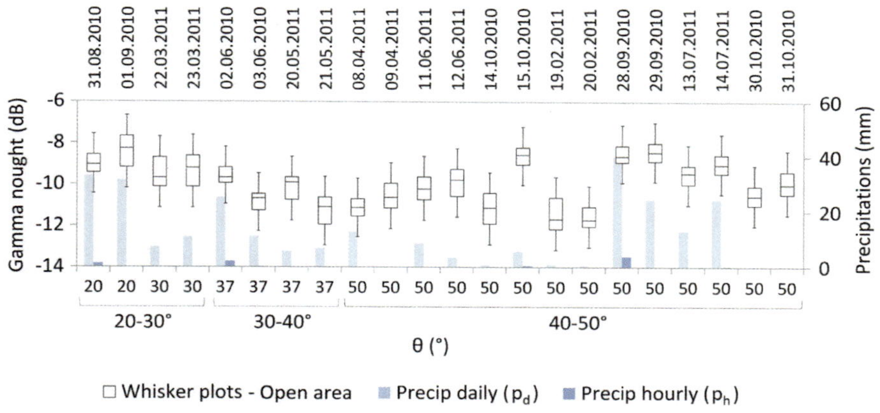

Fig. B.10 Incidence angle analysis with CSK SM Himage HH 20°–50° Asc. backscatter intensity (γ^0) (frame CSK1). The boxplots represent open areas with the sample minimum, *lower quartile* (Q1), median (Q2), *upper quartile* (Q3) and sample maximum. Weather information is depicted in *bright blue* and *dark blue* for daily (p_d) and hourly (p_h) precipitations, respectively

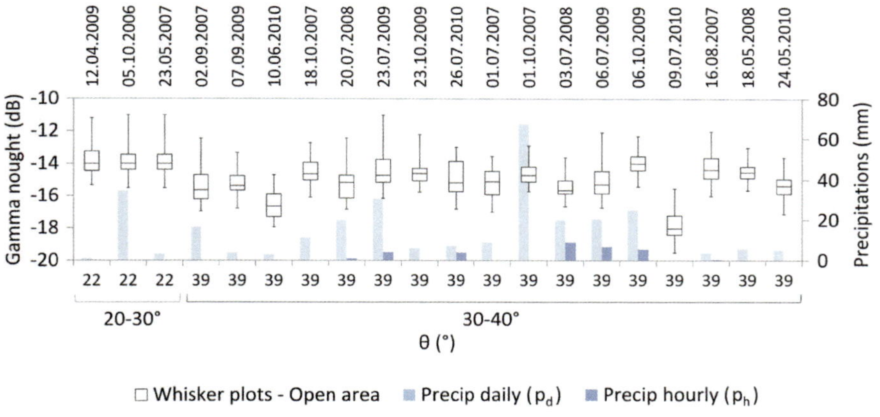

Fig. B.11 Incidence angle analysis with ALOS PALSAR FBD HV 39° Asc. and PLR HV 22° Asc. backscatter intensity (γ^0) (frames FBD7-8, PLR3). The boxplots represent open areas with the sample minimum, *lower quartile* (Q1), median (Q2), *upper quartile* (Q3), and sample maximum. Weather information is depicted in *bright blue* and *dark blue* for daily (p_d) and hourly (p_h) precipitations, respectively

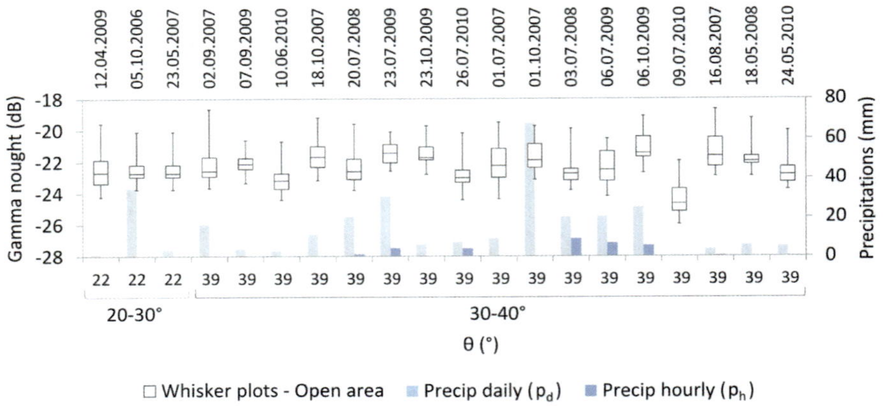

Fig. B.12 Incidence angle analysis with ALOS PALSAR FBD HV 39° Asc. and PLR HV 22° Asc. backscatter intensity (γ^0) (frames FBD7-8, PLR3). The boxplots represent open areas with the sample minimum, *lower quartile* (Q1), median (Q2), *upper quartile* (Q3), and sample maximum. Weather information is depicted in *bright blue* and *dark blue* for daily (p_{d}) and hourly (p_{h}) precipitations, respectively

B.2.2 Polarisation

See Figs. B.13, B.14, B.15, B.16, B.17, B.18, B.20 and B.21.

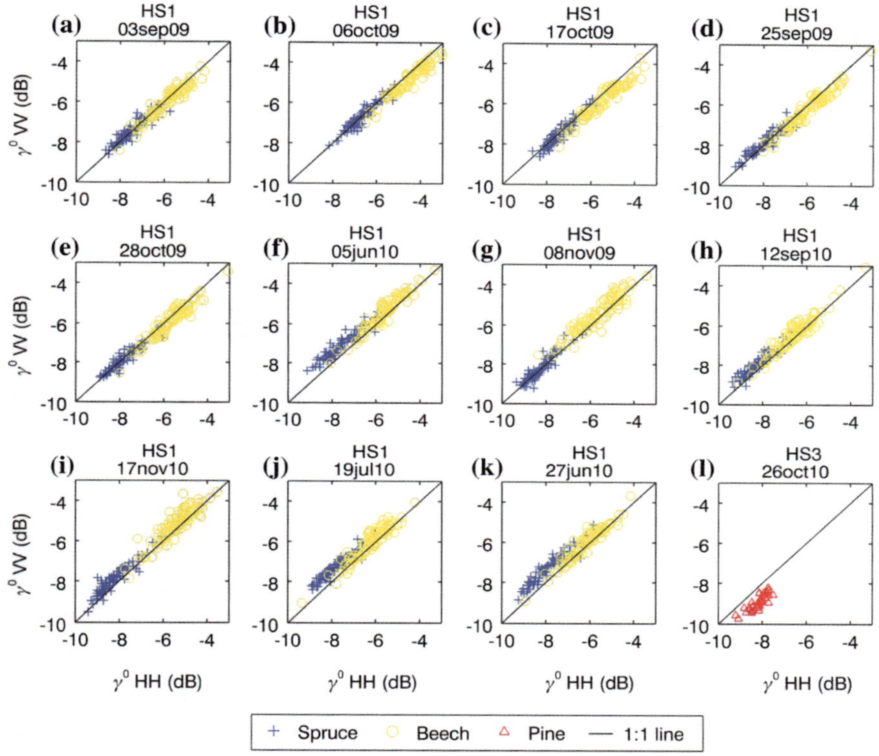

Fig. B.13 Polarisation analysis with TSX HS HH 30°–40° Asc. versus TSX HS VV 30°–40° Asc. backscatter intensity (γ^0) (frames HS1,3). *Blue*, *yellow* and *red* represent spruce, beech and pine, respectively

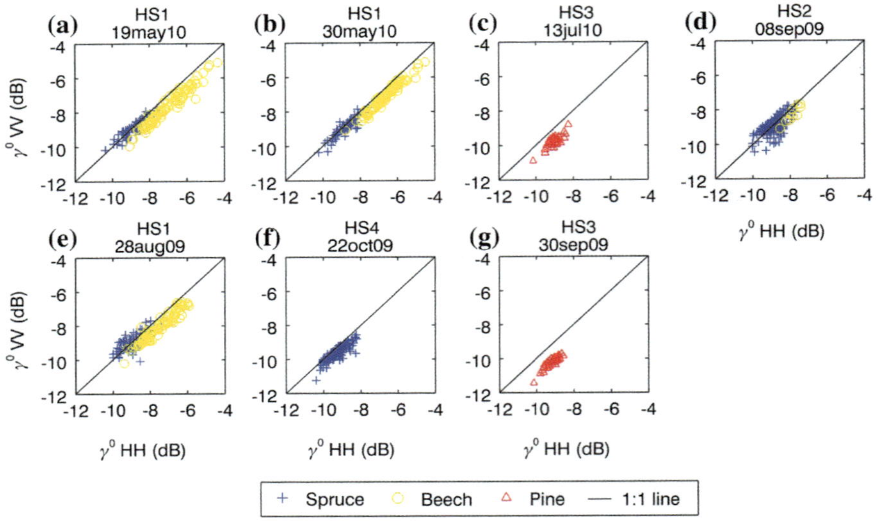

Fig. B.14 Polarisation analysis with TSX HS HH 40°–50° Asc. versus TSX HS VV 40°–50° Asc. backscatter intensity (γ^0) (frames HS1–4). *Blue*, *yellow* and *red* represent spruce, beech and pine, respectively

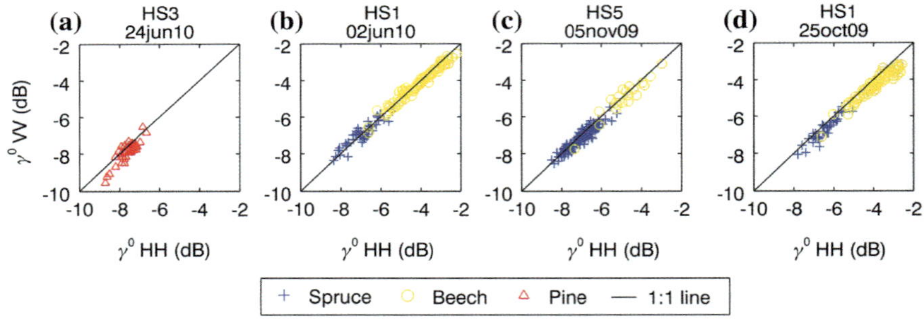

Fig. B.15 Polarisation analysis with TSX HS HH 30°–40° Desc. versus TSX HS VV 30°–40° Desc. backscatter intensity (γ^0) (frames HS1,3,5). *Blue*, *yellow* and *red* represent spruce, beech and pine, respectively

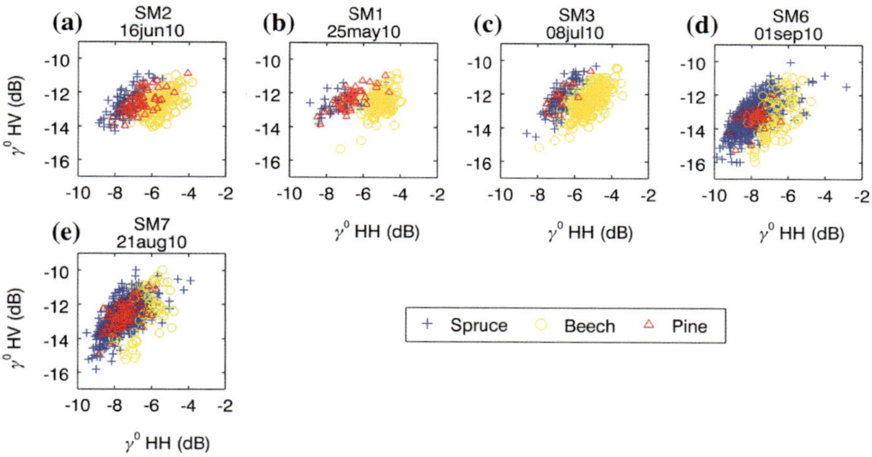

Fig. B.16 Polarisation analysis with TSX SM HH 30°–40° Asc. versus TSX SM HV 30°–40° Asc. backscatter intensity (γ^0) (frames SM1–3 and SM6,7). *Blue*, *yellow* and *red* represent spruce, beech and pine, respectively

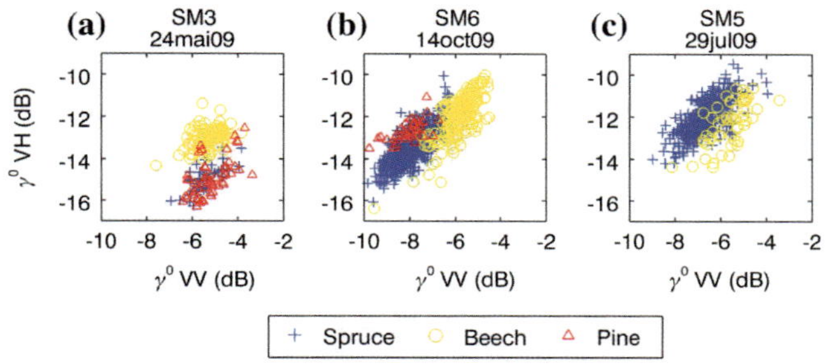

Fig. B.17 Polarisation analysis with TSX SM VV 30°–40° Desc. versus TSX SM VH 30°–40° Desc. backscatter intensity (γ^0) (frames SM3,5,6). *Blue*, *yellow* and *red* represent spruce, beech and pine, respectively

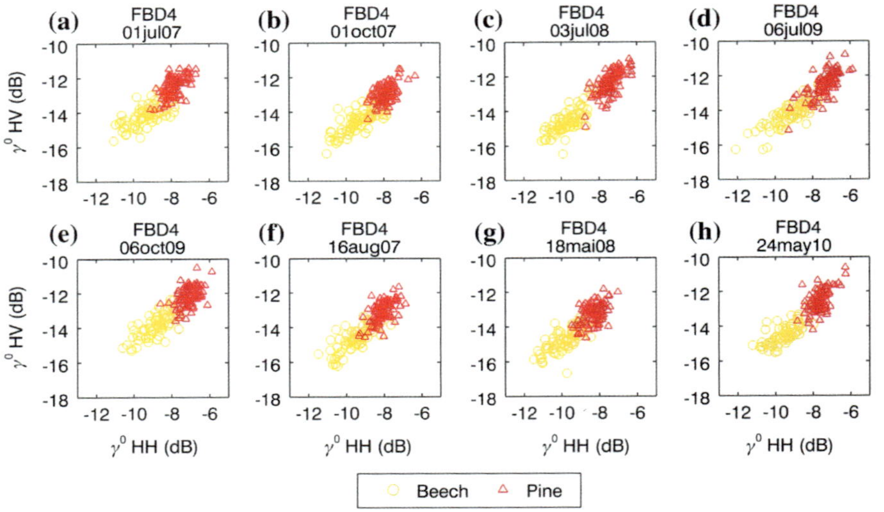

Fig. B.18 Polarisation analysis with PALSAR FBD HH 39° Asc. versus PALSAR FBD HV 39° Asc. backscatter intensity (γ^0) (frame FBD4). *Yellow* and *red* represent beech and pine, respectively

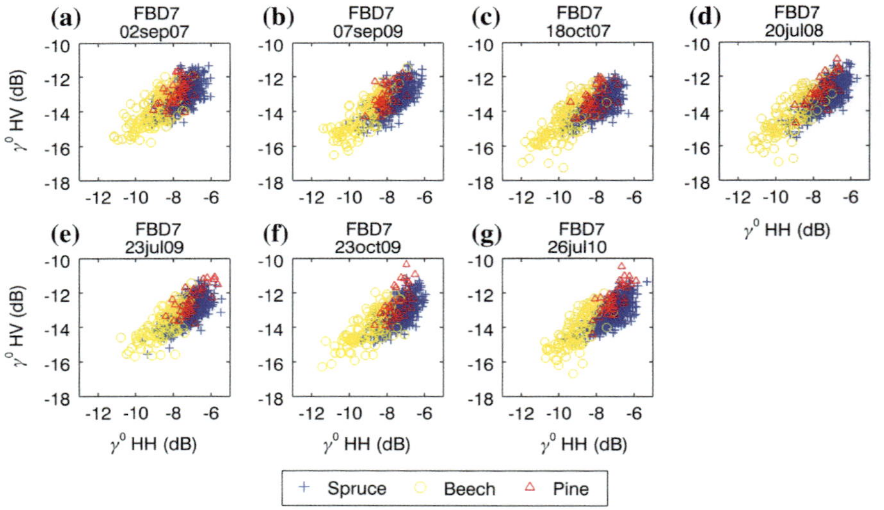

Fig. B.19 Polarisation analysis with PALSAR FBD HH 39° Asc. versus PALSAR FBD HV 39° Asc. backscatter intensity (γ^0) (frame FBD7). *Blue*, *yellow* and *red* represent spruce, beech and pine, respectively

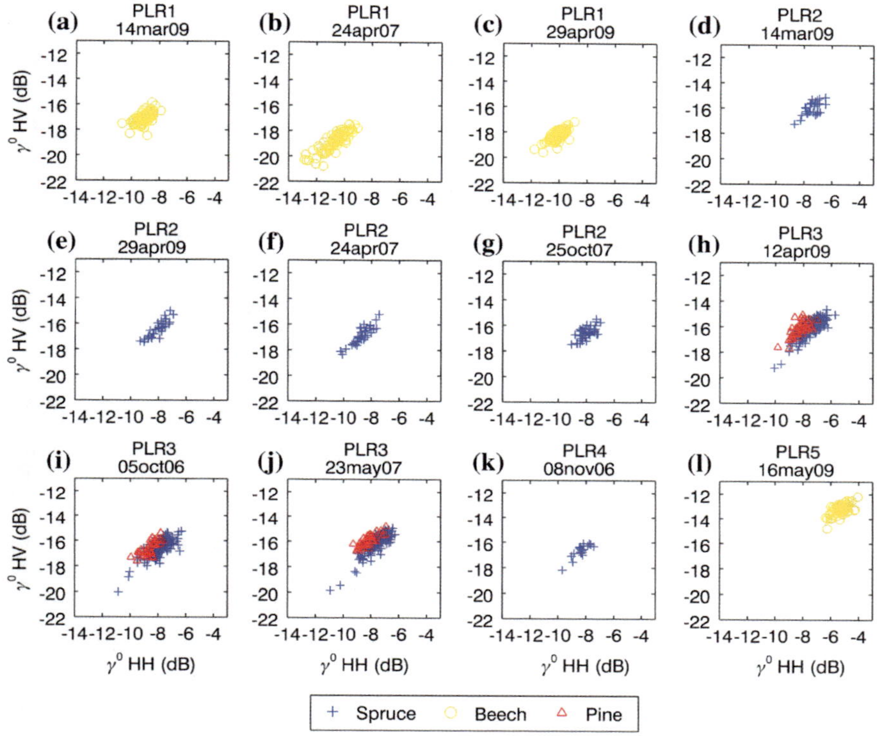

Fig. B.20 Polarisation analysis with PALSAR PLR HH 22° Asc. versus PALSAR PLR HV Asc. 22° backscatter intensity (γ^0) (frames PLR1–5). *Blue*, *yellow* and *red* represent spruce, beech and pine, respectively

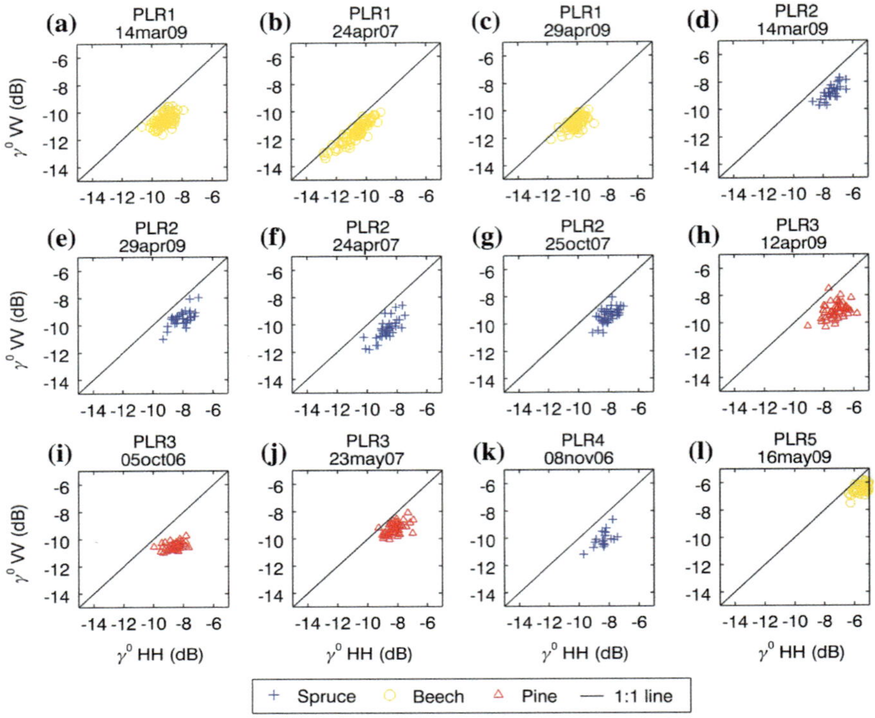

Fig. B.21 Polarisation analysis with PALSAR PLR HH 22° Asc. versus PALSAR PLR VV Asc. 22° backscatter intensity (γ^0) (frames PLR1–5). *Blue*, *yellow* and *red* represent spruce, beech and pine, respectively

B.2.3 GSV Sensitivity

See Figs. B.22, B.23, B.24, B.25, B.26 and B.27.

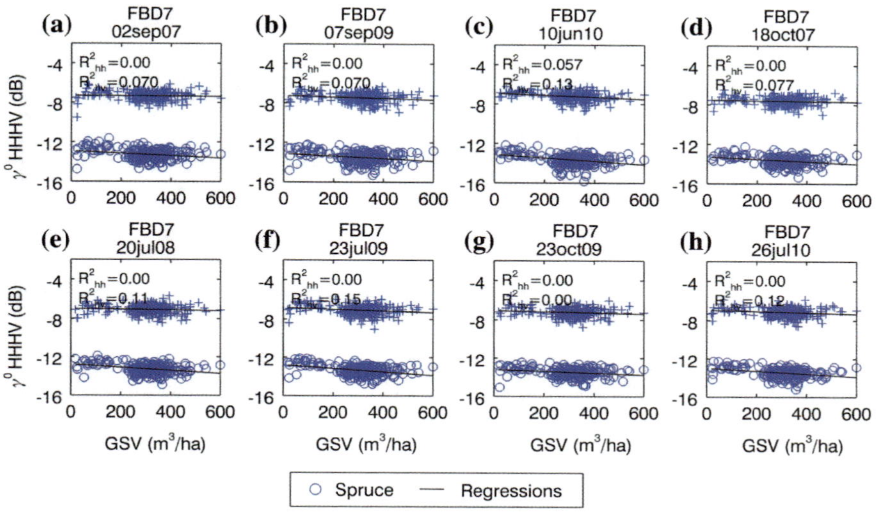

Fig. B.22 *GSV* sensitivity analysis with ALOS PALSAR FBD HH 39° Asc. backscatter intensity (γ^0) (frame FBD7). The forested areas represent Norway Spruce. Slopes <5° were selected to limit topographic effects

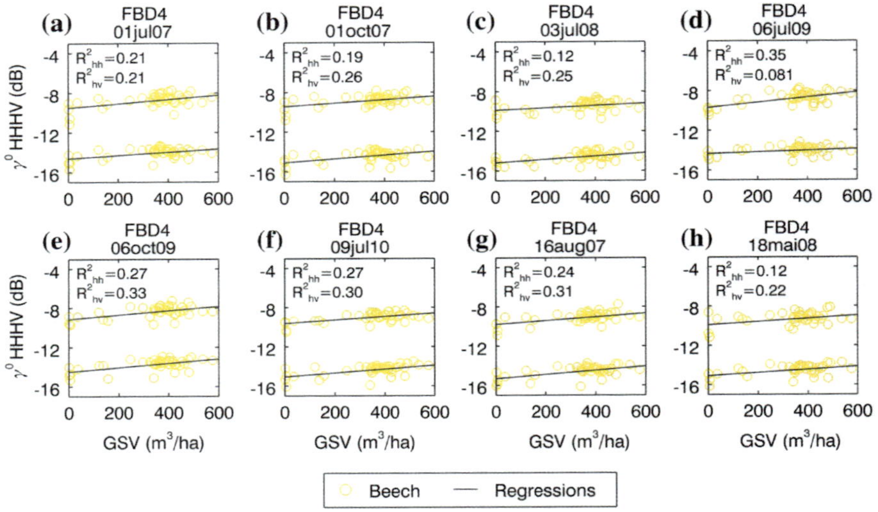

Fig. B.23 *GSV* sensitivity analysis with ALOS PALSAR FBD HH 39° Asc. backscatter intensity (γ^0) (frame FBD4). The forested areas represent European Beech. Slopes <5° were selected to limit topographic effects

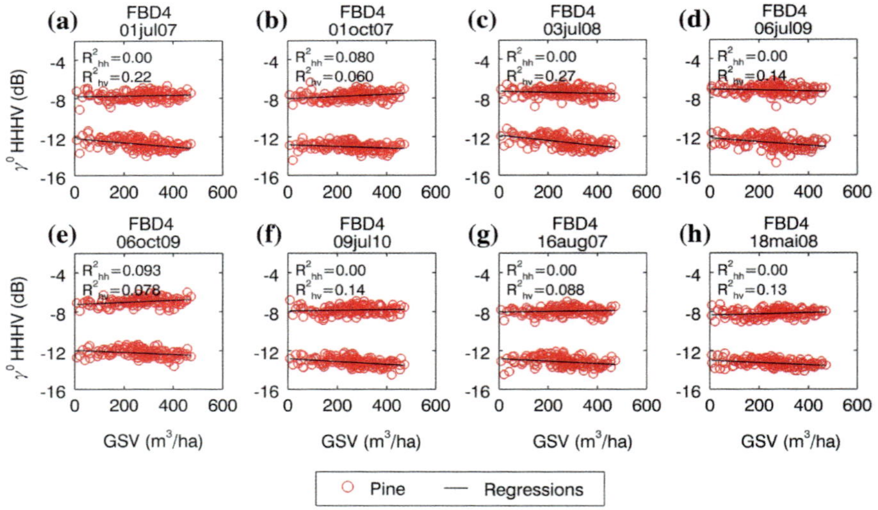

Fig. B.24 *GSV* sensitivity analysis with ALOS PALSAR FBD HH 39° Asc. backscatter intensity (γ^0) (frame FBD4). The forested areas represent Scots Pine. Slopes <5° were selected to limit topographic effects

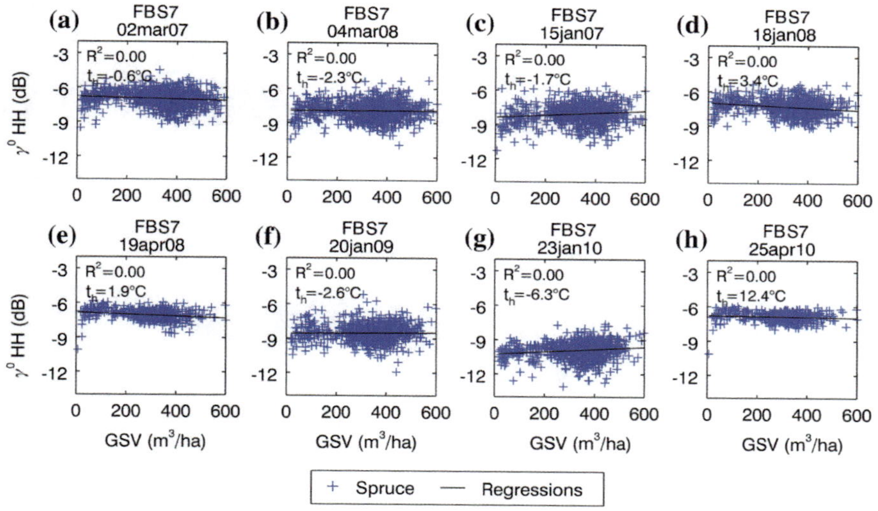

Fig. B.25 *GSV* sensitivity analysis for different air temperatures with ALOS PALSAR FBS HH 39° Asc. backscatter intensity (γ^0) (frame FBS7). The forested areas represent Norway Spruce. Slopes <5° were selected to limit topographic effects

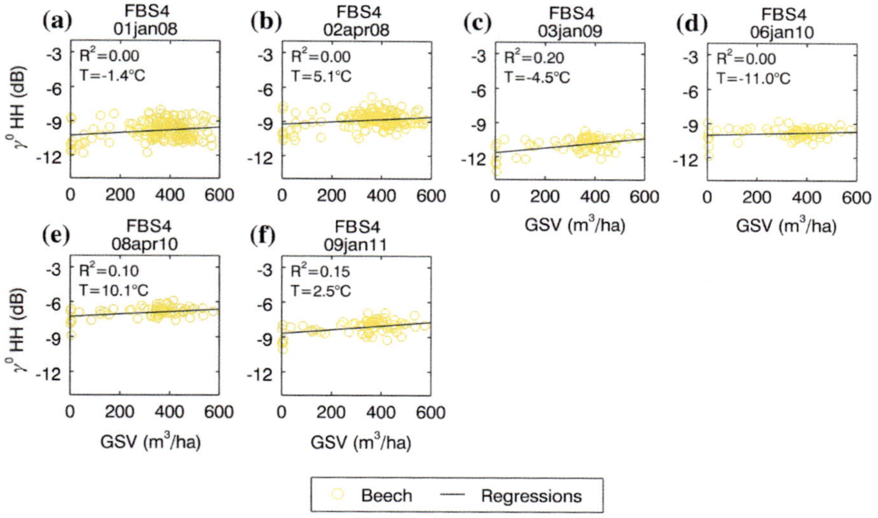

Fig. B.26 *GSV* sensitivity analysis for different air temperatures with ALOS PALSAR FBS HH 39° Asc. backscatter intensity (γ^0) (frame FBS4). The forested areas represent European Beech. Slopes <5° were selected to limit topographic effects

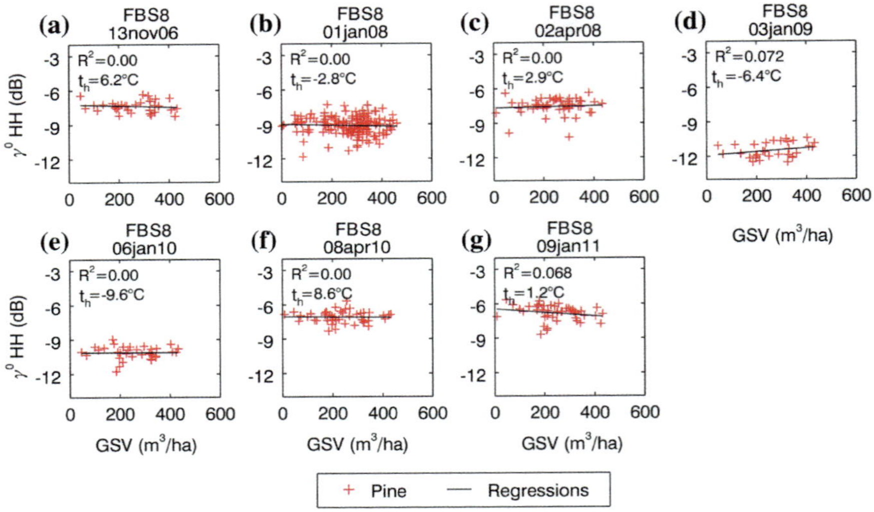

Fig. B.27 *GSV* sensitivity analysis for different air temperatures with ALOS PALSAR FBS HH 39° Asc. backscatter intensity (γ^0) (frame FBS8). The forested areas represent Scots Pine. Slopes <5° were selected to limit topographic effects

B.3 Interferometric Phase and Coherence Analysis

B.3.1 Comparison of Interferometric Systems

See Fig. B.28.

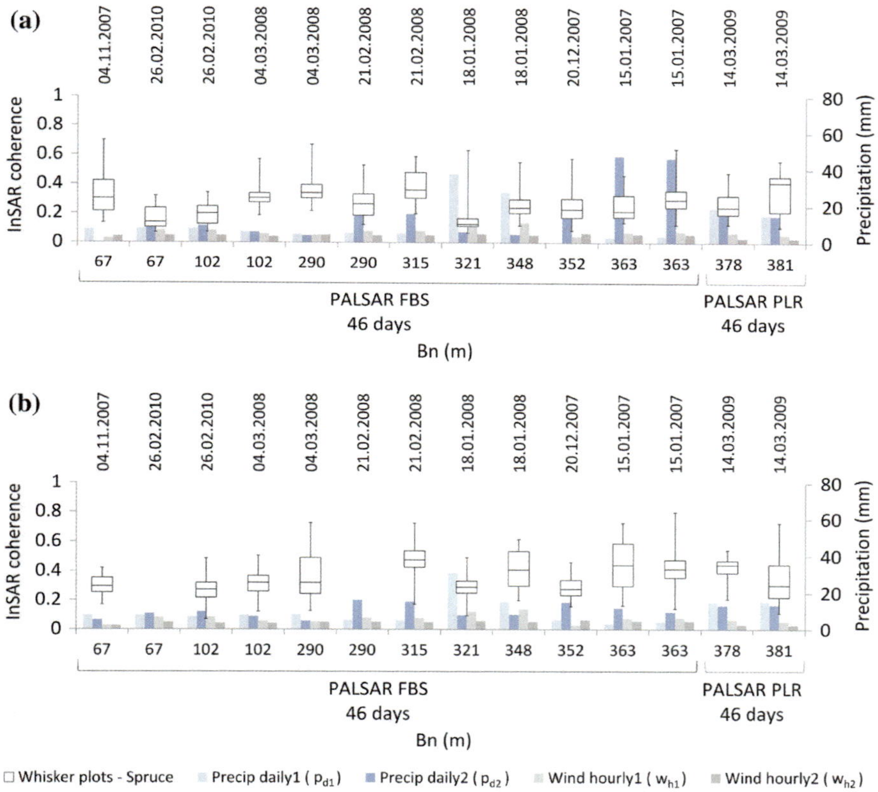

Fig. B.28 Comparisons of interferometric coherence (γ) between **a** forested areas and **b** open areas with PALSAR FBS HH 39° Asc. (frames FBS1,3 and FBS5–7) and PALSAR PLR HH 21° Asc. (frames PLR1,2) coherence. The boxplots represent spruce with the sample minimum, *lower quartile* (Q1), median (Q2), *upper quartile* (Q3) and sample maximum. Weather information is depicted in *bright blue*, *dark blue*, *bright grey* and *dark grey* for daily precipitations acquisition 1 (p_{d1}), daily precipitations acquisition 2 (p_{d2}), hourly wind acquisition 1 (w_{h1}) and hourly wind acquisition 2 (w_{h2}), respectively. Slopes of $<5°$ were selected to limit topographic effects

B.3.2 Polarisations

See Figs. B.29, B.30, B.31, B.32, B.33, B.34, B.35 and B.36.

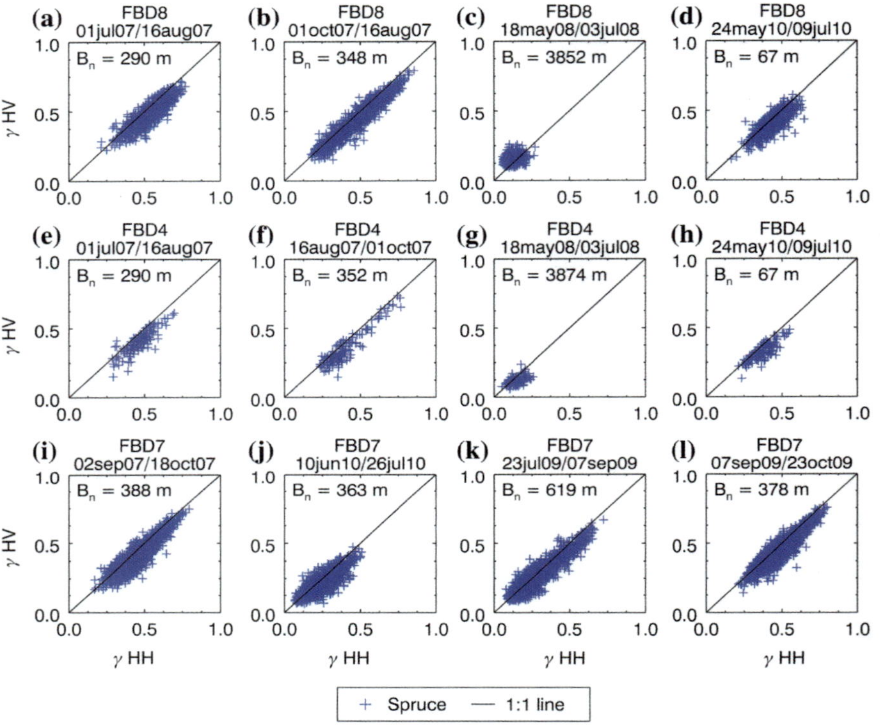

Fig. B.29 Polarisation analysis with PALSAR FBD HH 39° Asc. versus PALSAR FBD HV 39° Asc. interferometric coherence (γ) (frames FBD4 and FBD6–8). *Blue* points represent Norway Spruce forest stands. Slopes of <5° were selected to limit topographic effects

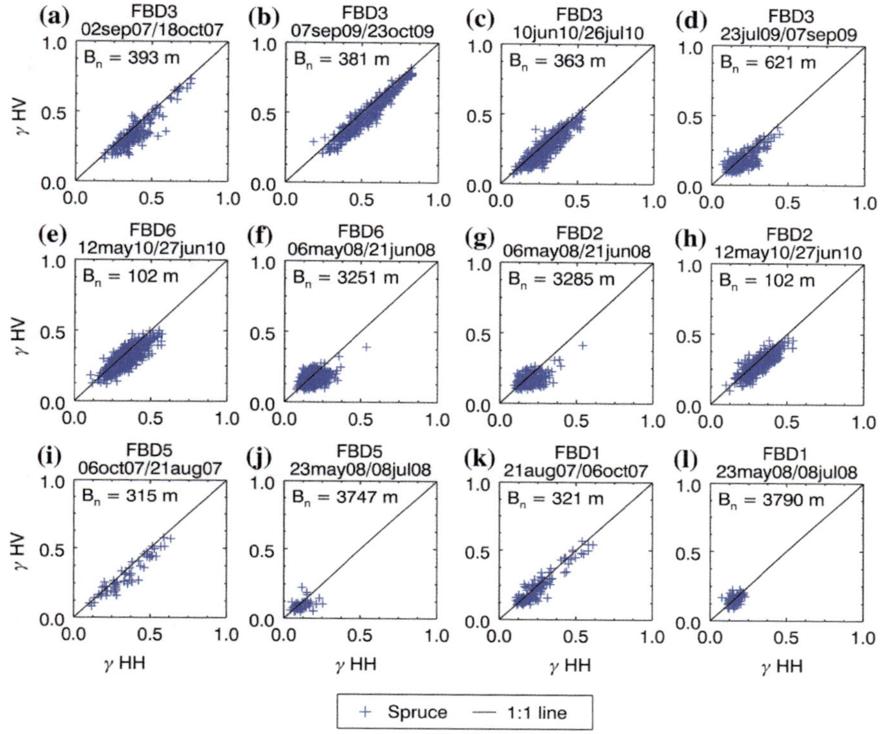

Fig. B.30 Polarisation analysis with PALSAR FBD HH 39° Asc. versus PALSAR FBD HV 39° Asc. interferometric coherence (γ) (frames FBD1–3 and FBD5,6). *Blue* points represent Norway Spruce forest stands. Slopes of <5° were selected to limit topographic effects

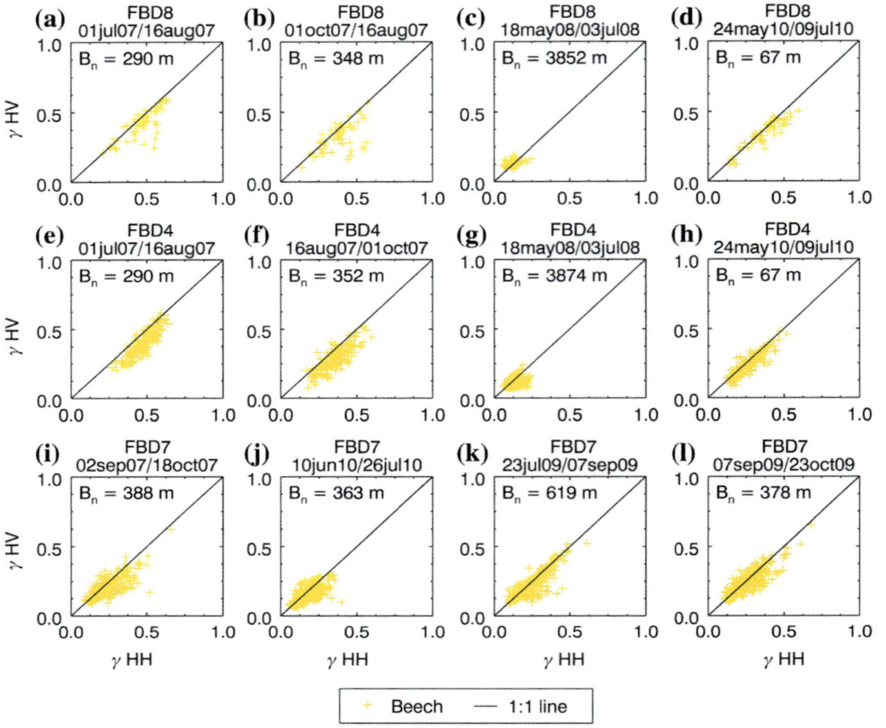

Fig. B.31 Polarisation analysis with PALSAR FBD HH 39° Asc. versus PALSAR FBD HV 39° Asc. interferometric coherence (γ) (frames FBD4 and FBD6–8). *Yellow* points represent European Beech forest stands. Slopes of $<10°$ were selected to limit topographic effects

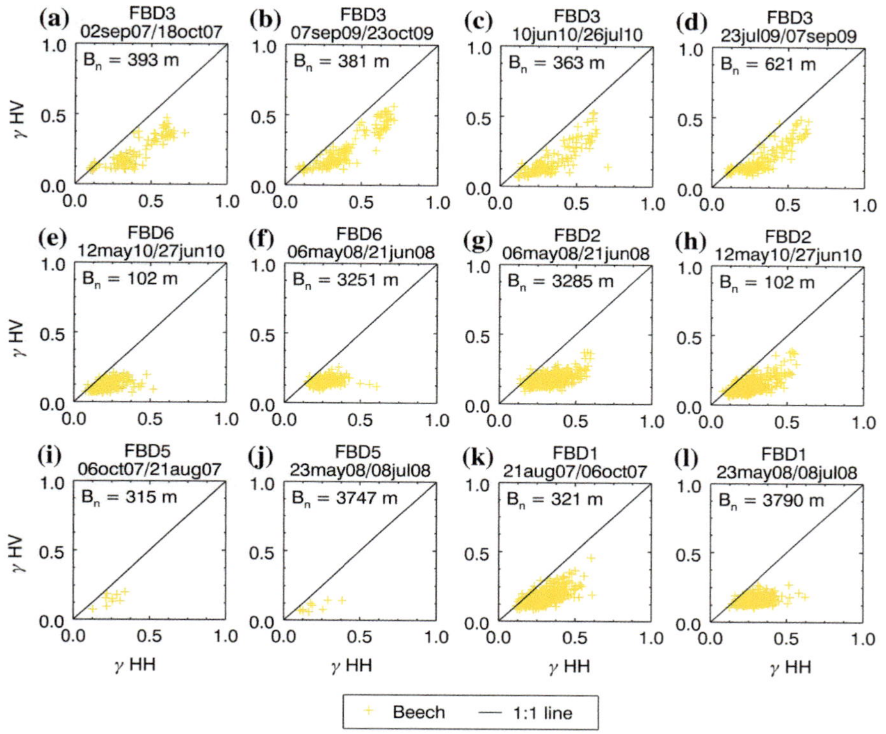

Fig. B.32 Polarisation analysis with PALSAR FBD HH 39° Asc. versus PALSAR FBD HV 39° Asc. interferometric coherence (γ) (frames FBD1–3 and FBD5,6). *Yellow* points represent European Beech forest stands. Slopes of $<10°$ were selected to limit topographic effects

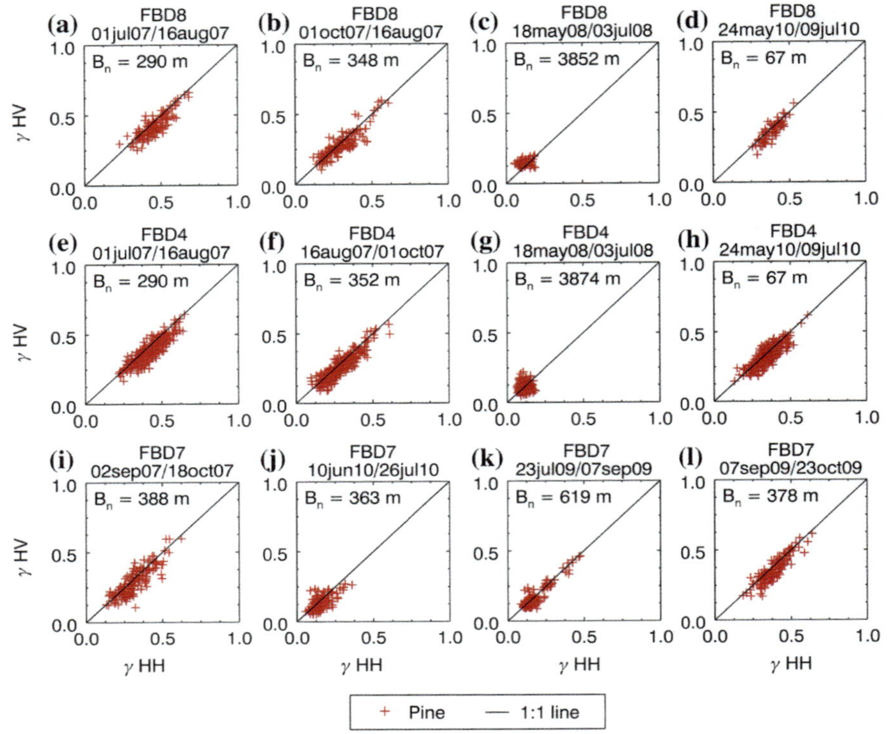

Fig. B.33 Polarisation analysis with PALSAR FBD HH 39° Asc. versus PALSAR FBD HV 39° Asc. interferometric coherence (γ) (frames FBD4 and FBD6–8). *Red* points represent Scots Pine forest stands. Slopes of <10° were selected to limit topographic effects

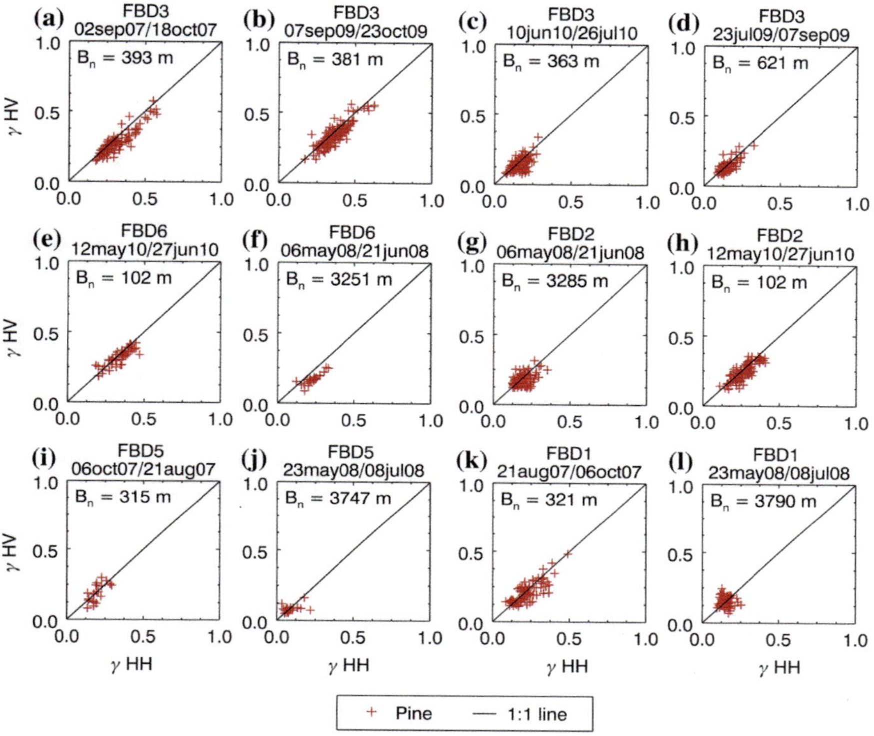

Fig. B.34 Polarisation analysis with PALSAR FBD HH 39° Asc. versus PALSAR FBD HV 39° Asc. interferometric coherence (γ) (frames FBD1–3 and FBD5,6). *Red* points represent Scots Pine forest stands. Slopes of <10° were selected to limit topographic effects

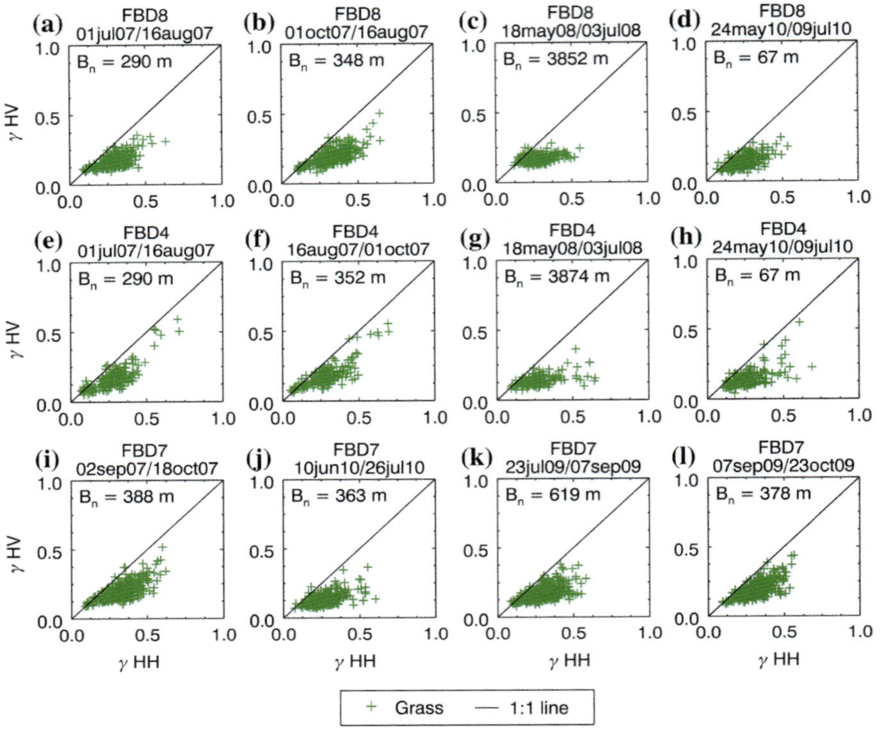

Fig. B.35 Polarisation analysis with PALSAR FBD HH 39° Asc. versus PALSAR FBD HV 39° Asc. interferometric coherence (γ) (frames FBD4 and FBD6–8). *Green* points represent open areas

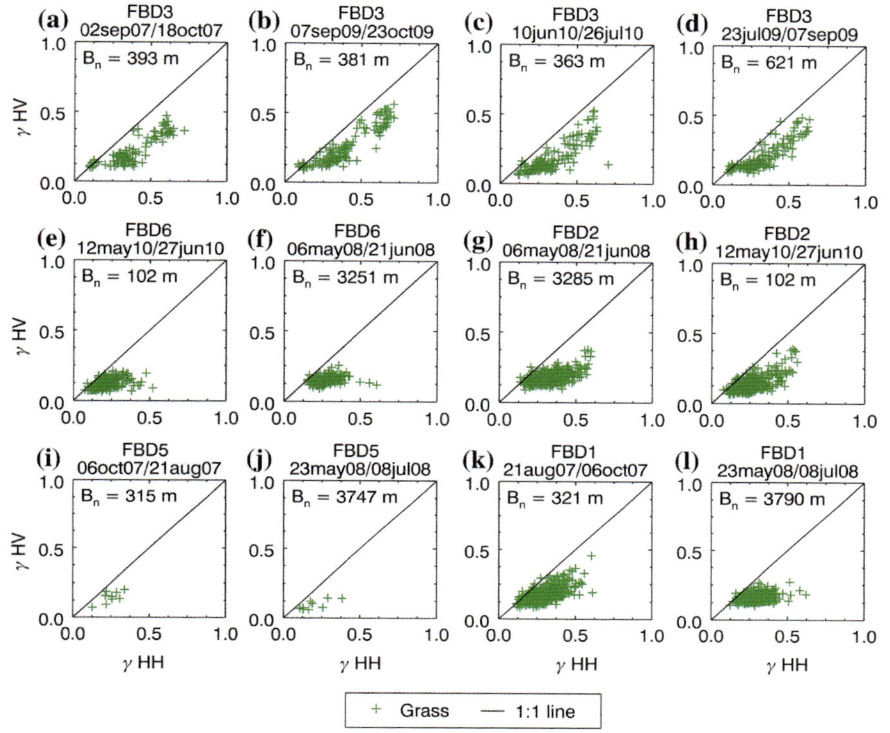

Fig. B.36 Polarisation analysis with PALSAR FBD HH 39° Asc. versus PALSAR FBD HV 39° Asc. interferometric coherence (γ) (frames FBD1–3 and FBD5,6). *Green* points represent open areas

B.3.3 GSV Sensitivity

See Figs. B.37, B.38, B.39, B.40, B.41, B.42, B.43, B.44, B.45 and B.46.

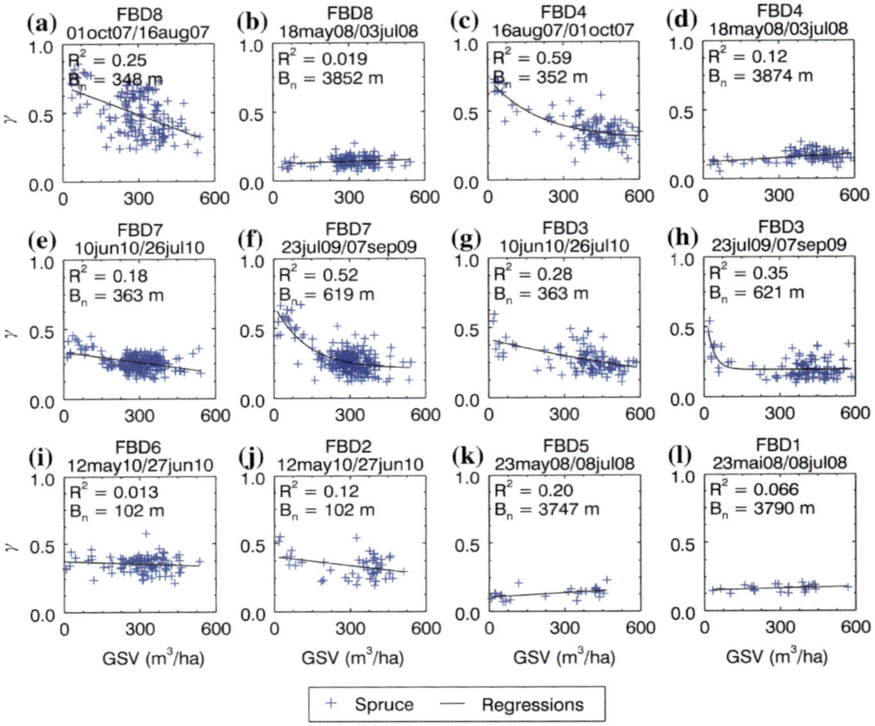

Fig. B.37 *GSV* sensitivity analysis with ALOS PALSAR FBD HH 39° Asc. interferometric coherence (γ) (frames FBD1–8). *Blue* points represent Norway Spruce. Slopes of <5° were selected to limit topographic effects

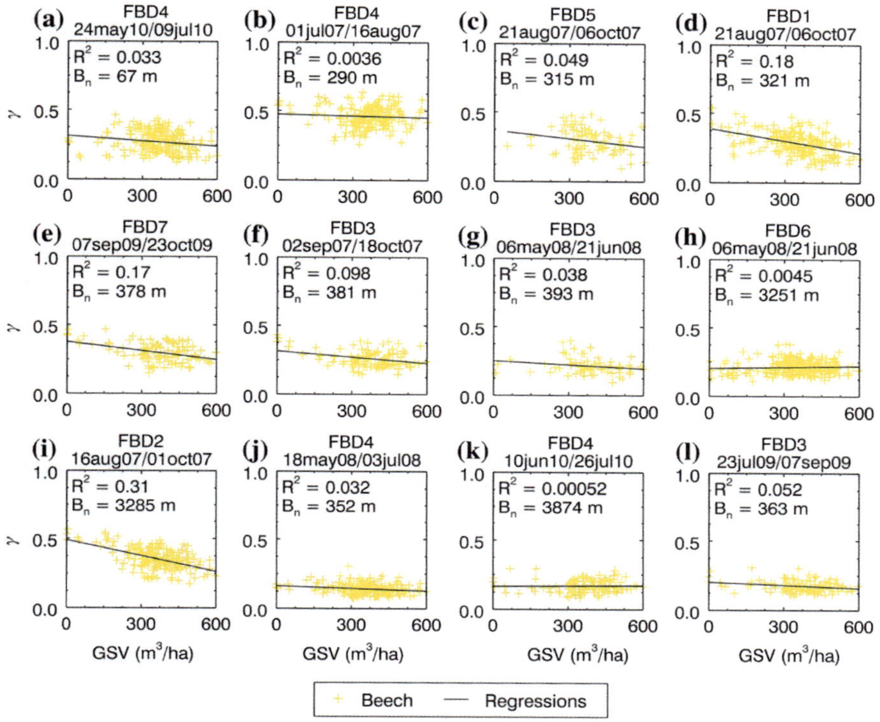

Fig. B.38 *GSV* sensitivity analysis with ALOS PALSAR FBD HH 39° Asc. interferometric coherence (γ) (frames FBD1–7). *Yellow* points represent European beech. Slopes of <10° were selected to limit topographic effects

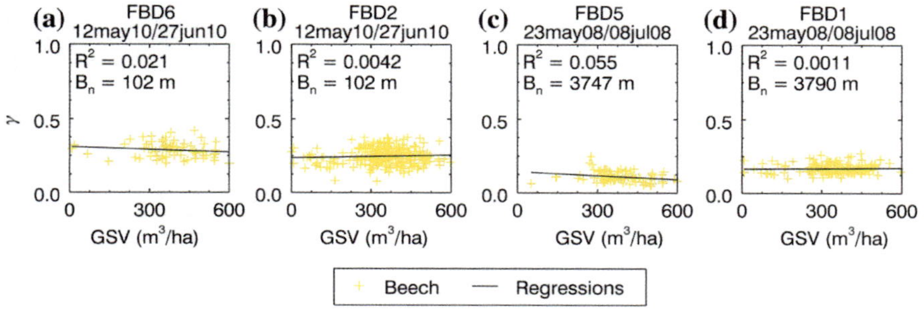

Fig. B.39 *GSV* sensitivity analysis with ALOS PALSAR FBD HH 39° Asc. interferometric coherence (γ) (frames FBD1,2 and FBD5,6). *Yellow* points represent European beech. Slopes of <10° were selected to limit topographic effects

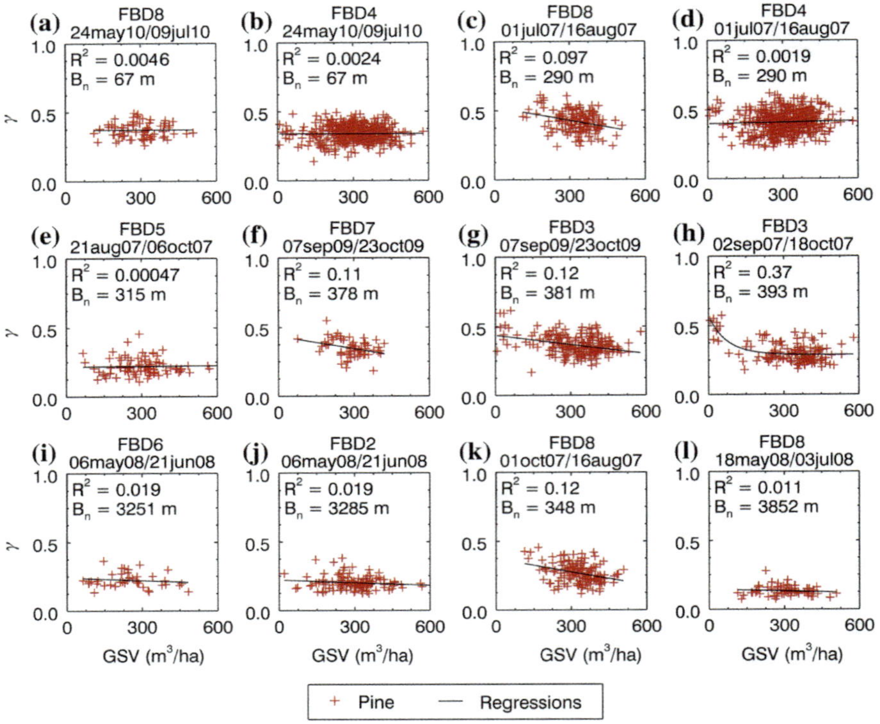

Fig. B.40 *GSV* sensitivity analysis with ALOS PALSAR FBD HH 39° Asc. interferometric coherence (γ) (frames FBD2–8). *Red* points represent Scots Pine. Slopes of $<10°$ were selected to limit topographic effects

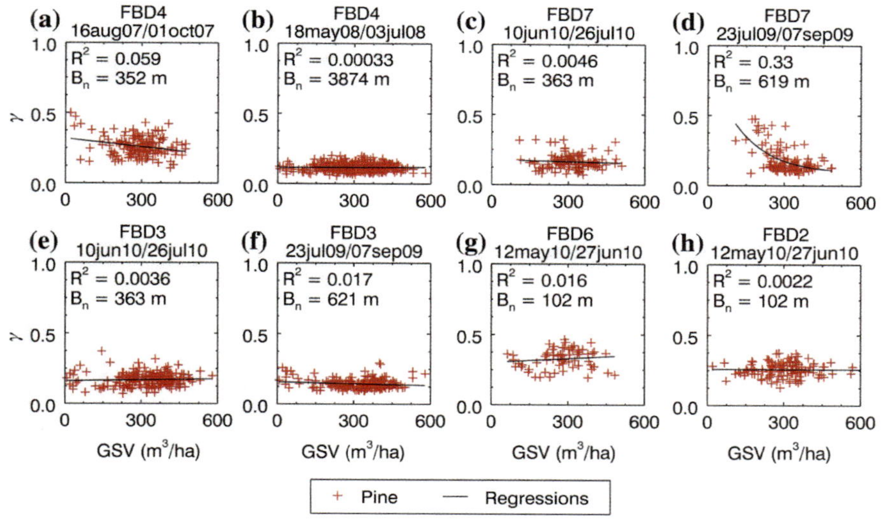

Fig. B.41 *GSV* sensitivity analysis with ALOS PALSAR FBD HH 39° Asc. interferometric coherence (γ) (frames FBD1–4 and FBD6,7). *Red* points represent Scots Pine. Slopes of <10° were selected to limit topographic effects

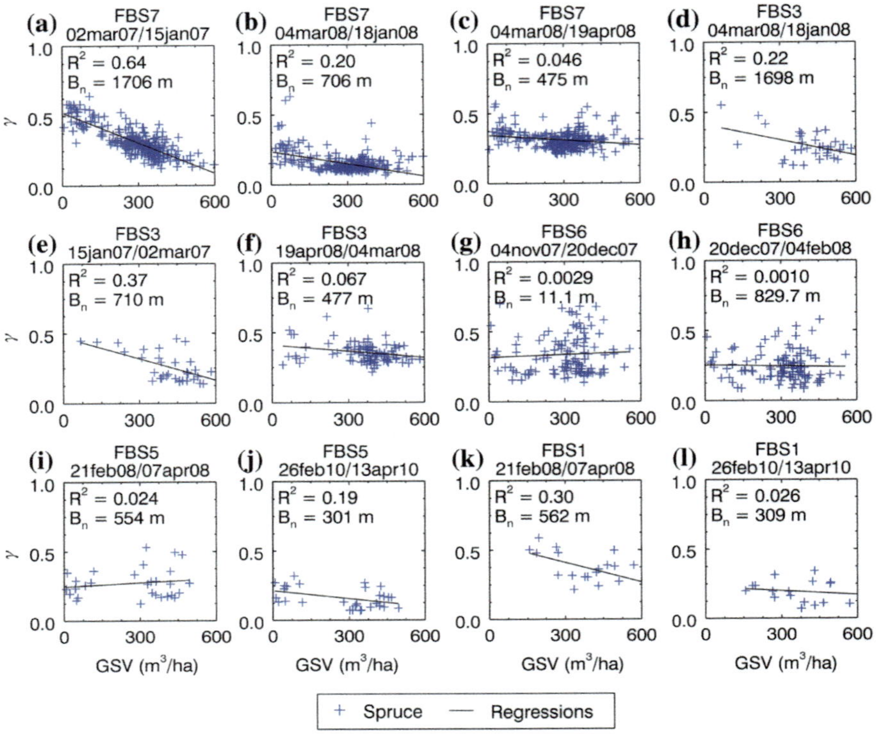

Fig. B.42 *GSV* sensitivity analysis with ALOS PALSAR FBS HH 39° Asc. interferometric coherence (γ) (frames FBS1,3 and FBS5–7). *Blue* points represent Norway Spruce. Slopes of <5° were selected to limit topographic effects

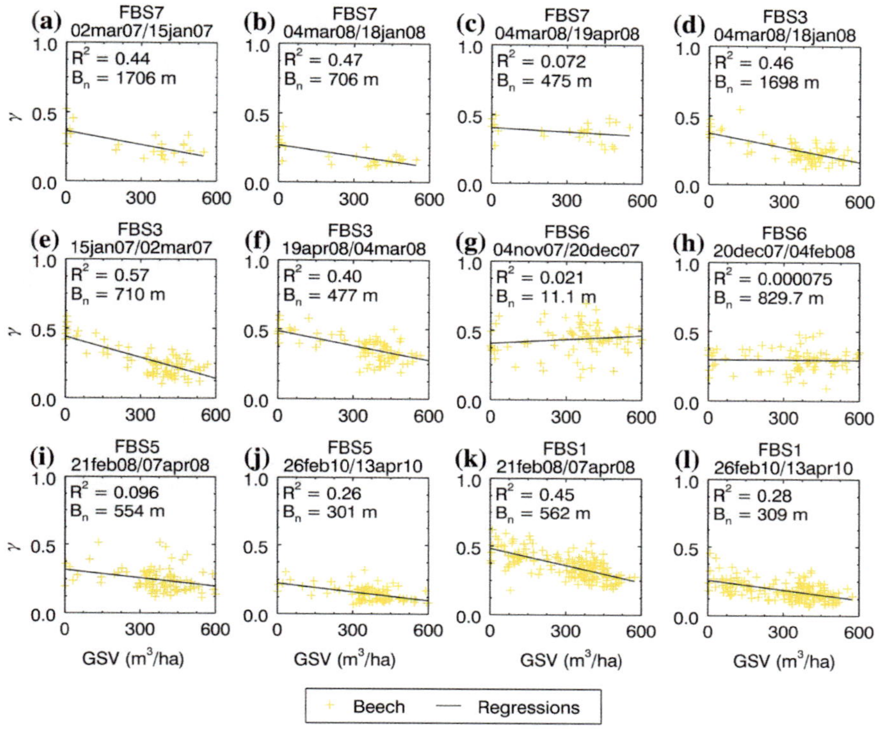

Fig. B.43 *GSV* sensitivity analysis with ALOS PALSAR FBS HH 39° Asc. interferometric coherence (γ) (frames FBS1,3 and FBS5–7). *Yellow* points represent European beech. Slopes of $<10°$ were selected to limit topographic effects

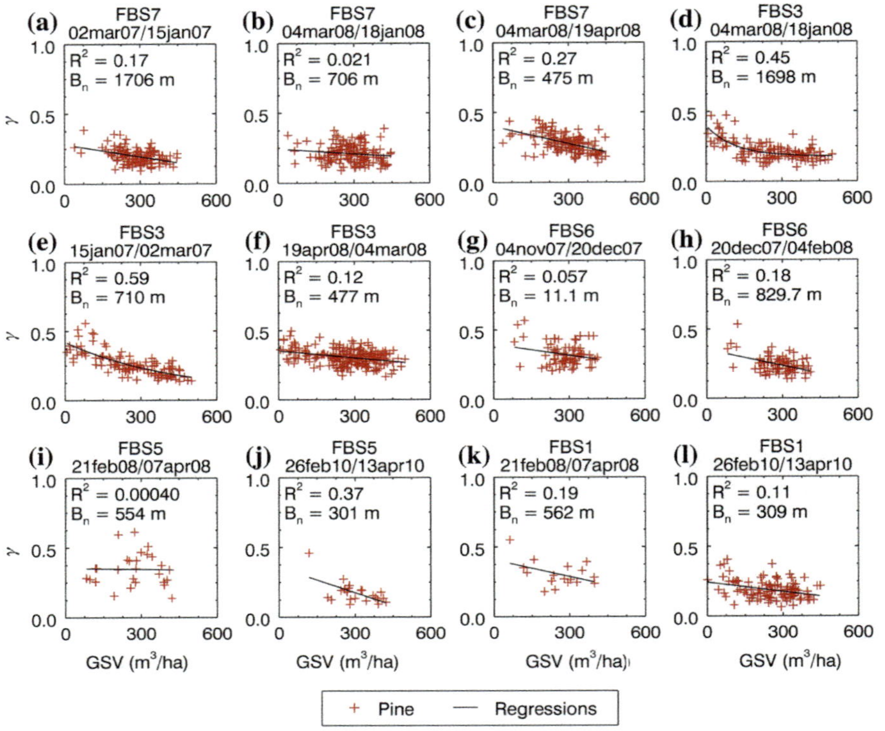

Fig. B.44 *GSV* sensitivity analysis with ALOS PALSAR FBS HH 39° Asc. interferometric coherence (γ) (frames FBS1,3 and FBS5–7). *Red* points represent Scots Pine. Slopes of $<10°$ were selected to limit topographic effects

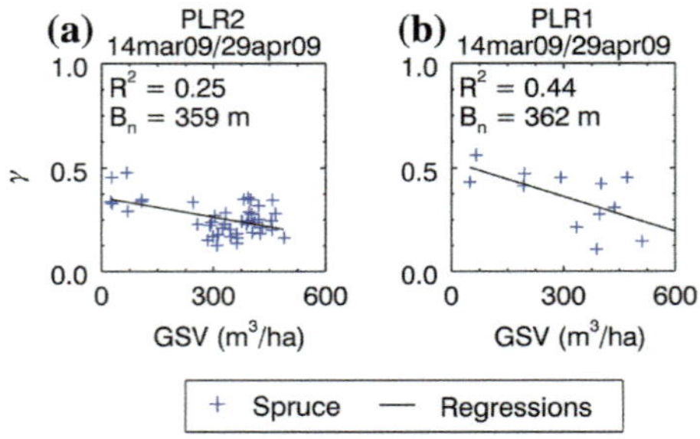

Fig. B.45 *GSV* sensitivity analysis with ALOS PALSAR PLR HH 39° Asc. interferometric coherence (γ) (frames PLR1,2). *Blue* points represent Norway Spruce. Slopes of $<5°$ were selected to limit topographic effects

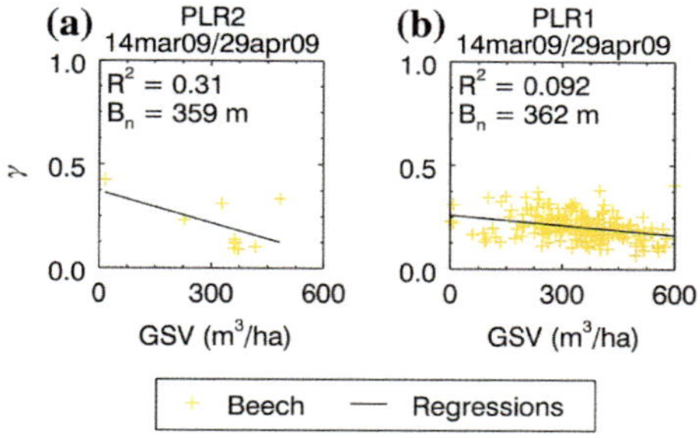

Fig. B.46 *GSV* sensitivity analysis with ALOS PALSAR PLR HH 39° Asc. interferometric coherence (γ) (frames PLR1,2). *Yellow* points represent European beech. Slopes of <10° were selected to limit topographic effects

B.3.4 LOCI Coherence

See Figs. B.47, B.48 and B.49.

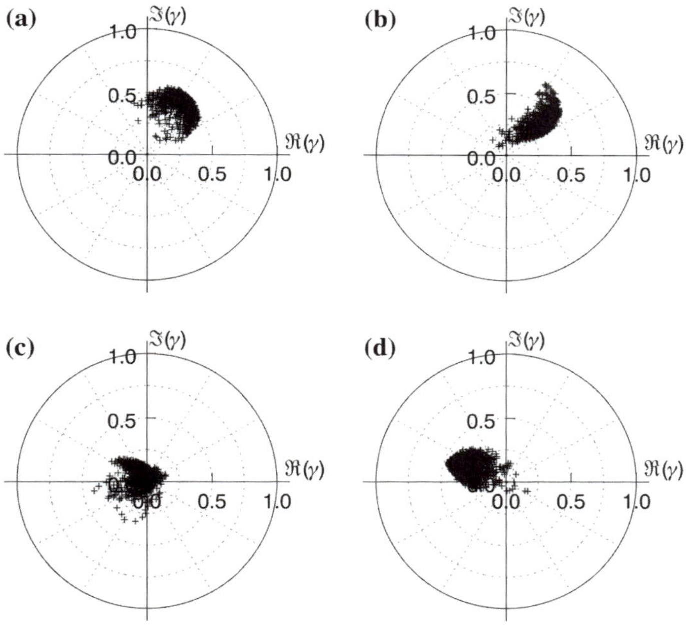

Fig. B.47 Loci coherence analysis for **a**, **b** mature dense and **c**, **d** mature sparse spruce forest with ALOS PALSAR PLR 22° Asc. $B_n = 359$ m complex interferometric coherence ($\tilde{\gamma}$) (frame PLR2). The PALSAR polarimetric interferometric scene was acquired on 14th March/19th April 2009. Each point describes the complex coherence for one polarisation state

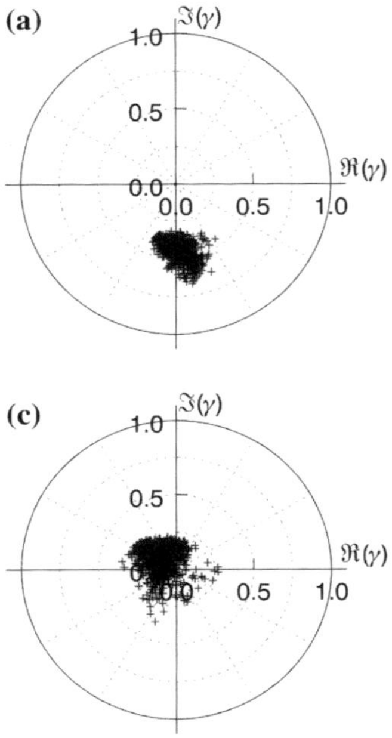

Fig. B.48 Loci coherence analysis for **a**, **b** young dense and **c**, **d** young sparse spruce forest with ALOS PALSAR PLR 22° Asc. $B_n = 359$ m complex interferometric coherence ($\tilde{\gamma}$) (frame PLR2). The PALSAR polarimetric interferometric scene was acquired on 14th March/19th April 2009. Each point describes the complex coherence for one polarisation state

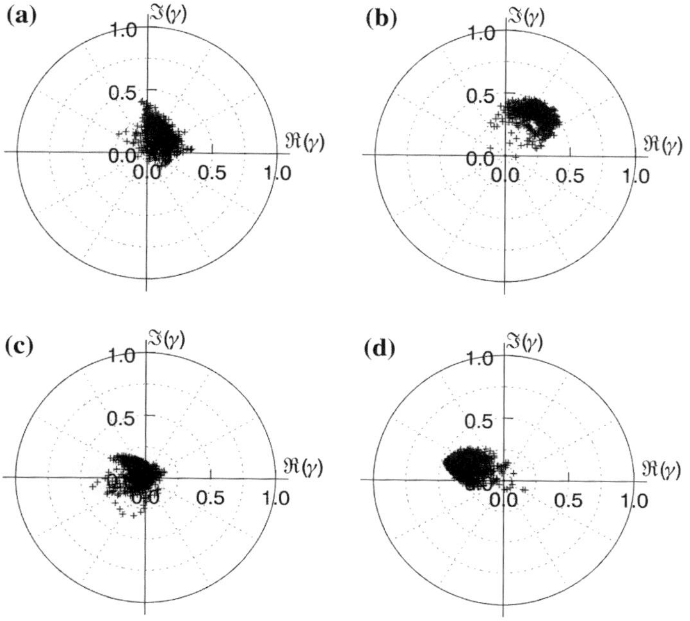

Fig. B.49 Loci coherence analysis for **a**, **b** mature dense beech forest and **c**, **d** open area with ALOS PALSAR PLR 22° Asc. $B_n = 359$ m complex interferometric coherence ($\tilde{\gamma}$) (frame PLR2). The PALSAR polarimetric interferometric scene was acquired on 14th March/19th April 2009. Each point describes the complex coherence for one polarisation state

B.4 Forest GSV Estimation

See Table B.1 and Figs. B.50 and B.51.

Table B.1 Statistics for modelling with parametric regressions of PALSAR FBS HH 39° Asc. (frames FBS1,3 and FBS5–7) and PALSAR PLR HH 21° Asc. (frame PLR1,2) coherence in the Thuringian Forest

Frame	Acquisition dates	Baseline (m)	p_{h1} (mm)	p_{h2} (mm)	R^2_{training}	RMSE (m^3 ha^{-1})	Bias (m^3 ha^{-1})
FBS1	21feb08–07apr08	562	0.4	0.1	0.01	246	39
FBS1	26feb10–13apr10	309	1.1	0.0	0.22	207	43
FBS3	15jan07–02mar07	1698	0.0	0.0	0.14	101	32
FBS3	04mar08–18jan08	710	3.5	0.1	0.72	183	5
FBS3	19apr08–04mar08	477	0.1	0.5	0.01	146	46
FBS5	21feb08–07apr08	554	0.4	0.2	0.10	284	7
FBS5	26feb10–13apr10	301	1.1	0.0	0.22	210	13
FBS6	04nov07–20dec07	11	0.0	0.0	–	–	–
FBS6	20dec07–04feb08	830	0.0	0.0	–	–	–
FBS7	15jan07–02mar07	−1706	0.0	0.0	0.80	126	6
FBS7	04mar08–18jan08	−706	4.8	0.0	0.45	163	11
FBS7	19apr08–04mar08	475	0.0	1.2	0.29	187	31
PLR1	14mar09–29apr09	362	15.1	14.7	0.23	195	6
PLR2	14mar09–29apr09	359	18.9	17.3	0.28	184	13

Three statistics were retained, namely R^2 for training, *RMSE* and bias Modelling was performed over spruce forest stands

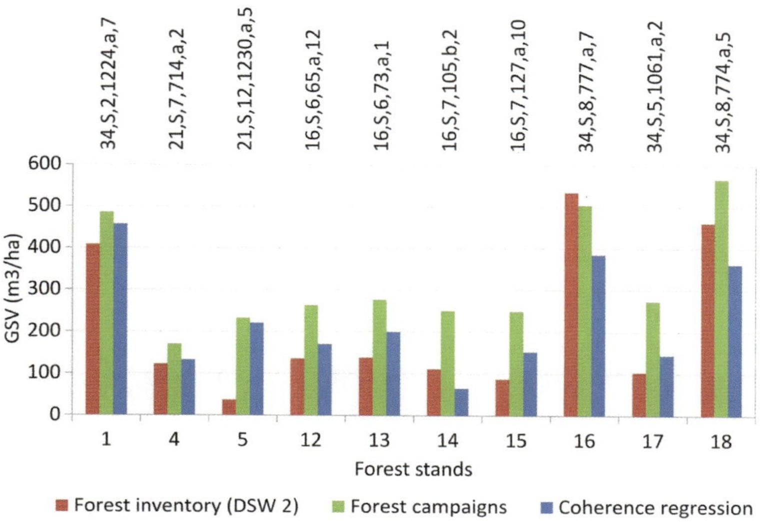

Fig. B.50 Comparison of the estimated GSV (regression approach) with GSV measured during forest campaigns and provided in the forest inventory. The estimated GSV is based on the average of ALOS PALSAR FBD 39° Asc. HH coherences (γ) available on frame FBD7

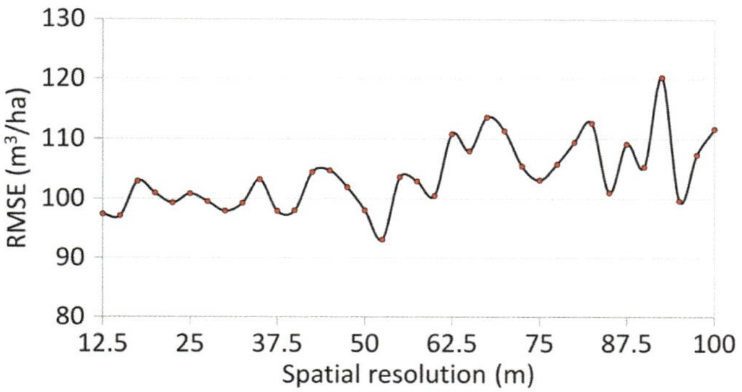

Fig. B.51 Influence of the radar spatial resolution on the GSV estimation accuracy (RMSE). The analysis is based on ALOS PALSAR FBS 39° Asc. HH (frames FBS7) coherence (γ). The interferometric acquisition was performed on 2nd March 2007/15th January 2007

Appendix C
Statistics Formulas

C.1 Statistics Formulas

$$RMSE = \sqrt{\frac{1}{N}\sum_{1}^{N}\left(\hat{y}_i - y_i\right)^2} \tag{C.1}$$

$$MAE = \frac{1}{N}\sum_{1}^{N}\left|\hat{y}_i - y_i\right| \tag{C.2}$$

$$Bias = \frac{1}{N}\sum_{1}^{N}\left(\hat{y}_i - y_i\right) \tag{C.3}$$

$$R^2 = 1 - \frac{SS_{\text{error}}}{SS_{\text{total}}} = 1 - \frac{\sum_i \left(y_i - \hat{y}_i\right)^2}{\sum_i \left(y_i - \overline{y}\right)^2} \tag{C.4}$$

N. Ackermann, *Growing Stock Volume Estimation in Temperate Forested Areas Using a Fusion Approach with SAR Satellites Imagery*, Springer Theses, DOI 10.1007/978-3-319-13138-2

Appendix D
TanDEM-X Height

See Fig. D.1.

Fig. D.1 Visualisation of the TanDEM-X derived interferometric height. The forested areas are located around Görsdorf

N. Ackermann, *Growing Stock Volume Estimation in Temperate Forested Areas Using a Fusion Approach with SAR Satellites Imagery*,
Springer Theses, DOI 10.1007/978-3-319-13138-2